Universitext

Springer
Berlin
Heidelberg
New York
Barcelona
Budapest
Hong Kong
London
Milan
Paris
Santa Clara
Singapore
Tokyo

Bruce P. Kitchens

Symbolic Dynamics

One-sided, Two-sided and
Countable State Markov Shifts

With 65 Figures

 Springer

Bruce P. Kitchens
Mathematical Sciences Department
IBM T. J.Watson Research Center
P.O. Box 557
Yorktown Heights, NY 10598
USA
e-mail: kitch@watson.ibm.com

Library of Congress Cataloging-in-Publication Data
Kitchens, Bruce P.
Symbolic dynamics: one-sided, two-sided, and countable state Markov shifts / Bruce P. Kitchens.
p. cm. – (Universitext). Includes bibliographical references (p. –) and index.
ISBN 3-540-62738-3 (softcover: acid-free paper)
1. Symbolic dynamics.   I. Title.   QA614.85.K57   1998   514'.74–dc21   97-39657   CIP

Mathematics Subject Classification (1991): 58F03, 34C35, 54H20, 28D20

ISBN 3-540-62738-3 Springer-Verlag Berlin Heidelberg New York

© Springer-Verlag Berlin Heidelberg 1998
Printed in Germany

Typesetting: Author's input files edited and reformatted by Jörg Steffenhagen, Königsfeld, using a Springer TEX macro package
Cover design: design & production GmbH, Heidelberg

SPIN: 10566561   41/3143 – 5 4 3 2 1 0 – Printed on acid-free paper

Preface

Nearly one hundred years ago Jacques Hadamard used infinite sequences of symbols to analyze the distribution of geodesics on certain surfaces. That was the beginning of symbolic dynamics. In the 1930's and 40's Arnold Hedlund and Marston Morse again used infinite sequences to investigate geodesics on surfaces of negative curvature. They coined the term symbolic dynamics and began to study sequence spaces with the shift transformation as dynamical systems. In the 1940's Claude Shannon used sequence spaces to describe information channels. Since that time symbolic dynamics has been used in ergodic theory, topological dynamics, hyperbolic dynamics, information theory and complex dynamics. Symbolic dynamical systems with a finite memory are studied in this book. They are the topological Markov shifts. Each can be defined by transition rules and the rules can be summarized by a transition matrix. The study naturally divides into two parts. The first part is about topological Markov shifts where the alphabet is finite. The second part is concerned with topological Markov shifts whose alphabet is countably infinite. The techniques used in the two cases are quite different. When the alphabet is finite most of the methods are combinatorial or algebraic. When the alphabet is infinite the methods are much more analytic.

This book grew from notes for a graduate course taught at Wesleyan University in the fall of 1994 and is intended as a graduate text and as a reference book for mathematicians working in related fields. Each chapter begins with an overview and then is divided into sections. At the end of each section are exercises. At the end of each chapter are historical notes and references.

The first section of the first chapter contains the beginning definitions of the spaces with a finite alphabet and the shift transformation. The second section is made up of examples. The examples show how topological Markov shifts arise in other areas of dynamics and in information theory. There are maps of the interval, Complex quadratic maps, the horseshoe, a toral automorphism and a channel code. The third section contains a short discussion of Perron-Frobenius theory for nonnegative, square matrices. The last section explains the basic dynamical properties of the shift transformation acting on a topological Markov shift. These properties include the topological entropy, the zeta function and a characterization of the continuous maps to be studied.

The second chapter is about homeomorphisms between topological Markov shifts which commute with the shift transformations. Such a map is called a topological conjugacy and when one exists between two topological Markov shifts they are said to be to be topologically conjugate. In the first section we see that a topological conjugacy can be decomposed into a sequence of elementary conjugacies. This leads to the development of an algorithm to determine when two one-sided topological Markov shifts are conjugate. We note that it is not known whether or not it is possible for such an algorithm to exist in the two-sided setting. In the second section shift-equivalence for two-sided topological Markov shifts is defined as an invariant for conjugacy. This is an algebraic relationship between transition matrices. This relationship is exploited to show relationships between the Jordan forms of the matrices and the dimension groups defined by the matrices.

Chapter three is about automorphisms of topological Markov shifts. An automorphism is a homeomorphism from a topological Markov shift to itself which commutes with the shift transformation. The automorphisms of a shift form a group under composition. The first section begins with a striking result which shows the automorphism group of the one-sided sequences on two symbols is isomorphic to $\mathbb{Z}/2\mathbb{Z}$ while the automorphism group of the two-sided sequences on two symbols contains every finite group. In the second section automorphisms are examined using the decomposition of conjugacies developed in chapter two. This allows us to show the automorphism groups of one-sided topological Markov shifts are generated by elements of finite order. It has recently been shown there are two-sided topological Markov shifts where the automorphism group is not generated by elements of finite order together with the shift transformation. The third section contains a discussion of subgroups of automorphism groups. This includes a discussion of finite subgroups and some observations about finitely presented groups. The fourth section contains observations about how the automorphisms act on the space and induce an automorphism of the dimension group. The decomposition of a conjugacy from chapter two is used to show how an automorphism of the space induces an automorphism of the dimension group.

The fourth chapter is about embeddings and factor maps. A factor map is a continuous map from one topological Markov shift onto another which commutes with the shift transformations. In the first section we see that factor maps fall into two distinct categories. They are either uniformly finite-to-one or uncountable-to-one on most points. Uniformly finite-to-one factor maps are examined in the second and third sections. A number of necessary conditions are developed for the existence of a uniformly finite-to-one factor map between two topological Markov shifts. These include conditions on the Jordan form of the transition matrices and conditions on some of the Bowen-Franks groups. In the fourth section necessary and sufficient conditions are developed for one topological Markov shift to be conjugate to a subsystem of another and for one topological Markov shift to be an unbounded-to-one factor of another.

Chapter five is about almost-topological conjugacy of topological Markov shifts. The first section is about reducible Markov shifts. The second section contains a description of the notion of almost-topological conjugacy and then develops a necessary and sufficient condition for two topological Markov shifts to be almost-topologically conjugate. The condition is that the topological entropy and the period of the two Markov shifts be the same.

Chapter six contains some additional topics. The first section contains a discussion of sofic systems. Sofic systems form a larger class of symbolic systems than the topological Markov shifts. A symbolic system on finitely many symbols is sofic if it can be obtained as the continuous image of a topological Markov shift. Markov measures on topological Markov shifts and the measure of maximal entropy are discussed in section two. Section three is about Markov subgroups. A Markov subgroup is a two-sided topological Markov shift with a group structure which makes the shift transformation an automorphism. There is a structure theorem for Markov subgroups and they are completely characterized up to topological conjugacy. A cellular automata is a continuous map from a topological Markov shift into itself. These are briefly discussed in section four. In the fifth section we examine a type of channel code used in data storage devices. These are called run-length limited codes. The constraints are described by finite transition rules and so the techniques previously developed are applicable. A method is developed to construct channel codes which meet certain encoding and decoding requirements.

The second part of this book is about topological Markov shifts with a countably infinite alphabet. There is far less known about these shifts than there is about finite state, topological Markov shifts. To examine countable state Markov shifts it is first necessary to extend the Perron-Frobenius theory for nonnegative, finite, square matrices to nonnegative, square, countably infinite matrices. This requires a considerable amount of work. First, the matrices are divided into three classes. Roughly speaking the classes correspond to how well the Perron- Frobenius Theorem can be approximated. This analytic division will also be mirrored in the dynamics of the Markov shifts the matrices define. Next the Perron-Frobenius Theorem is generalized and the differences in the classes of matrices is explored. The methods are very different than for finite matrices. Next there are a number of examples where the different types of behavior are illustrated.

In the second section countable state, topological Markov shifts are defined using countable transition matrices. The spaces for the Markov shifts are not compact and it leads to a number of difficulties. Topological entropy can be formulated in terms of the matrices, metrics, compactifications or invariant measures. Several of these formulations are given and compared. Finally, we examine some invariant measures and see that a form of the variational principle holds for a countable state topological Markov shift. We conclude by showing a countable state Markov shift has a maximal measure if and only if the transition matrix meets certain recurrence conditions.

Kinnakeet, North Carolina   *Bruce Kitchens*
May 2, 1997

Contents

Chapter 1. Background and Basics

We will be working with sequence spaces. There will be two types of sequence spaces. One is composed of one-sided infinite sequences and the other is composed of two-sided infinite sequences. These are metric spaces where two one-sided sequences are said to be close together if they agree for a long time at the beginning and two two-sided sequences are said to be close together if they agree for a long time around the center. In the first part of the book the sequences will have entries coming from a finite set and in the second part the entries will come from a countable set. When the entries are from a finite set the sequence space is usually homeomorphic to the standard Cantor set. The transformation we study will be the shift transformation which shifts each sequence once to the left. The richness of the dynamics does not arise from the topology of the space or the definition of the transformation but from the definition of which sequences belong to each space. We study subshifts of finite type. They are the sequence spaces characterized by having a finite rule which determines the sequences belonging to each space.

In this chapter we define subshifts of finite type, examine some of their basic properties and describe a number of examples where subshifts of finite type arise naturally.

In the first section we define the symbolic spaces, their topology and the shift operator. The second section is composed entirely of examples. The examples illustrate how subshifts of finite type arise in other types of dynamics and in information theory. The third section contains a quick review of Perron-Frobenius theory for nonnegative matrices. In the fourth section we examine some of the basic dynamical properties of subshifts of finite type. These include topological entropy, the zeta function of a dynamical system and continuous maps between subshifts of finite type.

§ 1.1 Subshifts of Finite Type

Begin with a finite set $\{1, \ldots, n\}$. It is the alphabet or state set for the sequences. The set is a metric space when we say two distinct points have distance one and a point is distance zero from itself. The metric is given by the formula $d(j,k) = 1 - \delta_{jk}$ where δ_{jk} is 0 if $j \neq k$ and 1 if $j = k$. Each point is both open and closed. In the topology defined by the metric the space is compact and has the discrete topology. We form two sequence spaces. The one-sided sequence space is $\{1, \ldots, n\}^{\mathbb{N}}$, where $\mathbb{N} = \{0, 1, 2, \ldots\}$. Its points are of the form $(.x_0 x_1 x_2 \ldots)$ where each $x_i \in \{1, \ldots, n\}$. The two-sided sequence space is $\{1, \ldots, n\}^{\mathbb{Z}}$. Its points are of the form $(\ldots x_{-2} x_{-1} . x_0 x_1 x_2 \ldots)$ where each $x_i \in \{1, \ldots, n\}$. We put the product topology on each space. By Tychonoffs Theorem for products of compact spaces both spaces are compact. A set of the form $[i_0, \ldots, i_\ell]_t = \{x : x_t = i_0, \ldots, x_{t+\ell} = i_\ell\}$ is both open and closed. Such a set is called a *cylinder set*. The cylinder sets form a countable basis for the topology on each space. Every open set is a countable union of cylinder sets. Since the cylinder sets are open-closed and generate the topology both spaces have topological dimension zero. Equivalently, they are totally disconnected. The only connected components are points. The product metric $d(x,y) = \sum_{i=0}^{\infty} \frac{1 - \delta_{x_i y_i}}{2^i}$ on $\{1, \ldots, n\}^{\mathbb{N}}$ generates the topology. An equivalent metric on the space (and the one we will use for simplicity) is $d(x,y) = 1/2^k$ where $k = max\{m : x_i = y_i \text{ for } i < m\}$. Similarly, the product metric $d(x,y) = \sum_{i=-\infty}^{+\infty} \frac{1 - \delta_{x_i y_i}}{2^{|i|}}$ on $\{1, \ldots, n\}^{\mathbb{Z}}$ generates the topology. The equivalent metric on this space is $d(x,y) = 1/2^k$ where $k = max\{m : x_i = y_i \text{ for } |i| < m\}$. All this says that both spaces are compact zero-dimensional metric spaces. Both (when $n > 1$) are homeomorphic to the standard middle thirds Cantor set (Exercise 1). Denote by X_n the space $\{1, \ldots, n\}^{\mathbb{N}}$ with the topology just discussed and denote by Σ_n the space $\{1, \ldots, n\}^{\mathbb{Z}}$ with the corresponding topology. Next define the shift transformation on each space. The shifts $\sigma : X_n \to X_n$ and $\sigma : \Sigma_n \to \Sigma_n$ are defined by $\sigma(x)_i = x_{i+1}$. On X_n the shift is a continuous, onto, n-to-1 transformation. On Σ_n the shift is a homeomorphism. The dynamical system (X_n, σ) is called the *one-sided shift on n symbols* or the *full one-sided n-shift*. Similarly the dynamical system (Σ_n, σ) is called the *two-sided shift on n symbols* or the *full n-shift*.

Next we will identify certain points with different types of dynamical behavior. First examine the one-sided shift. A *periodic point* is a point of the form $x = (.x_0 \ldots x_{p-1} x_0 \ldots x_{p-1} \ldots)$. It has the property that $\sigma^p(x) = x$. If p is the least such power of the shift (greater than zero) that fixes x, we say x has *period p*. There are countably many periodic points and they are dense in the space. A point of the form $x = (.x_0 \ldots x_{k-1} x_k \ldots x_{k+p-1} x_k \ldots x_{k+p-1} \ldots)$ is a *preperiodic* or *eventually periodic point*. It has the property that $\sigma^k(x) = \sigma^{k+p}(x)$.

The periodic points and preperiodic points are the points with finite orbits under σ. A point whose orbit under σ is dense is called a *transitive point*. The transitive points form a dense G_δ in the one-sided shift space (Exercise 4). Things are similar for the two-sided shift. The periodic points are of the form $x = (\ldots x_0 \ldots x_{p-1}.x_0 \ldots x_{p-1} \ldots)$. There are countably many of them. They are the points with finite orbits and are dense in the space. A point whose orbit is dense in both forward and backward time; that is, under both positive and negative powers of σ, is called a *doubly transitive point*. In the two-sided shift space the doubly transitive points form a dense G_δ.

A square $\{0,1\}$ matrix is *irreducible* if for every pair of indices i and j there is an $\ell > 0$ with $(A^\ell)_{ij} > 0$. A square $\{0,1\}$ matrix corresponds to a directed graph. The vertices of the graph are the indices for the rows and columns of A. There is an edge from vertex i to vertex j if and only if $A_{ij} = 1$. A directed graph is *strongly connected* if it is possible to get from any vertex to any other by traversing a sequence of edges as directed. A square $\{0,1\}$ matrix is irreducible if and only if the corresponding graph is strongly connected. Given a square $\{0,1\}$ matrix A, we will denote its corresponding graph by G_A. Sometimes it will be more convenient to think in terms of the matrix and sometimes the graph.

Example 1.1.1 The matrix A and the graph G_A are shown in Figure 1.1.1.

$$A = \begin{bmatrix} 1 & 1 \\ 1 & 0 \end{bmatrix}$$

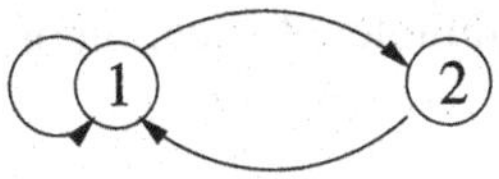

Figure 1.1.1

Let A be a square $\{0,1\}$ matrix with its rows and columns indexed by $\{1, \ldots, n\}$. The matrix A defines a closed, shift invariant subset of X_n. It is defined by selecting the sequences x in X_n with $A_{x_i x_{i+1}} = 1$ for all $i \in \mathbb{N}$. We can also think of these sequences as the infinite paths on G_A. This set of sequences is closed in X_n and so compact. The dynamical system consisting of this compact space with the subspace topology and the restriction of the shift transformation is the *one-sided subshift of finite type defined by A* or the *one-sided topological Markov shift defined by A*. It is denoted by (X_A, σ). Similarly, the matrix A defines a closed shift invariant subset of Σ_n. It is the set $\{x \in \Sigma_n : A_{x_i x_{i+1}} = 1, \forall i \in \mathbb{Z}\}$. Here, we think of the set of allowable sequences as the biinfinite paths from G_A. The dynamical system consisting of this space with the subspace topology and the restriction of the shift is called the *two-sided subshift of finite type defined by A* or the *two-sided topological Markov shift defined by A*. It is denoted by (Σ_A, σ). The matrix A is called the *transition matrix* for both X_A and Σ_A. It tells which transitions between states are allowed. A subshift of finite type (either one or two-sided) defined by an irreducible transition matrix is either one periodic orbit or is homeomorphic to the usual middle thirds Cantor set. It has dense periodic points. In the one-sided subshift of finite type the transitive points form a dense G_δ and in the two-sided subshift of finite type the doubly transitive points form a dense G_δ. As for the full shifts the cylinder sets form a countable basis for the topology and so any open set is a countable union of cylinder sets.

Exercises

Let A be an $n \times n$, irreducible, $\{0,1\}$ transition matrix with X_A and Σ_A the subshifts of finite type defined by A.

1. Show that X_A and Σ_A are either finite, consisting of a single σ orbit, or homeomorphic to the standard middle thirds Cantor set.
2. Show that there are countably many cylinder sets in X_A and Σ_A and any open set is a union of cylinder sets.
3. Suppose X_A and Σ_A are infinite. Show that each has countably many periodic points and they are dense.
4. Suppose X_A and Σ_A are infinite. Show that the transitive points form a dense G_δ in X_A and the doubly transitive points form a dense G_δ in Σ_A. (A G_δ is a countable intersection of open sets.)

§1.2 Examples

This section contains examples. They illustrate how subshifts of finite type arise in other areas of dynamics and information theory. They are meant to motivate the discussions which follow. The descriptions are geometric and heuristic. Many of the properties of the examples are stated but not proved. Some properties are proved, some are described and some are left as exercises.

Example 1.2.1 Let I be the unit interval $[0,1]$ and $f : I \to I$ be the function defined by the graph in in Figure 1.2.1. This function is usually referred to as the *full tent map*.

We will see that this map can be modelled by the one-sided 2-shift. Let P_0 be the closed interval $[0,1/2]$ and P_1 be the closed interval $[1/2,1]$. Then $P_0 \cap P_1 = \{1/2\}$. The sets P_0 and P_1 have disjoint interiors. Let $\mathcal{P} = \{P_0, P_1\}$. We will abuse terminology and refer to $\mathcal{P}$ as a "partition" of I. Let $z \in I$ be a point so that $f^k(z) \neq 1/2$ for any k. Then for every k, $f^k(z)$ lies in either P_0 or P_1. We associate with the point z a sequence $x \in \{0,1\}^{\mathbb{N}}$ by choosing x_i so

Figure 1.2.1

that $f^i(z) \in P_{x_i}$. This sequence is the $\mathcal{P}$-*name of* z. The point $1/2 \in I$ has two $\mathcal{P}$-names. One is $(.0100\dots)$ and the other is $(.1100\dots)$. This is because $1/2$ is in both P_0 and P_1, $f(1/2) = 1 \in P_1$ and $f^i(1/2) = 0 \in P_0$ for all $i > 1$. If $z \in I$ and $f^k(z) = 1/2$ then z also has two $\mathcal{P}$-names. They are $(.x_0 \dots x_{k-1}0100\dots)$ and $(.x_0 \dots x_{k-1}1100\dots)$ where x_0 through x_{k-1} are defined by the P_j which contains the corresponding iterate of z. This is similar to assigning to each point in I its diadic decimal expansion.

Observe that $f(P_j) = P_0 \cup P_1 = I, j = 1, 2$. The condition that an element of the partition is mapped by f onto a union of elements of the partition is the *Markov condition* on $\mathcal{P}$. A consequence of this condition is that every sequence in $\{0, 1\}^{\mathbb{N}}$ is the $\mathcal{P}$-name of some point in I. Next observe that for $n \geq 1$, $\mathcal{P} \cap f^{-1}(\mathcal{P}) \cap \dots \cap f^{-n+1}(\mathcal{P})$ is made up of 2^n closed intervals, each of length $(1/2)^n$, with disjoint interiors. This means that no two points in I have the same $\mathcal{P}$-name. We can define a map π from $\{0, 1\}^{\mathbb{N}}$ to I by sending a sequence in $\{0, 1\}^{\mathbb{N}}$ to the unique point in I having this sequence as its $\mathcal{P}$-name. The map π has the following properties:

1. π is continuous;
2. π is onto;
3. $f \circ \pi = \pi \circ \sigma$;
4. π is one-to-one on all transitive points (in fact on all but a countable number of points);
5. π is at most two-to-one.

The map π is continuous because a set of the form $P_{i_0} \cap f^{-1}(P_{i_1}) \cap \dots \cap f^{-n+1}(P_{i_{n-1}})$ in I has diameter $(1/2)^n$. This means two points in $\{0, 1\}^{\mathbb{N}}$ which agree in the first 2^n coordinates are mapped by π to within $(1/2)^n$ of each other in I. The map is onto since every point in I has at least one $\mathcal{P}$-name. The map commutes with f and σ by construction. The point $1/2$ and its preimages under f are the only points with two preimages in $\{0, 1\}^{\mathbb{N}}$. If a point has a dense f orbit in I it is not a preimage of $1/2$. As a result π is one-to-one on all transitive points. The partition $\mathcal{P}$ is an example of a *Markov partition*.

Example 1.2.2 Let I again be the unit interval $[0,1]$ and $f : I \to I$ be the function defined by the graph in Figure 1.2.2.

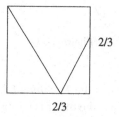

2/3

2/3

Figure 1.2.2

This time let P_0 be the closed interval $[0, 2/3]$, P_1 be the closed interval $[2/3, 1]$ and $\mathcal{P} = \{P_0, P_1\}$. Observe that f maps P_0 onto $P_0 \cup P_1$ and P_1 onto P_0. This is the Markov condition on $\mathcal{P}$. Here, the $\mathcal{P}$-names which occur for points in I never have adjacent ones. Instead of using a full shift to model f on I we use the subshift of finite type defined by the transition matrix

$$A = \begin{pmatrix} 1 & 1 \\ 1 & 0 \end{pmatrix}.$$

Define a map $\pi : X_A \to I$ by sending a sequence in X_A to the unique point in I having the sequence as its $\mathcal{P}$-name. Since the diameters of $\mathcal{P} \cap f^{-1}(\mathcal{P}) \cap \cdots \cap f^{-n}(\mathcal{P})$ go to zero as n goes to infinity, π is well-defined and continuous. The properties 1-5 of π in Example 1.2.1 hold for this π as well. The points with more than one preimage under π are now 2/3 and all its preimages under f.

Example 1.2.3 Let $f : \mathbb{R} \to \mathbb{R}$ be defined by the graph in Figure 1.2.3.

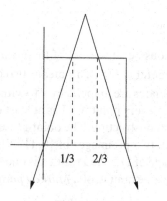

1/3   2/3

Figure 1.2.3

Let I be the unit interval $[0, 1]$ and observe that if x is outside of the unit interval then x goes to negative infinity under iteration by f. The only points that do not go to negative infinity under iteration are the points that always remain in I. We can identify these points. If $x \in (1/3, 2/3)$ then $f(x) \notin I$. If $x \in (1/9, 2/9) \cup (7/9, 8/9)$ then $f(x) \in (1/3, 2/3)$ and $f^2(x) \notin I$. If $x \in I$ but not in $(1/9, 2/9) \cup (1/3, 2/3) \cup (7/9, 8/9)$ then $x, f(x), f^2(x) \in I$. By this reasoning we see that the standard middle thirds Cantor set, $C \subsetneq I$, consists of all points in $\mathbb{R}$ which do not go to negative infinity under iteration by f. The map f takes C onto itself. Let $\mathcal{P} = \{P_0, P_1\}$ be the partition of C with $P_0 = [0, 1/2] \cap C$ and $P_1 = [1/2, 1] \cap C$. Observe that f maps each P_j homeomorphically onto C making $\mathcal{P}$ a Markov partition for f restricted to C. We define the map $\pi : \{0, 1\}^{\mathbb{N}} \to C$ as before. Now notice P_0 and P_1 are disjoint and so π is a homeomorphism.

Example 1.2.4 Let $\mathbb{C}$ be the complex plane. Define a map $f : \mathbb{C} \to \mathbb{C}$ by $f(z) = z^2$. If $z \in \mathbb{C}$ and $|z| > 1$ then $f^n(z) \to \infty$ as $n \to \infty$. If $|z| < 1$ then $f^n(z) \to 0$ as $n \to \infty$. The only interesting dynamics occur on the circle of modulus one, $\mathbb{S}^1 \subset \mathbb{C}$. Here, the points are of the form $e^{2\pi i \alpha}$ for $\alpha \in [0, 1]$ and $f(e^{2\pi i \alpha}) = e^{2\pi i (2\alpha)}$. The map f doubles the angle of a point. Define the partition $\mathcal{P} = \{P_0, P_1\}$ by $P_0 = \{e^{2\pi i \alpha} : \alpha \in [0, 1/2]\}$ and $P_1 = \{e^{2\pi i \alpha} : \alpha \in [1/2, 1]\}$. The partition $\mathcal{P}$ is shown in Figure 1.2.4. This is similar to Example 1.2.1.

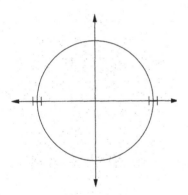

Figure 1.2.4

Now $f(P_0) = f(P_1) = P_0 \cup P_1 = \mathbb{S}^1$. The partition has the Markov condition. The model is the full one-sided two-shift. Define a map $\pi : \{0, 1\}^{\mathbb{N}} \to \mathbb{S}^1$ as before. The two points $(.000\ldots)$ and $(.111\ldots)$ are mapped by π to the point $1 = e^{2\pi i 0}$ on the circle. The two points $(.0111\ldots)$ and $(.1000\ldots)$ are mapped by π to $-1 = e^{2\pi i (1/2)}$ on the circle and all preimages under f of these points have two preimages under π. To be precise we can let $G = \{z \in \mathbb{S}^1 : f^n(z) \neq 1, \forall n \in \mathbb{N}\}$. The complement is a countable set. The $\mathcal{P}$-names of the points in G form a dense subset G' of $\{0, 1\}^{\mathbb{N}}$. The map

$\pi : G' \to G$ is well-defined. It is uniformly continuous so it extends to a map π from all of $\{0, 1\}^{\mathbb{N}}$ onto $\mathbb{S}^1$. Properties 1-5 of Example 1.2.1 hold for the map π here as well.

Example 1.2.5 Again let $\mathbb{C}$ be the complex plane and this time let $f : \mathbb{C} \to \mathbb{C}$ be defined by $f(z) = z^2 + 16$. Let D be the disk in $\mathbb{C}$ centered at the origin with radius 10 (10 is fairly arbitrary). Observe that $f(0) = 16$ and 16 is the only point in $\mathbb{C}$ with one preimage under f. Every other point has two preimages. Observe next that when a point z lies outside of D it goes to infinity under iteration by f. This means that the only points that do not go to infinity under iteration are the points which stay in D for all time. We want to identify these points. To do this notice that $f(4i) = f(-4i) = 0$ and $f^{-1}(D)$ consists of two disjoint disks (not round) D_0 and D_1 lying inside D as shown in Figure 1.2.5.

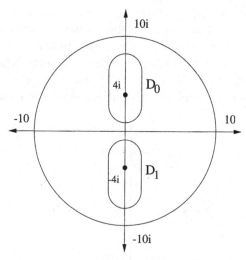

Figure 1.2.5

Next consider $f^{-1}(D_0)$. It consists of two disjoint disks, one lying inside D_0 and one lying inside D_1. This is shown in Figure 1.2.6a. Figure 1.2.6b shows D, $f^{-1}(D)$, and $f^{-2}(D)$.

We continue in this way to arrive at the set $J = \{z : f^n(z) \in D, \forall n \in \mathbb{N}\} = D \cap f^{-1}(D) \cap f^{-2}(D) \cap \cdots$. The set J is homeomorphic to the standard Cantor set. Define a partition $\mathcal{P} = \{P_0, P_1\}$ of J by $P_0 = D_0 \cap J$ and $P_1 = D_1 \cap J$. This partition of J has the Markov property. Each point of J has a unique $\mathcal{P}$-name and the map $\pi : \{0, 1\}^{\mathbb{N}} \to J$ is a homeomorphism with $\pi \circ \sigma = f \circ \pi$.

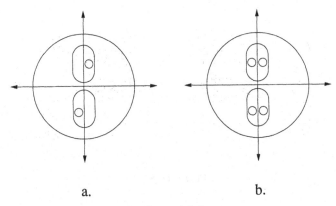

a.   b.

Figure 1.2.6

Example 1.2.6 This example is called *Smale's Horseshoe*. It is described by Figure 1.2.7. Begin with the region of $\mathbb{R}^2$ shown in Figure 1.2.7a labelled D. It is made up of four pieces. The pieces are labelled R_0, R_1, C_0 and C_1. We think of picking up the region D, stretching it vertically as shown in Figure 1.2.7b, bending it around and laying it down inside the original region as shown in Figure 1.2.7c. If this is done properly the map can be extended to a homeomorphism, f, of the entire plane, $\mathbb{R}^2$, to itself.

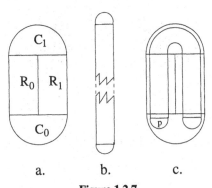

a.   b.   c.

Figure 1.2.7

There must be a fixed point for f in C_0 since C_0 is topologically a two dimensional disk and f maps C_0 inside itself. The fixed point is labelled p. Every point in C_0 and C_1 converges to p under forward iteration by f. The points in $R = R_0 \cup R_1$ which stay in R after one iteration of f are the points in $R \cap f^{-1}(R)$. The set $R \cap f^{-1}(R)$ is made up of two strips which are really four pieces, $R_i \cap f^{-1}(R_j)$ for $i,j \in \{0,1\}$. This is shown in Figure 1.2.8a. If we continue and examine $R \cap f^{-1}(R) \cap f^{-2}(R) \cap \cdots$ we see that it is a cross product of a horizontal line segment with a vertical Cantor set. This is pictured in Figure

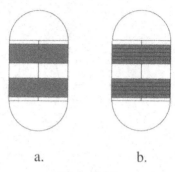

a.   b.

Figure 1.2.8

1.2.8b. Each horizontal line segment corresponds to an infinite intersection $R_{x_0} \cap f^{-1}(R_{x_1}) \cap f^{-2}(R_{x_2}) \cap \cdots$ for a sequence $x \in \{0,1\}^{\mathbb{N}}$.

Next we do the same for the map f^{-1}. Figure 1.2.9a shows the four regions which make up $R \cap f(R) \cap f^2(R)$. The set of points that stay in R for all time under iteration by f^{-1} is a cross product of a vertical line segment with a horizontal Cantor set. It is shown in Figure 1.2.9b.

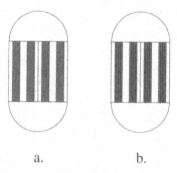

a.   b.

Figure 1.2.9

Each vertical line segment corresponds to an infinite intersection $\cdots \cap f^2(R_{x_2}) \cap f^1(R_{x_1}) \cap R_0$ for a sequence $x \in \{0,1\}^{\mathbb{N}}$. We see that the points that stay in R for all time, both forward and backward, make up the intersection of the two sets described. Denote this set by Ω. It is homeomorphic to the standard Cantor set. Define a partition of $\mathcal{P} = \{P_0, P_1\}$ of Ω by $P_0 = R_0 \cap \Omega$ and $P_1 = R_1 \cap \Omega$. Each point in Ω has a unique $\mathcal{P}$-name which is now a two-sided infinite sequence. The point $z \in \Omega$ has the $\mathcal{P}$-name $x \in \{0,1\}^{\mathbb{Z}}$ where $f^i(z) \in P_{x_i}$ for each $i \in \mathbb{Z}$. The map $\pi : \{0,1\}^{\mathbb{Z}} \to \Omega$ that sends each point x in $\{0,1\}^{\mathbb{Z}}$ to the point in Ω with its $\mathcal{P}$-name x is a homeomorphism and satisfies $\pi \circ \sigma = f \circ \pi$.

Example 1.2.7 This is an example of a toral automorphism. The two dimensional torus, $\mathbb{T}^2$, can be thought of as the unit square in $\mathbb{R}^2$ with opposite edges identified or as $\mathbb{R}^2$ modulo the integer lattice, $\mathbb{Z}^2$. The matrix

$$A = \begin{pmatrix} 2 & 1 \\ 1 & 1 \end{pmatrix}$$

acting on $\mathbb{R}^2$ induces a homeomorphism of the torus to itself because it has integer entries and determinant one. Let f_A denote the induced map on the torus. Figure 1.2.10a shows the image of the unit square in $\mathbb{R}^2$ under A and Figure 1.2.10b shows the pieces outside the unit square translated back into the unit square. This is the image modulo $\mathbb{Z}^2$. It demonstrates that the identifications in the unit square for the torus are preserved by A. So, A does in fact define a homeomorphism, f_A, from the torus to itself.

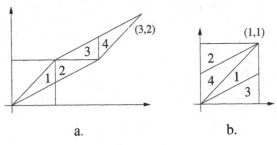

a.   b.

Figure 1.2.10

The matrix A has two eigenvalues $\frac{3+\sqrt{5}}{2} > 1$ and $0 < \frac{3-\sqrt{5}}{2} < 1$. The eigenline for $\frac{3+\sqrt{5}}{2}$ passes through the first quadrant with slope $\frac{-1+\sqrt{5}}{2}$. This will be called the *expanding eigenline*. The eigenline for $\frac{3-\sqrt{5}}{2}$ passes through the fourth quadrant and is perpendicular to the expanding eigenline. It will be called the *contracting eigenline*. These lines are shown in Figure 1.2.11. Figure 1.2.11 also shows five rectangles in $\mathbb{R}^2$, labelled R_1 through R_5. They are constructed so that their edges are parallel to the eigenlines and so that their union is a fundamental region in $\mathbb{R}^2$ for the torus. When projected onto the torus their images cover the torus and do not intersect in their interiors. Let $\mathcal{R} = \{R_1, R_2, R_3, R_4, R_5\}$ be the partition of the torus into the images of the rectangles from $\mathbb{R}^2$. We are using R_i to denote both the rectangle in $\mathbb{R}^2$ and its image in the torus. The boundary of each rectangle is made up of two line segments parallel to the expanding eigenline and two line segments parallel to the contracting eigenline. For each rectangle, we will refer to the first two line segments as the expanding boundary and the other two as the contracting boundary of the rectangle. The boundary of R_i is the expanding boundary of R_i union the contracting boundary of R_i.

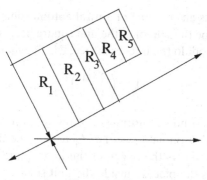

Figure 1.2.11

Figure 1.2.12 shows the images of the rectangles in $\mathbb{R}^2$ under A. Observe that when the image of a rectangle is translated back into the fundamental region, $R_1 \cup \cdots \cup R_5$, the image of each contracting boundary is contained in the contracting boundary of one of the original R_i's.

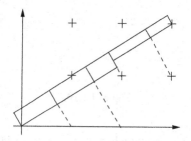

Figure 1.2.12

A similar diagram would show that when A^{-1} is applied to the R_i's the image of each expanding boundary is contained in one of the original expanding boundaries. This is a *Markov condition* on the partition. It is the two-sided analogue of the one-sided condition in Examples 1.2.1 to 1.2.5. For noninvertible maps (as in Examples 1.2.1 to 1.2.5) we require that the image of an element of the partition be equal to the union of elements of the partition. For invertible maps the Markov condition is more subtle. The point of the condition on the images of the boundaries of the elements of the partition is that if $R_{i_0} \cap f_A^{-1}(R_{i_1}) \cap \cdots \cap f_A^{-n}(R_{i_n}) \neq \phi$ and $R_{i_n} \cap f_A^{-1}(R_{i_{n+1}}) \cap \cdots \cap f_A^{-m}(R_{i_{n+m}}) \neq \phi$ then

$$R_{i_0} \cap \cdots \cap f_A^{-n}(R_{i_n}) \cap f_A^{-(n+1)}(R_{i_{n+1}}) \cdots \cap f_A^{-(n+m)}(R_{i_{n+m}}) \neq \phi.$$

The Markov partition, $\mathcal{R}$, allows us to model f_A by a two-sided subshift of finite type. The transition matrix is determined by the way f_A acts on $\mathcal{R}$. The images of R_1, R_2 and R_4 each stretch all the way across R_1, R_2 and R_3. The images of R_3 and R_5 stretch all the way across R_4 and R_5. This means that the transition matrix we need is

$$B = \begin{bmatrix} 1 & 1 & 1 & 0 & 0 \\ 1 & 1 & 1 & 0 & 0 \\ 0 & 0 & 0 & 1 & 1 \\ 1 & 1 & 1 & 0 & 0 \\ 0 & 0 & 0 & 1 & 1 \end{bmatrix}.$$

As before, a point $z \in \mathbb{T}^2$ with $f_A^n(z)$ not in the boundary of an R_i for any n has a well defined $\mathcal{R}$-name. The name will obey the transition rules of B. If a point $z \in \mathbb{T}^2$ lands in the boundary under some iterate of A the point will have more than one $\mathcal{R}$-name obeying the transition rules of B. There are a number of things we need to check, but as in the other examples, the map $\pi : \Sigma_B \to \mathbb{T}^2$ which sends a point $x \in \Sigma_B$ to the unique point $z \in \mathbb{T}^2$ having x as its $\mathcal{R}$-name is well-defined and has following properties:

1. π is continuous;
2. π is onto;
3. $A \circ \pi = \pi \circ \sigma$;
4. π is one-to-one on all doubly transitive points;
5. π is uniformly bounded-to-one.

Example 1.2.8 This example is used for storing information on a magnetic tape or disk. The track on a magnetic tape or disk is a sequence of magnets. The magnets are magnetized to record information and the information can be read back. The information is a sequence of zeros and ones. The zeros and ones are stored at the gaps between the magnets. Two adjacent magnets may either be magnetized in the same or in opposite directions. See Figure 1.2.13. An arrow head points to the north end of a magnet.

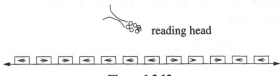

reading head

Figure 1.2.13

The reading head is a coil. When two adjacent magnets are lined up south to south or north to north and the transition passes under the reading head the changing magnetic field induces a current in the coil. The current counts as a one. When two adjacent magnets are lined up north to south and the transition passes under the reading head the magnetic field is unchanging so there is no current induced. This counts as a zero. It is illustrated in Figure 1.2.13. There is also a clock which tells when to look at the signal to see if there is a zero or one. The clock must constantly correct itself so that the signal is sampled at the correct spot. If it drifts too much the signal will be read incorrectly. The clock corrects itself by looking for the peaks in the current. This means it is necessary that the recorded data does not have very long stretches of zeros. There is also a physical constraint on how close the ones can occur on the track. This is due to the properties of the circuitry. If the ones are too close together the circuit will not have time to recover and the peaks will interfere with each other. Let this minimal distance be δ. Think of the raw data as $\{0,1\}^{\mathbb{Z}}$. One way to record information is to make each magnet have length δ. This keeps the ones far enough apart. Then take an arbitrary data sequence and add in a one every so many bits. Then record the new sequence. The added ones allow the clock to keep in phase. When it is read back the added bits can be discarded and the result is the original information. See Figure 1.2.14.

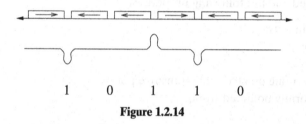

$$1 \qquad 0 \qquad 1 \qquad 1 \qquad 0$$

Figure 1.2.14

There is a more efficient way to record data. It is the method currently used in magnetic tape, magnetic disks and optical disk drives. Instead of making the magnets with length δ make them with length one-half δ. Now we must insure that the recorded string of zeros and ones does not have two consecutive ones. That keeps the ones a distance of at least δ apart on the track. We take an arbitrary data string, from $\{0,1\}^{\mathbb{Z}}$, and recode it. It is recoded so there are never two adjacent ones and so there are never more than seven consecutive zeros. See Figure 1.2.15.

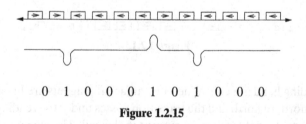

$$0 \quad 1 \quad 0 \quad 0 \quad 0 \quad 1 \quad 0 \quad 1 \quad 0 \quad 0 \quad 0$$

Figure 1.2.15

This is called a *run-length limited code*. We denote the constraint by $(1,7)$. It gives the minimum and maximum number of zeros allowed between adjacent ones. The graph in Figure 1.2.16 describes this constraint.

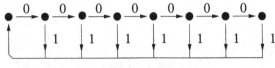

Figure 1.2.16

The graph defines a subshift of finite type with eight states. The allowable strings of zeros and ones are described by reading the labels on the edges. We can pass back and forth between sequences from the subshift of finite type and the constrained sequences of zeros and ones. We are really dealing with a subshift of finite type. We must do the coding in a way that can be decoded. Such codes will be discussed in Section 5.7. There are (1,7) codes with rate 2/3. Rate 2/3 means the code takes two bits of unconstrained data and turns it into three bits of (1,7) constrained data. So a string of raw data of length ℓ gets turned into a string of length 3/2 times ℓ. We have a longer string but we also have shorter magnets. If we store the unconstrained data we get d/δ bits of information into a distance d on the track. If we use the (1,7) code with rate 2/3 and magnets of length 1/2 times δ we have $2d/\delta$ magnets in a distance d on the track and we can store $(2/3)(2d/\delta)$ information bits on the same length of track. It means that we can store 4/3 times the amount of information we could store before. We have increased the capacity of the disk by 33 percent. We will consider this problem more carefully in Section 5.7.

Exercises

1. Show that conditions 1-5 on π in Example 1.2.1 hold for the maps π in Examples 1.2.1, 1.2.2 and 1.2.4.
2. Verify that the maps π in Examples 1.2.3 and 1.2.5 are homeomorphisms. Conclude by showing the two sets C and J are homeomorphic.
3. Show that the map $\pi : \{0,1\}^{\mathbb{Z}} \to \Omega$ in Example 1.2.6 is a well defined homeomorphism and that Ω is homeomorphic to the standard middle thirds Cantor set.
4. Define a map $f : \mathbb{R}^2 \to \mathbb{R}^2$ by $f(x, y) = (y, 1 - ay^2 + bx)$. When $b \neq 0$ the map is a homeomorphism. It is called the Henon mapping. Verify that for sufficiently large a and negative b sufficiently near zero there is a horseshoe (as in example 1.2.6) in $\mathbb{R}^2$.
5. In Example 1.2.7 show that if $R_{i_0} \cap f_A^{-1}(R_{i_1}) \cap \cdots \cap f_A^{-n}(R_{i_n}) \neq \phi$ and $R_{i_n} \cap f_A^{-1}(R_{i_{n+1}}) \cap \cdots \cap f_A^{-m}(R_{i_{n+m}}) \neq \phi$ then

$$R_{i_0} \cap \cdots \cap f_A^{-n}(R_{i_n}) \cap f_A^{-(n+1)}(R_{i_{n+1}}) \cdots \cap f_A^{-(n+m)}(R_{i_{n+m}})$$

is not empty. Verify that if $x \in \Sigma_B$ then $f_A^n(R_{x_{-n}}) \cap \cdots \cap R_{x_0} \cap \cdots \cap f_A^{-n}(R_{x_n})$ is nonempty and the diameter goes to zero as n goes to infinity. Show that $\pi : \Sigma_B \to \mathbb{T}^2$ is well defined and continuous. Conclude by showing that properties 1–5 hold for π.

6. For Example 1.2.8 show that there is a one-to-one correspondence between the points in the two-sided subshift of finite type defined by the directed graph in Figure 1.2.14 and the biinfinite sequences of zeros and ones that obey the (1,7) run-length constraint.

§ 1.3 Perron-Frobenius Theory

In this section we state and discuss the Perron-Frobenius Theorem and then discuss the structure of an arbitrary square, nonnegative matrix. This matrix theory is central to what follows. This is classical matrix theory so proofs are not included. Excellent references are Gantmacher's *Theory of Matrices* and Seneta's *Nonnegative Matrices and Markov Chains*.

In Section 1.1 we associated to a square $\{0, 1\}$ matrix a directed graph. Now let A be a real, nonnegative, square matrix. Associate with it a directed graph with labelled edges. The vertices of the graph are the indices for the rows and columns of the matrix. If $A_{ij} > 0$ there is an edge from vertex i to vertex j labelled A_{ij}. If $A_{ij} = 0$ there is no edge from vertex i to vertex j. Denote the graph by G_A. We also extend the definition of irreducible to real, nonnegative, square matrices. We say such a matrix is *irreducible* if for every pair of indices i and j there is an $\ell > 0$ with $(A^\ell)_{ij} > 0$. Equivalently, the matrix A is irreducible when the graph G_A is strongly connected.

Let A be a real, nonnegative, square matrix. Fix an index i and let $p(i) = gcd\{\ell : (A^\ell)_{ii} > 0\}$. This is the *period of the index i*. When A is irreducible the period of every index is the same and is called the *period of A*. When A is irreducible the period is also the greatest common divisor of the lengths of the closed, directed paths in G_A.

Example 1.3.1 The matrix

$$A = \begin{pmatrix} 1 & 1 \\ 1 & 0 \end{pmatrix}$$

is irreducible and has period one. The matrix

$$B = \begin{bmatrix} 0 & 0 & 0 & 1 & 0 \\ 0 & 0 & 0 & 0 & 1 \\ 0 & 0 & 0 & 1 & 1 \\ 1 & 1 & 0 & 0 & 0 \\ 0 & 0 & 1 & 0 & 0 \end{bmatrix}$$

is irreducible with period two.

If the matrix has period one it is said to be *aperiodic*. Observe that if the matrix A is irreducible and aperiodic there will be a k so that for all $\ell \geq k$ the matrix A^ℓ is strictly positive. First we state the Perron-Frobenius Theorem for aperiodic irreducible matrices and then generalize to arbitrary irreducible matrices.

Aperiodic Perron-Frobenius Theorem 1.3.2 *Suppose A is a nonnegative, square matrix. If A is irreducible and aperiodic there exists a real eigenvalue $\lambda > 0$ such that:*

(a) *λ is a simple root of the characteristic polynomial;*
(b) *λ has strictly positive left and right eigenvectors;*
(c) *the eigenvectors for λ are unique up to constant multiple;*
(d) *$\lambda > |\mu|$, where μ is any other eigenvalue;*
(e) *if $0 \leq B \leq A$ (entry by entry) and β is an eigenvalue for B then $|\beta| \leq \lambda$ and equality can occur if and only if $B = A$;*
(f) *$\lim_{n \to \infty} A^n / \lambda^n = r\ell$ where ℓ and r are the left and right eigenvectors for A normalized so that $\ell \cdot r = 1$.*

The special eigenvalue, λ, is the Perron value *of the matrix A. A positive eigenvector corresponding to λ is called a* Perron eigenvector.

The crucial step in proving this theorem is to show there is a single eigenline passing through the positive quadrant and every nonnegative vector gets closer and closer to this line under iteration by A. It is instructive to think of this fact about the line as a fixed point theorem. Denote the unit simplex in $\mathbb{R}^n$, $\{x \in \mathbb{R}^n : x_i \geq 0, \sum_i x_i = 1\}$, by Δ. Since A is aperiodic we can choose an ℓ so that A^ℓ is strictly positive and define a map α from Δ to itself by $\alpha(x) = \frac{A^\ell x}{|A^\ell x|}$, where $|y| = \Sigma|y_i|$ is the norm. Since A^ℓ is strictly positive α maps Δ strictly inside itself. The image of the boundary is in the interior. We know immediately that there is at least one fixed point for α. A fixed point for α corresponds to an eigenline for A^ℓ. There is in fact a metric which makes α a contraction mapping. For the correct metric there is a δ, $0 < \delta < 1$, with $d(\alpha(x), \alpha(y)) \leq \delta d(x, y)$. The contraction mapping principle then asserts that there is a unique fixed point for α and every point in Δ approaches the fixed point asymptotically. Since $\frac{Ax}{|Ax|}$ will also map Δ into itself the unique fixed point for α corresponds to a unique eigenline for A in the first quadrant. The existence of this line and the asymptotic properties of the other nonnegative vectors guarantee the eigenvalue associated with this line is strictly larger than any other eigenvalue for A. The other statements in the theorem follow.

Example 1.3.3 Let A be the matrix from Example 1.2.7.

$$A = \begin{pmatrix} 2 & 1 \\ 1 & 1 \end{pmatrix}$$

It is irreducible and aperiodic. We have already seen the matrix A has two eigenvalues $\frac{3+\sqrt{5}}{2} > 1$ and $0 < \frac{3-\sqrt{5}}{2} < 1$. The eigenline for $\frac{3+\sqrt{5}}{2}$ passes through the first quadrant with slope $\frac{-1+\sqrt{5}}{2}$ while the eigenline for $\frac{3-\sqrt{5}}{2}$ passes through the fourth quadrant and is perpendicular to the eigenline for $\frac{3+\sqrt{5}}{2}$.

Next we examine arbitrary irreducible matrices. Let A be a real, nonnegative, square, irreducible matrix with period p. Fix an index i and define the *cyclic subsets*, $C_0, \ldots, C_{p-1}$ by

$$C_k = \{j : (A^{np+k})_{ij} > 0, \text{ some } n \in \mathbb{N}\}.$$

Since A is irreducible every index is in some C_k and since A has period p the C_k are disjoint.

Suppose A is an irreducible $\{0,1\}$ transition matrix with period p and G_A is its associated directed graph. Let $C_1, \ldots, C_{p-1}$ be the cyclic subsets of the indices. We also think of the C_k as cyclic subsets of the vertices of G_A. Observe that for $i \in C_k$ every edge which begins at i ends at a vertex in $C_{k+1 \mod p}$. Every cycle in the graph will have length np for some $n \in \mathbb{N}$. This emphasizes the periodic structure of A and G_A.

Example 1.3.4 Let B be the matrix in Example 1.3.1. It has period two with $C_0 = \{1, 2, 3\}$ and $C_1 = \{4, 5\}$. The graph G_B is shown in Figure 1.3.1. All cycles have even length.

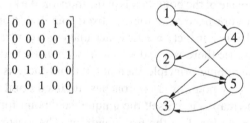

$$\begin{bmatrix} 0 & 0 & 0 & 1 & 0 \\ 0 & 0 & 0 & 0 & 1 \\ 0 & 0 & 0 & 0 & 1 \\ 0 & 1 & 1 & 0 & 0 \\ 1 & 0 & 1 & 0 & 0 \end{bmatrix}$$

Figure 1.3.1

Now we state the Perron-Frobenius Theorem for irreducible matrices.

Perron-Frobenius Theorem 1.3.5 Suppose A is a nonnegative, square matrix. If A is irreducible there exists a real eigenvalue $\lambda > 0$ such that:

(a) λ is a simple root of the characteristic polynomial;
(b) λ has strictly positive left and right eigenvectors;
(c) the eigenvectors for λ are unique up to constant multiple;
(d) $\lambda \geq |\mu|$, where μ is any other eigenvalue;
(e) if $0 \leq B \leq A$ and β is an eigenvalue for B then $|\beta| \leq \lambda$ and equality occurs if and only if $B = A$.

The special eigenvalue, λ, is the *Perron value* of the matrix A. A positive eigenvector corresponding to λ is called a *Perron eigenvector*.

This should be compared to the statement for aperiodic matrices in Theorem 1.3.2. Observe (d) of Theorem 1.3.2 is weakened and strict inequality for the modulus of the eigenvalues is no longer asserted. This will be discussed in Theorems 1.3.6 and 1.3.7. Also notice, statement (f) of Theorem 1.3.2 is dropped. The asymptotic behavior of the powers of A will be described in Theorem 1.3.8. The geometric structure of the eigenvalues is described in the next two observations.

Theorem 1.3.6 *Suppose A is a nonnegative, square, irreducible matrix with period p and Perron value λ. Then A has exactly p complex eigenvalues with modulus λ. Each is a simple root of the characteristic polynomial. They are: $\lambda \exp(i2\pi k/p)$ for $k = 0, \ldots, p-1$ (that is, they are the p roots of $x^p - \lambda^p = 0$).*

Theorem 1.3.7 *Suppose A is a nonnegative, square, irreducible matrix with period p. If μ is an eigenvalue for A then $\mu \exp(i2\pi k/p)$ for $k = 0, \ldots, p-1$ are also eigenvalues. The spectrum of A in the complex plane is invariant under rotation by $2\pi/p$.*

Finally we examine the asymptotic behavior of the powers of A.

Theorem Theorem 1.3.8 *Suppose A is a nonnegative square, irreducible matrix with period p and Perron value λ. Let $C_0, \ldots, C_{p-1}$ be the cyclic subsets of the indices. Then for $i, j \in C_k$*

$$\lim_{n \to \infty} (A^{np}/\lambda^{np})_{ij} = r_i \ell_j$$

where ℓ and r are the left and right eigenvectors for A normalized so that $\ell \cdot r = p$.

Next we describe the structure of an arbitrary nonnegative, square matrix. Except for a few results in Section 1.4 we will not deal with reducible matrices until Chapter 5. Given a nonnegative, square matrix A let I_A denote the set of indices for the rows and columns of A. Define a relation on I_A by saying i is related to j if there are $k, \ell > 0$ with $(A^k)_{ij} > 0$ and $(A^\ell)_{ji} > 0$. Notice that some indices may not be related to themselves or any other index. In terms of G_A the vertex i is related to the vertex j if there is a path from i to j and from j to i. A vertex is not related to itself if there is no path from it to itself.

An index i is *transient* if it is not related to itself. If an index i is not transient it belongs to a relation class. The classes are disjoint. A relation class determines a submatrix of A. The submatrix is composed of all A_{ij} with i and j belonging to the class. This submatrix is called an *irreducible component* of A. It corresponds to a maximal strongly connected subgraph of G_A.

Example 1.3.9 The matrix

$$A = \begin{bmatrix} 1 & 1 & 1 & 0 & 0 \\ 1 & 0 & 0 & 0 & 0 \\ 0 & 0 & 0 & 1 & 1 \\ 0 & 0 & 0 & 0 & 1 \\ 0 & 0 & 0 & 1 & 0 \end{bmatrix}$$

has two irreducible components and one transient state.

Theorem 1.3.10 *Let A be a square nonnegative matrix. Then there is a permutation matrix P so that $P^{-1}AP$ has the form*

$$P^{-1}AP = \begin{bmatrix} A_1 & * & * & \cdots & * \\ 0 & A_2 & * & \cdots & * \\ 0 & 0 & A_3 & \cdots & * \\ \vdots & \vdots & \vdots & \ddots & \vdots \\ 0 & 0 & 0 & \cdots & A_m \end{bmatrix}$$

where A_i is either a nonnegative, square, irreducible matrix corresponding an irreducible component of A or the one by one matrix $[0]$ corresponding to a transient state.

This is proved by examining G_A.

Let A be a nonnegative square matrix. Define the *spectral radius of A* to be $\max\{|\mu| : \mu$ is an eigenvalue for $A\}$.

Corollary 1.3.11 *Let A be a nonnegative square matrix. Then the spectral radius of A is equal to $\max_i\{\lambda_i : \lambda_i$ is the Perron value for the ith irreducible component of $A\}$.*

We say that an irreducible component of A whose Perron value is equal to the spectral radius of A is a *maximal irreducible component*.

Exercises

1. Let $A = \begin{pmatrix} 2 & 1 \\ 1 & 1 \end{pmatrix}$ be the matrix from Example 1.3.3. Define a norm and a metric on $\mathbb{R}^2$ by $||u|| = |u_1| + |u_2|$ and $d(u,v) = ||u - v||$. Let $\Delta = \{u \geq 0 : ||u|| = 1\}$ be the unit simplex in $\mathbb{R}^2$. Think of Δ as $\{\begin{pmatrix} x \\ 1-x \end{pmatrix} : 0 \leq x \leq 1\}$. Define a map $\alpha : \Delta \to \Delta$ by $\alpha(u) = \frac{Au}{||Au||}$. Show that in this metric α is a contraction mapping.

* For the next 5 exercises suppose A is a strictly positive matrix and we have established that the map α described in Exercise 1 is a contraction mapping.

2. Show that A has a positive real eigenvalue, λ, with strictly positive left and right eigenvalues.

3. Show that $\lambda > |\mu|$ where μ is any other eigenvalue for A. Hint: Consider the boundary of the positive quadrant.

4. Show that λ is a simple root of the characteristic polynomial and that the left and right eigenvectors for λ are unique up to constant multiple.

5. Show that $\lim_{n \to \infty} A^n / \lambda^n = r\ell$ where ℓ and r are the left and right eigenvectors for A normalized so that $\ell \cdot r = 1$.

6. Prove that if $0 \leq B \leq A$ (entry by entry) and β is an eigenvalue for B then $|\beta| \leq \lambda$ and equality can occur if and only if $B = A$.

7. Suppose B is a nonnegative, irreducible, square matrix with period p. Assume that B^p is a block diagonal matrix with the indices of the kth diagonal block indexed by the indices in the kth cyclic subset C_i. Apply the Aperiodic Perron-Frobenius Theorem 1.3.2 to each of the diagonal blocks of B^p and use the results to prove the Perron-Frobenius Theorem 1.3.5.

8. Prove Theorem 1.3.10.

§1.4 Basic Dynamics

In this section we will discuss some of the basic dynamical properties of sub-shifts of finite type. These include topological entropy, periodic points, block maps and an alternative definition of subshifts of finite type.

Let A be an $n \times n$, $\{0,1\}$ transition matrix. In Section 1.1 we defined X_A to be the one-sided subshift of finite type defined by A and Σ_A to be the two-sided subshift of finite type defined by A. The *alphabet* of X_A and Σ_A is the n element set which indexes the rows and columns of A. We denote this set by L_A. Its elements are *states* or *symbols*. Let G_A denote the directed graph associated with A, V_A denote the vertices of G_A and E_A denote the edges of G_A. An allowable *word* or *block* of length ℓ is $[i_0, \ldots, i_{\ell-1}]$ where $A_{i_r i_{r+1}} = 1$ for $r = 0, \ldots, \ell - 2$. Observe that $(A^\ell)_{ij}$ is the number of allowed words of length $\ell + 1$ beginning with i and ending with j. Define $\mathcal{W}(A, \ell)$ to be the set of allowable words of length ℓ. So $\mathcal{W}(A, 1) = L_A$ and the cardinality of $\mathcal{W}(A, \ell)$ is $\sum_{i,j} (A^{\ell-1})_{ij}$. Define $\mathcal{W}(A)$ to be the set of all allowable words. It is the union of all the $\mathcal{W}(A, \ell)$. A cylinder set for X_A or Σ_A is $[i_0, \ldots, i_{\ell-1}]_t = \{x : x_t = i_0, \ldots, x_{i+\ell-1} = i_{\ell-1}\}$ where $[i_0, \ldots, i_{\ell-1}]$ is in $\mathcal{W}(A)$ and t is in either $\mathbb{N}$ or $\mathbb{Z}$. The cylinder sets form a countable open-closed basis for the topology and every open set is a union of cylinder sets.

In Section 1.1 we defined a point in X_A or Σ_A to be transitive if its orbit under σ is dense and a point in Σ_A to be doubly transitive if its forward and backward σ orbits are dense. We say a subshift of finite type is *transitive* if it contains a transitive point. A point x in a subshift of finite type is *nonwandering* if for every open set $\mathcal{U}$ containing x there is an $n \neq 0$ with $\sigma^n(\mathcal{U}) \cap \mathcal{U} \neq \phi$. A subshift of finite type is said to be *nonwandering* if every point is nonwandering. Our first result in this section ties these dynamical ideas to the irreducibility of the transition matrix.

Theorem 1.4.1 *Let A be a $\{0, 1\}$ transition matrix. Then*

(i) X_A *is transitive if and only if A is irreducible,*
(ii) Σ_A *is transitive and nonwandering if and only if A is irreducible and*
(iii) Σ_A *contains a doubly transitive point if and only if A is irreducible.*

Proof. (*i*) Consider the graph G_A. If X_A is transitive then it is possible to go from any vertex of G_A to any other so G_A is strongly connected and A is irreducible. The converse holds because we can list all words which occur in X_A and when G_A is strongly connected we can construct a point in X_A that contains every word in the list.

(*ii*) Suppose Σ_A is transitive and nonwandering. Let $i, j \in L_A$. We want to see that in G_A there is a path from i to j. Transitivity means that there is a path from i to j or from j to i. Suppose there is a path $[j, x_1, \ldots, x_k, i]$. Since Σ_A is nonwandering there is an n with arbitrarily large absolute value and $\sigma^n([j, x_1, \ldots, x_k, i]) \cap [j, x_1, \ldots, x_k, i] \neq \phi$. This means that there is a path from i to j. If A is irreducible the argument for (*i*) shows that Σ_A is transitive. If A is irreducible Σ_A is nonwandering by the same type of reasoning.

(*iii*) Suppose Σ_A contains a doubly transitive point. Then G_A is strongly connected and A is irreducible. If A is irreducible then we can construct a doubly transitive point for Σ_A just as we constructed a transitive point for X_A in the proof of (*i*). □

If A is an irreducible transition matrix we say X_A and Σ_A are *irreducible*.

The topological entropy of a subshift of finite type measures the exponential growth rate of the number of allowable words. From the dynamics viewpoint it is a measure of the randomness of the dynamical system. From the information theory point of view it is a measure of how much information can be stored in the set of allowed sequences. Define the *topological entropy* of X_A and Σ_A to be

$$\lim_{\ell \to \infty} \frac{1}{\ell} \log |\mathcal{W}(A, \ell)|,$$

where $|\cdot|$ denotes the cardinality of the set. Denote by $h(X_A, \sigma)$ the topological entropy of the shift acting on X_A and by $h(\Sigma_A, \sigma)$ the topological entropy of the shift acting on Σ_A. The Perron-Frobenius Theorem 1.3.5 allows us to compute this limit.

Observation 1.4.2 *Let A be a transition matrix. Then $h(X_A, \sigma) = h(\Sigma_A, \sigma) = \log \lambda$, where λ is the spectral radius of A. If A is irreducible, then λ is the Perron value of A.*

Proof. First suppose A is irreducible. By the Perron-Frobenius Theorem 1.3.5 there exist constants c_1, c_2 with

$$c_1 \lambda^\ell \leq |\mathcal{W}(A, \ell)| \leq c_2 \lambda^\ell$$

for all sufficiently large ℓ. This gives the desired result in the irreducible case.

Next suppose A is reducible. Since $|\mathcal{W}(A, \ell)| = \Sigma(A^{\ell-1})_{ij}$ we can use the result for irreducible matrices and Corollary 1.3.11 to see that the entropy is greater than or equal to $\log \lambda$ where λ is the spectral radius of A. We can then apply Theorem 1.3.10 and the Perron-Frobenius Theorem 1.3.5 to see that there is a constant c such that $|\mathcal{W}(A, \ell)| \leq c \lambda^\ell$ for all sufficiently large ℓ. This shows that the entropy is less than or equal to $\log \lambda$. $\quad\square$

In Section 1.1 we saw that the periodic points are dense in X_A and Σ_A. Now we obtain more information about their numbers. We know that A_{ii}^k is the number of allowable words of length $k + 1$ beginning and ending at i. Each block of length $k + 1$ beginning and ending with the same symbol defines a point fixed by the kth power of the shift in both X_A and Σ_A. The converse is also true: a point fixed by the kth power of the shift in X_A or in Σ_A determines a block in $\mathcal{W}(A, k + 1)$ beginning and ending with the same symbol. There is a one-to-one correspondence between the points fixed by the kth power of the shift in X_A, the points fixed by the kth power of the shift in Σ_A and the blocks in $\mathcal{W}(A, k + 1)$ beginning and ending with the same symbol. If $N(A, k)$ is the number of points fixed by the kth power of the shift in X_A and Σ_A then it is equal to the trace of A^k. If we let $tr A^k$ denote the trace of A^k and $\{\mu_1, \dots, \mu_n\}$ be the roots of the characteristic polynomial of A, with multiplicity, then $N(A, k) = tr A^k = \sum \mu_j^k$. The Perron-Frobenius Theorem 1.3.5 yields the following:

Observation 1.4.3 *If A is an irreducible transition matrix then the number of points fixed by the kth power of the shifts in X_A and Σ_A, called $N(A, k)$, has the property that*

$$\lim_{k\to\infty} \frac{N(A, kp)}{\lambda^{kp}} = \lim_{k\to\infty} \frac{trA^{kp}}{\lambda^{kp}} = p,$$

where λ is the Perron value of A and p is the period of A.

We can recover the number of points of each period from the sequence $\{N(A, k)\}$. The zeta function provides a convenient way to keep track of the $N(A, k)$. Define the *zeta function* of X_A and Σ_A to be

$$\zeta_A(t) = \exp\left[\sum_{k=1}^{\infty} \frac{N(A, k)}{k} t^k\right].$$

Notice that all the $N(A, k)$ can be recovered from the function $\zeta_A(t)$.

Observation 1.4.4 *If A is an irreducible transition matrix then the zeta function of X_A and Σ_A is $\zeta_A(t) = \det(I - tA)^{-1}$.*

Proof. This is a computation. We already know $N(A, k) = trA^k = \sum_j \mu_j^k$ where $\{\mu_1, \dots, \mu_n\}$ are the roots of the characteristic polynomial of A, with multiplicity. Then

$$\zeta_A(t) = \exp\left[\sum_{k=1}^{\infty} \frac{N(A, k)}{k} t^k\right] = \exp\left[\sum_{k=1}^{\infty}\sum_{j=1}^{n} \frac{(\mu_j t)^k}{k}\right]$$

$$= \exp\left[\sum_{j=1}^{n}\sum_{k=1}^{\infty} \frac{(\mu_j t)^k}{k}\right]$$

$$\overset{*}{=} \exp\left[-\sum_{j=1}^{n} \log(1 - \mu_j t)\right]$$

$$= \prod_{j=1}^{\infty}(1 - \mu_j t)^{-1} = \det(I - tA)^{-1}.$$

The step $(*)$ follows because

$$\log_e(1 - x) = \int_x^0 \frac{1}{1-t} dt = \int_x^0 \sum_{k=0}^{\infty} t^k dt = \sum_{k=1}^{\infty} \frac{t^k}{k}\bigg]_x^0 = -\sum_{k=1}^{\infty} \frac{x^k}{k}. \qquad \square$$

Let $c_A(x)$ be the characteristic polynomial of A. Define $c_A^*(x)$ to be $c_A(x)$ with all powers of x factored out. The polynomial $c_A^*(x)$ has the same nonzero roots with the same multiplicities as $c_A(x)$ but zero is not a root. The polynomial $c_A^*(x)$ is called the *characteristic polynomial of A away from zero*. Suppose the degree of $c_A^*(x)$ is d and observe that $det(I - tA) = \prod_{j=1}^{n}(1 - t\mu_j) = t^d c_A^*(1/t)$. By our formula for the zeta function we obtain the following:

Lemma 1.4.5 *The number of points of each period in X_A and Σ_A is determined by the characteristic polynomial of A away from zero.*

Example 1.4.6 Let A be as in Example 1.1.1 and B be as in Example 1.2.7.

$$A = \begin{bmatrix} 1 & 1 \\ 1 & 0 \end{bmatrix} \qquad B = \begin{bmatrix} 1 & 1 & 1 & 0 & 0 \\ 1 & 1 & 1 & 0 & 0 \\ 0 & 0 & 0 & 1 & 1 \\ 1 & 1 & 1 & 0 & 0 \\ 0 & 0 & 0 & 1 & 1 \end{bmatrix}.$$

Then

$$c_A^*(x) = x^2 - x - 1, \qquad \zeta_A(t) = \frac{1}{1 - t - t^2}$$

and

$$c_B^*(x) = x^2 - 3x + 1, \qquad \zeta_B(t) = \frac{1}{1 - 3t + t^2}.$$

Suppose X_A and X_B are two one-sided subshifts of finite type. A *block map* from X_A to X_B is determined by a map $\varphi_0 : \mathcal{W}(A, \ell) \to L_B$ that is *consistent with A and B*. The map φ_0 is consistent with A and B if for every word of length $\ell + 1$, $[i_0, \dots, i_\ell] \in \mathcal{W}(A, \ell + 1)$, $[\varphi_0([i_0, \dots, i_{\ell-1}]), [\varphi_0([i_1, \dots, i_\ell])] \in \mathcal{W}(B, 2)$. The map φ_0 determines a map $\varphi : X_A \to X_B$ by $(\varphi(x))_i = \varphi_0([x_i, \dots, x_{i+\ell-1}])$. This maps X_A into X_B since φ_0 is consistent with A and B. We'll say such a map is an ℓ-*block map*. In practice, we will abuse notation and use φ for both the map on blocks and on points in X_A. A block map between two subshifts of finite type is continuous and commutes with the shifts.

Example 1.4.7 Let $X_A = X_B = \{0, 1\}^{\mathbb{N}}$ and $\varphi_0 : \mathcal{W}(A, 2) \to L_B$ be $\varphi_0([j, k]) = j + k \mod 2$. So $(\varphi(x))_i = x_i + x_{i+1} \mod 2$, φ maps X_A onto X_B and is exactly two-to-one.

If Σ_A and Σ_B are two-sided subshifts of finite type we have another degree of freedom in defining a block map. As in the one-sided case we start with a map on blocks $\varphi_0 : \mathcal{W}(A, \ell) \to L_B$ that is consistent with A and B. Now choose two integers m and a with $m + a + 1 = \ell$. Define $\varphi : \Sigma_A \to \Sigma_B$ by $(\varphi(x))_i = \varphi_0([x_{i-m}, \dots, x_{i+a}])$. We think of m as the memory and a as the anticipation of φ. This seemingly minor difference will have profound consequences when we examine shift-commuting homeomorphisms between subshifts of finite type. Another way to view this is to define $\varphi' : \Sigma_A \to \Sigma_B$ by $(\varphi'(x))_i = \varphi_0([x_i, \ldots, x_{i+\ell-1}])$ and then let $\varphi = \sigma^{-m} \circ \varphi'$. This cannot be done for one-sided shifts because σ is not invertible on one-sided shifts.

Example 1.4.8 Let $\Sigma_A = \Sigma_B = \{0, 1\}^{\mathbb{Z}}$, and φ_0 be the same as in in Example 1.4.7. Let $m = -2$ and $a = 3$. Then $\varphi(x)_i = x_{i+2} + x_{i+3} \mod 2$ and φ is exactly two-to-one and onto. We could also let $\varphi'(x)_i = x_i + x_{i+1} \mod 2$ and get $\varphi = \sigma^2 \circ \varphi'$. Look at the one-sided case. The φ' here is exactly the φ of Example 1.4.7. The map $\sigma^2 \circ \varphi'$ can be applied to $\{0, 1\}^{\mathbb{N}}$ but it is an onto eight-to-one map.

Next we have the most basic theorem of symbolic dynamics. It is true for both one- and two-sided subshifts of finite type.

Theorem 1.4.9 *If φ is a continuous, shift-commuting map between two subshifts of finite type then it is a block map.*

Proof. The proof is essentially the same for both the one- and two-sided shifts so we will prove only the one-sided version. Let X_A and X_B be subshifts of finite type and φ a continuous shift-commuting map between them. Let $j \in L_B$. The cylinder set $[j]_0$ is open and closed. Since φ is continuous, $\varphi^{-1}([j]_0)$ is both open and closed in X_A. Since it is open it is a countable union of cylinder sets. Since it is closed it is compact and so is equal to a finite union of these cylinder sets. By splitting these cylinder sets into longer cylinder sets we can assume that $\varphi^{-1}([j]_0)$ is a disjoint union of cylinder sets, all of the same length.

$$\varphi^{-1}([j]_0) = \bigcup_{r=1}^{N(j)} [i_0^r, \ldots, i_{\ell(j)-1}^r]_0$$

for some $N(j)$ and $l(j)$. This is true for each $j \in L_B$. We take $\ell = max\{\ell(j)\}$ and define $\varphi_0 : \mathcal{W}(A, \ell) \to L_B$ by saying $\varphi_0([i_0, \ldots, i_{\ell-1}]) = j$, where j is the unique element of L_B with

$$[i_0, \ldots, i_{\ell-1}]_0 \subseteq \varphi^{-1}([j]_0).$$

This partitions $\mathcal{W}(A, \ell)$. The map on words produces the original φ. □

There is a natural "relabelling" of a subshift of finite type. Let A be a transition matrix with alphabet L_A. We define a new transition matrix $A^{[\ell]}$ with alphabet $L_A^{[\ell]} = L_{A^{[\ell]}} = \mathcal{W}(A, \ell)$. The transitions for $A^{[\ell]}$ are defined by

$$(A^{[\ell]})_{[i_0, \ldots, i_{\ell-1}][j_0, \ldots, j_{\ell-1}]} = 1$$

<   2 1 3 2 1 1 2 . 1 1 2 3 1 2 3 1 3 3

<   2 1 3 2 1 1 2 . 1 1 2 3 1 2 3 1 3 3
         1 3 2 1 1 2 1   1 2 3 1 2 3 1 3 3
         3 2 1 1 2 1 1   2 3 1 2 3 1 3 3

Figure 1.4.1

if and only if $i_r = j_{r-1}$ for $r = 1, \ldots, \ell - 1$. This means that the two blocks can overlap when j_0 is placed on i_1. Consider the two one-sided shifts X_A and $X_{A^{[\ell]}}$. There is a natural correspondence between their points. It is illustrated in Figure 1.4.1 where the sequence above the line is a point in X_A and the triple sequence below the line is a point in $X_{A^{[3]}}$.

On the two-sided shifts there is also a natural correspondence but once more there is one more degree of freedom. We can shift where we place the blocks by composing with a power of the shift. This means that the transition matrix is still canonical but the map is canonical only up to a power of the shift. This extra degree of freedom will be used when we examine conjugacies in the next section. These are called *higher block presentations*. When the new subshift of finite type is made using ℓ-blocks we say it is the *ℓ-block presentation*.

There is an alternative definition of a subshift of finite type that is often useful. Start with a full n-shift, either one- and two-sided. Let B be a finite collection of words in the full n-shift. Define $X_{\bar{B}}$ to be the set $\{x \in X_n :$ no subword of x is in $B\}$. The set B is a list of forbidden words. It is easily seen that $X_{\bar{B}}$ can be identified with a subshift of finite type defined by a transition matrix. First, we can assume that all the words in B are the same length, say ℓ. Let $L_A = \{1, \ldots, n\}^\ell - B$. Think of these blocks as the "good" blocks. L_A is the set of allowed words of length ℓ. The space $X_{\bar{B}}$ consists of the points with every block of length ℓ in L_A. Define a transition matrix, A, with indices L_A by

$$A_{[i_0, \ldots, i_{\ell-1}][j_0, \ldots, j_{\ell-1}]} = 1.$$

if and only if $i_r = j_{r-1}$ for $r = 1, \ldots, \ell - 1$. Then X_A is the ℓ-block presentation of $X_{\bar{B}}$. Exactly the same is true for the two-sided shifts.

Example 1.4.10 Begin with the full two-shift. Let $B = \{[1,1], [0,0,0]\}$ then $L_A \subseteq \mathcal{W}([1^{(2)}], 3)$ is $\{[0,0,1], [0,1,0], [1,0,0], [1,0,1]\}$ and

$$A = \begin{bmatrix} 0 & 1 & 0 & 0 \\ 0 & 0 & 1 & 1 \\ 1 & 0 & 0 & 0 \\ 0 & 1 & 0 & 0 \end{bmatrix}.$$

Exercises

1. Show that the limit in the definition of entropy always exists and is equal to the infimum of $\{\frac{1}{\ell}\log|\mathcal{W}(A,\ell)|\}$. Use the fact that $|\mathcal{W}(A,k+\ell)| \leq |\mathcal{W}(A,k)||\mathcal{W}(A,\ell)|$.
2. Compute the zeta function for the matrix A in Example 1.4.10.
3. There is a product formula for the zeta function. The relationship between the sum and product formulas is the reason the function is called the zeta function. Fix a subshift of finite type X_A or Σ_A. Let $\mathcal{O}(A)$ be the set of all its periodic orbits. For an orbit $o \in \mathcal{O}(A)$ let $\ell(o)$ denote its length; that is, the number of points in the orbit. Define

$$\zeta(t) = \left[\prod_{o \in \mathcal{O}(A)} (1 - t^{\ell(o)})\right]^{-1}$$

Prove that the sum and product formulas define the same function. Hint: Use the * identity used in the proof of Observation 1.4.4.
4. Prove Theorem 1.4.9 for two-sided subshifts of finite type.
5. Show that the map φ in Example 1.4.7 is onto and exactly two-to-one.
6. Prove that the map from X_A to its ℓ-block presentation, $X_{A^{[\ell]}}$, is a shift-commuting homeomorphism.
7. Show that the maps $\varphi_0, \psi_0 : L_{A^{[\ell]}} \to L_A$ defined by $\varphi_0([i_0, \ldots, i_{\ell-1}]) = i_0$ and $\psi_0([i_0, \ldots, i_{\ell-1}]) = i_{\ell-1}$ produce two different maps $\varphi, \psi : \Sigma_{A^\ell} \to \Sigma_A$ that are both shift-commuting homeomorphisms from the ℓ-block presentation of Σ_A to Σ_A.
8. In the alternative definition of subshift of finite type show that the identification between points in $X_{\tilde{B}}$ and points in X_A defines a shift-commuting homeomorphism between the spaces.

Notes

Section 1.1

Infinite symbolic sequences were first used in dynamics by J. Hadamard in 1898 [Hd] to study geodesics on a family of surfaces with constant negative curvature. Over the next forty years mathematicians used symbolic sequences to examine similar geometric problems, questions about trajectories arising from differential equations, and questions about the combinatorics of the sequences themselves. The term *Symbolic Dynamics* was first used as the title of a paper by G.A. Hedlund and M. Morse in 1938 [MH]. There they examined the shift transformation restricted to a closed, invariant subset of a full shift as a dynamical system. It was the first time the shift transformation acting on a subshift was explicitly considered in this light.

Subshifts of finite type are the combinatorial or topological version of the finite state Markov shifts in classical probability theory [Fe]. C. Shannon in 1948 [S] considered an information channel as a subshift of finite type supporting Markov probability measures and investigated some of the asymptotic combinatorial properties of the subshift of finite type. In 1964 W. Parry [P1] used the notion of a $\{0, 1\}$ transition matrix and considered subshifts of finite type from the ergodic theory point of view as the underlying space for shift invariant probability measures. He investigated many of the dynamical properties of these measures. In 1968 and 1969 G.A. Hedlund wrote two papers [H1] and [H2] describing results about continuous, shift commuting maps from a full shift to itself. The results were obtained by him and others while working on a contract with the Institute for Defense Analysis in Princeton, New Jersey over a period of ten years. Following this, subshifts of finite type were found to arise as models for maps of the interval, as models for hyperbolic automorphisms of the two-torus, as invariant subsets of differentiable dynamical systems and as dynamical systems of interest for their own sake.

Section 1.2

Markov partitions for maps of the interval and circle were first used by W. Parry in the mid-1960's [P2]. He set up the correspondence between noninvertible maps of the interval or circle that admit Markov partitions and one-sided subshifts of finite type, and then used this correspondence to examine dynamical properties of the maps.

The circle in Example 1.2.4 and the set J in Example 1.2.5 are examples of *Julia sets*. Both sets are dynamically defined by quadratic maps of the complex plane with the form $f(z) = z^2 + c$. The dynamics of these maps were investigated by P. Fatou [Fa] and G. Julia [Ju] before 1920 and they continue to be a subject of active research.

The horseshoe of Example 1.2.6 was defined by S. Smale in the early-1960's and a complete description of it can be found in 1965 [Sm] where he showed that f restricted to the set Ω can be identified with two-sided two-shift. The set Ω together with the fixed point p is the *nonwandering set* for the map f.

Markov partitions for automorphisms of the two dimensional torus as in Example 1.2.7 were first discovered by K. Berg in his Ph.D thesis [Be] in 1967. He found the Markov partitions and set up the correspondence between the automorphism acting on the torus and a two-sided subshift of finite type. R. Adler and B. Weiss in 1967 [AW1] and [AW2] also found Markov partitions for automorphisms of the two dimensional torus. Following this, Y. Sinai [Si1],

R. Bowen [Bo1] and D. Fried [Fd] found Markov partitions for increasingly more general classes of maps.

Subshifts of finite type were used by C. Shannon [S] in 1948 to model finite memory information channels and the formulation for run-length limited codes as presented in Example 1.2.8 was known to electrical engineers by 1970 [Fz]. We will examine this type coding problem carefully in Chapter 5.

Section 1.3

An historical discussion of the Perron-Frobenius Theorem and references to the original papers can be found on page 25 of E. Seneta's *Non-negative Matrices and Markov Chains, 2nd ed.* [Se].

Section 1.4

Topological entropy was defined for an information channel and computed for subshifts of finite type by C. Shannon in 1948 [S], where he called it the *capacity* of a channel. W. Parry [P2] in 1966 defined topological entropy for the shift transformation acting on any closed, shift invariant subset of a full shift. He called it the *absolute entropy* and had already computed it for subshifts of finite type in 1964 [P1]. R. Adler, A. Konheim and M. McAndrew defined topological entropy for any continuous map of a compact metric space to itself in 1965 [AKM] and computed it for several examples. Their definition was the topological analog of the definition of measure-theoretic entropy (see Chapter 5) which was formulated by A. Kolmogorov [Ko] and Y. Sinai [Si2] in the late 1950's. The Perron-Frobenius Theorem shows that the entropy of an irreducible and aperiodic subshift of finite type must be the logarithm of an algebraic integer greater than or equal to one and it must be strictly larger than the absolute value of its other conjugates. In 1983 D. Lind [Li1], [Li2] showed that any such number can occur as the largest eigenvalue for of a transition matrix.

The zeta function for a dynamical system was introduced by M. Artin and B. Mazur in 1965 [ArM]. In 1970 R. Bowen and O. Lanford [BL] proved the formula for the zeta function which appears in Observation 1.4.4.

Theorem 1.4.9 was proved by M. Curtis, G.A. Hedlund and R. Lyndon sometime before 1967 and can be found in [H1] and [H2].

References

[AKM] R. Adler, A. Konheim and M.H. MacAndrew, *Topological Entropy*, Transactions of the American Mathematical Society no. 114 (1965).

[AW1] R. Adler and B. Weiss, *Entropy, a Complete Metric Invariant for Automorphisms of the Torus*, Proceedings of the National Academy of Sciences, USA no. 57 (1967), 1573–1576.

[AW2] R. Adler and B. Weiss, *Similarity of Automorphisms of the Torus*, Memoirs of the American Mathematical Society no. 98 (1970).

[ArM] M. Artin and B. Mazur, *On Periodic Points*, Annals of Mathematics **81** (1965).

[Be]   K. Berg, *On the Conjugacy Problem for K-systems*, Ph. D Thesis, University of Minnesota, 1967.

[Bo1] R. Bowen, *Markov Partitions for Axiom A Diffeomorphisms*, American Journal of Mathematics **92** (1970) 725–747.

[BL]   R. Bowen and O. Lanford, *Zeta Functions of Restrictions of the Shift Transformation*, in Global Analysis, Proceedings of Symposia in Pure and Applied Math (S-S. Chern and S. Smale, eds.), vol. XIV, American Mathematical Society, 1970, pp. 43–49.

[Fa]   P. Fatou, *Sur les Equations Fonctionnelles*, Bulletin de la Société Mathématique de France **47** (1919), 161–247.

[Fe]   W. Feller, *An Introduction to Probability Theory and Its Applications*, 3rd. edition, Wiley, 1968.

[Fz]   P. Franaszek, *Sequence State Methods for Run-Length Limited Codes*, IBM Journal of Research and Development **14** (1970), 376–383.

[Fd]   D. Fried, *Finitely Presented Dynamical Systems*, Ergodic Theory and Dynamical Systems **7** (1987), 489–507.

[G]   W. Gantmacher, *The Theory of Matrices*, vol. 1, Chelsea Publishing Co., 1959.

[Hd]   J. Hadamard, *Les surfaces à courbures opposées et leurs lignes géodésiques*, Journal de Mathématiques Pures et Appliquées **4** (1898), 27–73.

[H1]   G.A. Hedlund, *Transformations Commuting with the Shift*, Topological Dynamics (J. Auslander and W. Gottschalk, eds.), W.A. Benjamin, 1968.

[H2]   G.A. Hedlund, *Endomorphisms and Automorphisms of the Shift Dynamical System*, Mathematical Systems Theory **3** no. 4 (1969), 320–375.

[Ju]   G. Julia, *Iteration des Applications Fonctionnelles*, Journal de Mathématiques Pures et Appliquées **8** (1918), 47–245, reprinted in *Oeuvres de Gaston Julia*, Gauthier-Villars, volume I, 121–319.

[Ko]   A. Kolmogorov, *A New Metric Invariant for Transient Dynamical Systems*, Academiia Nauk SSSR, Doklady **119** (1958), 861–864. (Russian)

[Li1] D. Lind, *Entropies and Factorizations of Topological Markov Shifts*, Bulletin of the American Mathematical Society **9** (1983), 219–222.

[Li2] D. Lind, *The Entropies of Topological Markov Shifts and a Related Class of Algebraic Integers*, Ergodic Theory and Dynamical Systems **4** (1984), 283–300.

[MH]   M. Morse and G.A. Hedlund, *Symbolic Dynamics*, American Journal of Mathematics **60** (1938), 815–866.

[P1]   W. Parry, *Intrinsic Markov Chains*, Transactions of the American Mathematical Society **112** (1964), 55–66.

[P2]   W. Parry, *Symbolic Dynamics and Transformations of the Unit Interval*, Transactions of the American Mathematical Society **122** (1966), 368–378.

[Se]   E. Seneta, *Non-negative Matrices and Markov Chains*, Springer-Verlag, 1981.

[S]   C. Shannon, *A Mathematical Theory of Communication*, Bell System Technical Journal (1948), 379–473 and 623–656, reprinted in *The Mathematical Theory of Communication* by C. Shannon and W. Weaver, University of Illinois Press, 1963.

[Si1]   Y. Sinai, *Construction of Markov Partitions*, Funkcional'nyi Analiz i Ego Prilozheniya **2** no. 3 (1968), 70–80 (Russian); English transl. in Functional Analysis and Its Applications **2** (1968), 245–253.

[Si2]   Y. Sinai, *On the Concept of Entropy for a Dynamical System*, Academiia Nauk SSSR, Doklady **124** (1959), 768–771. (Russian)

[Sm]   S. Smale, *Diffeomorphisms with Many Periodic Points, Differential and Combinatorial Topology*, Princeton University Press, 1965, pp. 63–80.

Chapter 2. Topological Conjugacy

In this chapter we investigate the problem of determining when two subshifts of finite type are the "same". Two subshifts of finite type are dynamically the same if there is a homeomorphism between them which commutes with the shifts. We examine this problem for both one and two-sided subshifts of finite type. In the one-sided setting we develop a simple algorithm which allows us to determine when two one-sided subshifts of finite type are the same. In the two-sided setting we will see that it is not known whether such an algorithm can exist.

In the first section we show a homeomorphism between two subshifts of finite type which commutes with the shifts can be decomposed into a sequence of elementary maps. The decomposition is mirrored in a sequence of elementary matrix equations between the defining transition matrices. These equations allow us to develop an algorithm which will determine when two one-sided subshifts of finite type are the same. The equations do not lead to such an algorithm for two-sided subshifts of finite type. In the second section we see how the equations developed in section one for two-sided subshifts of finite type lead to the notion of shift equivalence. Then we examine some of the algebraic consequences of shift equivalence between two transition matrices. These include the relationship between the Jordan forms away from zero, the Bowen-Franks groups, some inverse limit spaces and the dimension groups.

§ 2.1 Decomposition of Topological Conjugacies

Two subshifts of finite type are *topologically conjugate* or simply *conjugate* if there is a homeomorphism between the two which commutes with the shifts. The homeomorphism is called a *conjugacy*. The problem of determining when two subshifts of finite type are conjugate is quite different in the one and two-sided settings.

For a subshift of finite type defined by the transition matrix A with alphabet L_A the *predecessor set* of a symbol $j \in L_A$ is $p(j) = \{i \in L_A : [i,j] \in \mathcal{W}(A,2)\}$.

The *follower set* or *successor set* of a symbol $j \in L_A$ is $f(j) = \{k \in L_A : [j,k] \in W(A,2)\}$. Sometimes we use $p_A(j)$ and $f_A(j)$ to avoid confusion when working with more than one subshift of finite type.

First we concentrate on one-sided subshifts of finite type and then move to the two-sided ones. We begin with an example:

Example 2.1.1 Let X_3 be the full 3-shift with alphabet $L_3 = \{1,2,3\}$. We will construct another conjugate subshift of finite type. It is done by "splitting" the state 1 into two new states, 1_1 and 1_2. The new subshift of finite type is X_B with alphabet $L_B = \{1_1, 1_2, 2, 3\}$. The transitions are defined by $f(1_1) = \{1_1, 1_2\}$, $f(1_2) = \{2,3\}$ and $f(2) = f(3) = \{1_1, 1_2, 2, 3\}$. Then $p(1_1) = p(1_2) = \{1_1, 2, 3\}$ and $p(2) = p(3) = \{1_2, 2, 3\}$. The conjugacy from X_3 to X_B is a two-block map defined by $2 \to 2, 3 \to 3, [1,1] \to 1_1$ and $[1,2], [1,3] \to 1_2$. The inverse map simply drops the subscripts on 1_1 and 1_2. The new states, 1_1 and 1_2, have the same predecessors and disjoint successors.

The operation we used in the example is called a *state* or *symbol splitting (by followers)*. To define a general state splitting start with a subshift of finite type X_A with alphabet $L_A = \{1, \dots, n\}$. Split the symbol 1 by partitioning its followers into two nonempty sets, f_1 and f_2. So $f(1) = f_1 \cup f_2$ with $f_1 \cap f_2 = \phi$. Define a new subshift of finite type X_B with alphabet $L_B = \{1_1, 1_2, 2, \dots, n\}$. The transitions are defined for $j \in L_B$ in four cases. If

(1) $j \neq 1_1$ or 1_2 and $1 \notin f_A(j)$ then $f_B(j) = f_A(j)$.
(2) $j \neq 1_1$ or 1_2 and $1 \in f_A(j)$ then $f_B(j) = (f_A(j) - \{1\}) \cup \{1_1, 1_2\}$,
(3) the state is 1_r, for $r = 1$ or 2 and $1 \notin f_r$ then $f_B(1_r) = f_r$,
(4) the state is 1_r and $1 \in f_r$ then $f_B(1_r) = (f_r - \{1\}) \cup \{1_1, 1_2\}$.

The new states, 1_1 and 1_2, have the same predecessors and disjoint successors.

A state splitting has an inverse operation. It is called an *amalgamation (by followers)*. If a subshift of finite type, X_B, has two states with the same predecessors and disjoint successors we can "glue" them together to obtain a new subshift of finite type. The alphabet of the new subshift of finite type will contain one less symbol than L_B. Suppose $L_B = \{1_1, 1_2, 2, \dots, n\}, p_B(1_1) = p_B(1_2)$ and $f_B(1_1) \cap f_B(1_2) = \phi$. Define $L_A = \{1, \dots, n\}$ and define the transitions for $j \in L_B$ in four cases. If

(1) $j \neq 1, 1_1$ and $1_2 \notin f_B(j)$ then $f_A(j) = f_B(j)$,
(2) $j \neq 1, 1_1$ and $1_2 \in f_B(j)$ then $f_A(j) = (f_A(j) - \{1_1, 1_2\}) \cup \{1\}$.
(3) 1_1 and $1_2 \notin f_B(1_r)$ for $r = 1, 2$ then $f_A(1) = f_B(1_1) \cup f_B(1_2)$,
(4) 1_1 and $1_2 \in f_B(1_r)$ for $r = 1$ or 2 then $f_A(1) = (f_B(1_1) \cup f_B(1_2) - \{1_1, 1_2\}) \cup \{1\}$.

This process simply undoes a state splitting. At this point it should be noted that the ℓ-block presentation of a subshift of finite type can be obtained from the original subshift of finite type by a finite sequence of state splittings.

Next we prove a lemma with important computational consequences.

Lemma 2.1.2 *Suppose* X_A *is a subshift of finite type and* X_{B_1}, X_{B_2} *are subshifts of finite type obtained from* X_A *by amalgamations. Then there is a subshift of finite type* X_C *that can be obtained from both* X_{B_1} *and* X_{B_2} *by amalgamations.*

Proof. Suppose states 1_1 and 1_2 are amalgamated to obtain X_{B_1} and states 2_1 and 2_2 are amalgamated to obtain X_{B_2}. There are two case. The first case is when 1_1, 1_2, 2_1 and 2_2 are all distinct. The second case is when the two amalgamations have a state in common. We prove the result in the first case, leaving the proof in the second case as Exercise 4. Since two amalgamations are possible, $p_A(1_1) = p_A(1_2)$, $f_A(1_1) \cap f_A(1_2) = \phi$, $p_A(2_1) = p_A(2_2)$ and $f_A(2_1) \cap f_A(2_2) = \phi$. Consider 2_1 and 2_2 in X_{B_1}. Either the "new" state 1 will be in both $p_{B_1}(2_1)$ and $p_{B_1}(2_2)$ or it will be in neither since either 1_r is in both $p_A(2_1)$ and $p_A(2_2)$ or neither. Consequently $p_{B_1}(2_1) = p_{B_1}(2_2)$. At most one of $f_{B_1}(2_1)$ and $f_{B_1}(2_2)$ can contain 1 because if $f_A(2_r)$ contains 1_1 it must contain 1_2 and the other $f_A(2_s)$ contains neither. So $f_{B_1}(2_1) \cap f_{B_1}(2_2) = \phi$. In X_{B_1} it is possible to amalgamate 2_1 and 2_2. The new subshift of finite type does not depend on the order of the amalgamations. This is described in Figure 2.1.1 □

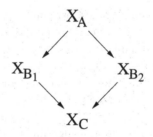

Figure 2.1.1

Example 2.1.3 Let X_A be the subshift of finite type defined by G_A in Figure 2.1.2. The alphabet of X_A is $L_A = \{1_1, 1_2, 2_1, 2_2, 3, 4\}$. Then X_{B_1} and X_{B_2} are defined by G_{B_1} and G_{B_2} and X_C is defined by G_C in Figure 2.1.2

Given a subshift of finite type, X_A, it is easy to check whether or not it is possible to amalgamate any states. After an amalgamation we can recheck and amalgamate again if possible. We continue until it is not possible to amalgamate any further states. The resulting subshift of finite type is called the *total column amalgamation* of X_A and is denoted by X_{A_c}. A consequence of Lemma 2.1.2 is the next lemma.

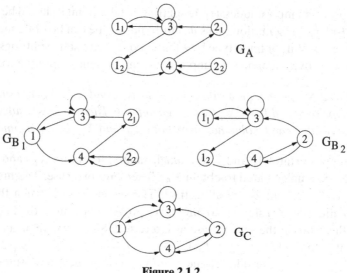

Figure 2.1.2

Lemma 2.1.4 *If* X_{B_1} *and* X_{B_2} *can both be obtained from* X_A *by a sequence of amalgamations then they have a common amalgamation. If* X_A *is a subshift of finite type then it has a unique, well-defined total column amalgamation (up to relabelling states).*

Proof. By Lemma 2.1.2 when there are two amalgamations we can complete the diagram in Figure 2.1.1. If there are two chains of amalgamations that begin with A we can extend each chain until we arrive at a common amalgamation. We do this one step at a time. Starting at the top of the diagram in Figure 2.1.3 we work down finding a common amalgamation at each step until we arrive at the bottom of the diagram in Figure 2.1.3. ☐

Next we examine the way state splittings and amalgamations are reflected in the transition matrices. Before proceeding we slightly generalize the definition of transition matrix. Let A be a square nonnegative integer matrix. It is identified with a directed graph G_A. The vertices of G_A are the indices of the rows and columns of A and the number of edges from vertex i to vertex j is A_{ij}. Denote the vertices of G_A by V_A and the edges by E_A. Define a subshift of finite type whose states are the edges of G_A. And say there is a transition from state e to state e' if the edge e ends at the vertex where the edge e' begins. This defines a subshift of finite type X_A with $L_A = E_A$. It is the same for one or two-sided subshifts of finite type. Let A be a $\{0, 1\}$ transition matrix. It defines two subshifts of finite type. The first subshift of finite type has for its alphabet $L_A = V_A$ as described in Section 1.1. The second subshift of finite type has

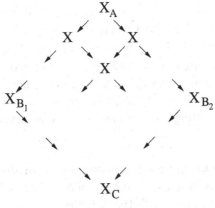

Figure 2.1.3

for its alphabet $L_A = E_A$ as just described. The subshift of finite type with $L_A = E_A$ is the 2-block presentation of the one with $L_A = V_A$. In practice this ambiguity is not really confusing.

Example 2.1.5 The graph of the matrix $A = \begin{pmatrix} 2 & 1 \\ 1 & 1 \end{pmatrix}$ from Example 1.2.7 is shown in Figure 2.1.4.

Figure 2.1.4

Observe that the $\{0, 1\}$ transition matrix describing X_A is the matrix B from Example 1.2.7. Both matrices A and B describe the same subshift of finite type.

Suppose A is a $\{0, 1\}$ transition matrix with alphabet $L_A = \{1_1, 1_2, \dots, n\}$ where 1_1 and 1_2 have the same predecessors and disjoint successors. The columns for 1_1 and 1_2 are the same and the rows have no ones in a common column. We amalgamate 1_1 and 1_2 to get a new transition matrix B. The matrix B can be described in two ways. One way is to define two $\{0, 1\}$ matrices R and S. The matrix R has rows indexed by L_A and columns indexed by L_B. For $j \in L_B, j \neq 1$ the column indexed by j in R is the same as the column indexed by j in A. The column indexed by 1 in R is the same as either of the columns indexed by 1_1 or 1_2 in A. The matrix S has rows indexed by L_B and columns indexed by L_A. For $j \in L_A, j \neq 1_1, 1_2$ the jth column of S has a 1 in the jth (now $j \in L_B$) entry and 0's elsewhere. The 1_1^{th} and 1_2^{th} columns of S have a 1 in the 1th entry and 0's elsewhere. Compute to see $A = RS$. To see $SR = B$

consider $\sum_k S_{ik}R_{kj}$. For $i \neq 1$ there is a single one in the ith row of S. It is in the ith position. So the ith row of SR is the same as the ith row of A with the 1_1^{th} entry removed. This gives $f_B(i)$. Now consider $\sum_k S_{1k}R_{kj}$. The 1th row of S has two 1's in it. They are in the 1_1^{th} and 1_2^{th} entries. So the 1th row of SR will be the sum of the 1_1^{th} and 1_2^{th} rows with the 1_1^{th} entry removed. This gives $f_B(1)$. We have $A = RS$ and $SR = B$. Another way to think of obtaining B is to add the 1_1^{th} and 1_2^{th} rows putting the sum in the row for 1_2 and then delete the 1_1^{th} row and column.

Example 2.1.6 Consider the transition matrices A, B_1 and B_2 below. The matrix B_1 is obtained from A by amalgamating 1_1 and 1_2. The matrix B_2 is obtained from A by amalgamating 2_1 and 2_2 (see Exercise 7).

$$A = \begin{bmatrix} 0 & 0 & 0 & 0 & 1 & 0 \\ 0 & 0 & 0 & 0 & 0 & 1 \\ 0 & 0 & 0 & 0 & 1 & 0 \\ 0 & 0 & 0 & 0 & 0 & 1 \\ 1 & 1 & 0 & 0 & 1 & 0 \\ 0 & 0 & 1 & 1 & 0 & 0 \end{bmatrix}$$

$$B_1 = \begin{bmatrix} 0 & 0 & 0 & 1 & 1 \\ 0 & 0 & 0 & 1 & 0 \\ 0 & 0 & 0 & 0 & 1 \\ 1 & 0 & 0 & 1 & 0 \\ 0 & 1 & 1 & 0 & 0 \end{bmatrix} \qquad B_2 = \begin{bmatrix} 0 & 0 & 0 & 1 & 0 \\ 0 & 0 & 0 & 0 & 1 \\ 0 & 0 & 0 & 1 & 1 \\ 1 & 1 & 0 & 1 & 0 \\ 0 & 0 & 1 & 0 & 0 \end{bmatrix}$$

Next suppose the transition matrix A is a nonnegative integer matrix with a repeated column. Let the repeated columns be indexed by 1_1 and 1_2. Looking at the graph G_A we see the vertices 1_1 and 1_2 have the same number of incoming edges from each vertex. If i is a vertex then A_{i1_1} is the number of edges from i to 1_1. It is the same as A_{i1_2}, the number of edges from i to 1_2. Pair each edge from i to 1_1 with an edge from i to 1_2. Make this pairing for each vertex of G_A. Each incoming edge to 1_1 is uniquely paired with an incoming edge to 1_2 and both begin at the same vertex. The edges in each pair have the same predecessors since they begin at the same vertex. The edges in each pair have disjoint successors since one ends at 1_1 and one ends at 1_2. We have pairs of edges with the same predecessors and disjoint successors. Identify the two symbols of X_A that occur in each of the edge pairings. This carries out several amalgamations simultaneously. In G_A the vertices 1_1 and 1_2 are identified. The edges in each pair are identified. The transition matrix changes as before. The new transition matrix, B, has one less row and column than A. We can write down R and S or we can add the 1_1^{th} row to the 1_2^{th} row putting the sum in the 1_2^{th} row and then delete the 1_1^{th} row and column.

Example 2.1.7 Let A and B be the matrices described below.

$$A = \begin{bmatrix} 2 & 2 & 1 \\ 0 & 0 & 2 \\ 1 & 1 & 0 \end{bmatrix} \qquad B = \begin{bmatrix} 2 & 3 \\ 1 & 0 \end{bmatrix}$$

Then X_B is obtained from X_A by amalgamating three pairs of edges. The matrices R and S are

$$R = \begin{bmatrix} 2 & 1 \\ 0 & 2 \\ 1 & 0 \end{bmatrix} \qquad S = \begin{bmatrix} 1 & 1 & 0 \\ 0 & 0 & 1 \end{bmatrix}.$$

It is demonstrated graphically in Figure 2.1.5.

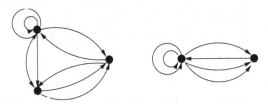

Figure 2.1.5

Given a matrix A we define the *total column amalgamation* and denote it by A_c. It is defined by repeatedly removing duplicated columns as just described. By the same reasoning as used in Lemma 2.1.4 we obtain the next lemma.

Lemma 2.1.8 *If A is a square nonnegative integer matrix its total column amalgamation is well-defined and unique (up to conjugation by a permutation matrix).*

If A is the transition matrix for X_A then A_c is the transition matrix for the total amalgamation of X_A. We have proven the following lemma.

Lemma 2.1.9 *If A and B are transition matrices and A_c is the total column amalgamation of A then X_A and X_{A_c} are topologically conjugate. If A and B have the same total column amalgamations then X_A and X_B are topologically conjugate.*

The final step is to show this condition is also necessary.

Theorem 2.1.10 *If A and B are irreducible, square, nonnegative integer matrices then the one-sided subshifts of finite type they define, X_A and X_B, are topologically conjugate if and only if A and B have the same total column amalgamations.*

Proof. We have already proven that if two matrices have the same total column amalgamation then the one-sided subshifts of finite type they define are topologically conjugate. Suppose $\varphi : X_A \to X_B$ is a conjugacy. We define a new subshift of finite type, X_C, which can be obtained from both X_A and X_B by a sequence of state splittings. Then Lemma 2.1.4 will show both A and B have the same total column amalgamation. Suppose both φ and φ^{-1} are ℓ-block maps. The alphabet of X_C is composed of pairs of ℓ-blocks. Define the new alphabet by

$$L_C = \{([x_0, \ldots, x_{\ell-1}], [y_0, \ldots, y_{\ell-1}]) : \varphi(x) = y\}.$$

The transitions are defined by saying $([z_0, \ldots, z_{\ell-1}], [w_0, \ldots, w_{\ell-1}])$ can follow $([x_0, \ldots, x_{\ell-1}], [y_0, \ldots, y_{\ell-1}])$ when $z_{i-1} = x_i$ and $w_{i-1} = y_i$ for $i = 1, \ldots, \ell-1$. It is the same overlapping idea as the transition rule for higher block presentations. Define a one-block map, π_A, from X_C to X_A by

$$\pi_A(([x_0, \ldots, x_{\ell-1}], [y_0, \ldots, y_{\ell-1}])) = x_0.$$

We want to show that X_C can be obtained from X_A by a sequence of state splittings. We know $X_{A^{[\ell]}}$ can be obtained from X_A by a sequence of state splittings. For $k = 0, \ldots, \ell - 1$ define X_{A_k} to be the subshift of finite type with alphabet

$$L_{A_k} = \{([x_0, \ldots, x_{\ell-1}], [y_0, \ldots, y_k]) : \varphi(x) = y\}.$$

The transitions are defined by saying the symbol $([z_0, \ldots, z_{\ell-1}], [w_0, \ldots, w_k])$ can follow the symbol $([x_0, \ldots, x_{\ell-1}], [y_0, \ldots, y_k])$ when $z_{i-1} = x_i$ for $i = 1, \ldots, \ell - 1$ and $w_{i-1} = y_i$ for $i = 1, \ldots, k$. Notice $X_{A^{[\ell]}}$ and X_{A_0} are the same since $[x_0, \ldots, x_{\ell-1}]$ determines y_0. Next observe that it is possible to go from X_{A_k} to $X_{A_{k+1}}$ by state splittings. Since $X_{A_{\ell-1}}$ is X_C we see X_C can be obtained from X_A by a sequence of state splittings. By Lemma 2.1.4 X_A and X_C have the same total column amalgamation. Everything is symmetric so X_A, X_B and X_C all have the same total column amalgamation. $\square$

For one-sided subshifts of finite type we have shown that there is a simple algorithm to determine when two transition matrices determine topologically conjugate one-sided subshifts of finite type. We will see that the situation for two-sided subshifts of finite type is completely different. At present it is not known whether or not the problem of determining when two transition matrices determine topologically conjugate two-sided subshifts of finite type is decidable.

At the beginning of this section we defined a state splitting by followers. It was done by partitioning the followers of a symbol into two nonempty disjoint sets. A state splitting by followers defines a conjugacy between two one-sided subshifts of finite type. In the two-sided case we have another degree of freedom. We can partition the predecessors of a symbol and let this define a *state* or *symbol splitting by predecessors*. This is illustrated in Example 2.1.11.

Example 2.1.11 Let Σ_3 be the full 3-shift with alphabet $L_3 = \{1, 2, 3\}$. We will "split" the state 1 into two new states and obtain a topologically conjugate subshift of finite type. Here we partition the predecessors of $1, p(1) = \{1, 2, 3\}$ into two disjoint sets $p_1 = \{1\}$ and $p_2 = \{2, 3\}$. The new subshift of finite type, Σ_B, has alphabet $L_B = \{1_1, 1_2, 2, 3\}$ and transitions defined by $f(1_1) = f(1_2) = \{1_1, 2, 3\}$ and $f(2) = f(3) = \{1_2, 2, 3\}$. This means that $p(1_1) = \{1_1, 1_2\}$, $p(1_2) = \{2, 3\}$ and $p(2) = p(3) = \{1_1, 1_2, 2, 3\}$. The map from Σ_3 to Σ_B is a two-block map with memory one and anticipation zero. It is $[1, 1] \to 1_1$, $[2, 1], [3, 1] \to 1_2$, $2 \to 2$ and $3 \to 3$. The two new states have the same followers and disjoint predecessors. This is demonstrated by the graphs in Figure 2.1.6.

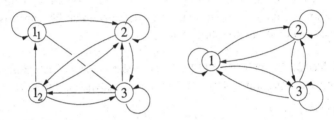

Figure 2.1.6

The one-sided subshifts of finite type determined by these graphs are not conjugate since their matrices do not have the same total column amalgamations. There is a factor map from X_B to X_3 defined by dropping the subscripts on the ones but it is not invertible. Points in X_3 beginning with 1 have two preimages.

In general a state splitting by predecessors and its inverse operation, an amalgamation of two states with disjoint predecessors and identical successors are defined exactly as a splitting by followers and an amalgamation of states with identical predecessors and disjoint successors but with the roles of predecessors and successors reversed.

On the matrix level this difference is reflected in the elimination of repeated rows. Suppose A is a nonnegative integer transition matrix with a repeated row. Let the repeated rows be indexed by 1_1 and 1_2. Looking at the graph, G_A, we see that the vertices 1_1 and 1_2 have the same number of outgoing edges to each vertex. This means $A_{1_1 i}$ is the number of edges from 1_1 to i. It is the same as the number of edges from 1_2 to i, $A_{1_2 i}$. We pair each edge from 1_1 to i with an edge from 1_2 to i. Make this pairing for each vertex. Every outgoing edge from 1_1 is uniquely paired with an outgoing edge from 1_2 and both edges in the pair end at the same vertex. The edges in each pair have the same successors since they end at the same vertex. The edges in each pair have disjoint predecessors since they begin at different vertices. We identify the two symbols of Σ_A that occur in each of our edge pairings. This carries out several amalgamations simultaneously. In the graph the vertices 1_1 and 1_2 are identified. The two edges in each pair are also identified. The new transition matrix, B, has one less row and column than A. We can write down R and S as follows. The rows of R are indexed by L_A and the columns by L_B. For $i \neq 1_1$ or 1_2, the ith row of R has a one in the ith entry and zeros elsewhere. For $i = 1_1$ or 1_2 the ith row of R has a one in the 1th entry and zeros elsewhere. The rows of S are indexed by L_B and the columns by L_A. For $i \neq 1$ the ith row of S is the same as the ith row of A. The 1th row of S is the same as the 1_1^{th} and the 1_2^{th} row of A. We see that $A = RS$ and $SR = B$. The matrix B can also be obtained from A by adding the 1_1^{th} column to the 1_2^{th} column, placing the sum in the 1_2^{th} column and eliminating the 1_1^{th} row and column. This would be the same as taking the transpose of the equations for a column amalgamation.

Example 2.1.12 Let A, B, R and S be as described below. Then B is obtained from A by identifying one pair of vertices and three pairs of edges.

$$A = \begin{pmatrix} 2 & 0 & 0 & 1 \\ 2 & 0 & 0 & 1 \\ 1 & 1 & 0 & 0 \\ 1 & 0 & 1 & 0 \end{pmatrix} = \begin{pmatrix} 1 & 0 & 0 \\ 1 & 0 & 0 \\ 0 & 1 & 0 \\ 0 & 0 & 1 \end{pmatrix} \begin{pmatrix} 2 & 0 & 0 & 1 \\ 1 & 1 & 0 & 0 \\ 1 & 0 & 1 & 0 \end{pmatrix} = RS$$

$$B = \begin{pmatrix} 2 & 0 & 1 \\ 2 & 0 & 0 \\ 1 & 1 & 0 \end{pmatrix} = \begin{pmatrix} 2 & 0 & 0 & 1 \\ 1 & 1 & 0 & 0 \\ 1 & 0 & 1 & 0 \end{pmatrix} \begin{pmatrix} 1 & 0 & 0 \\ 1 & 0 & 0 \\ 0 & 1 & 0 \\ 0 & 0 & 1 \end{pmatrix} = SR$$

Note that the ℓ-block presentation of a two-sided subshift of finite type can be obtained from the original by either a sequence of state splittings by predecessors or a sequence of state splittings by followers (Exercise 9). We can define the *total row amalgamation of A* in an analogous manner to the way in which the total column amalgamation was defined and obtain the analogue of Lemma 2.1.8. Denote the total row amalgamation of A by A_r.

Lemma 2.1.13 *If A is a square nonnegative integer matrix, its total row amalgamation is well-defined and unique (up to conjugation by a permutation matrix).*

Next we prove the two-sided analogue of Theorem 2.1.10.

Theorem 2.1.14 *If A and B are irreducible, square, nonnegative integer matrices then the two-sided subshifts of finite type they define, Σ_A and Σ_B, are topologically conjugate if and only if there exists a square, nonnegative integer matrix, C, so that A and C have the same total column amalgamation and B and C have the same total row amalgamation.*

Proof. It is clear that if such a matrix C exists then Σ_A and Σ_B are topologically conjugate. The converse is proved as in Theorem 2.1.10, by producing the matrix C. Suppose $\varphi : \Sigma_A \to \Sigma_B$ is a conjugacy. Choose a $k > 0$ so that we may write φ and φ^{-1} as $2k+1$ block maps with memory and anticipation k. Let $\ell = 2k + 1$ and replace φ with the map $\psi = \sigma^{-k} \circ \varphi$. Now $\psi(x)_0$ is a function of $[x_{-\ell+1}, \dots , x_0]$, ψ has memory $\ell - 1$ and anticipation 0. On the other hand $\psi^{-1}(y)_0$ is a function of $[y_0, \dots , y_{\ell-1}]$, and it has memory 0 and anticipation $\ell - 1$. Define the alphabet of Σ_C to be

$$L_C = \{([x_{-\ell+1}, \dots , x_0][y_0, \dots , y_{\ell-1}]) : \psi(x) = y\}.$$

The transitions for Σ_C are defined by the usual overlapping rule. The symbol $([z_{-\ell+1}, \dots , z_0][w_0, \dots , w_{\ell-1}])$ can follow $([x_{-\ell+1}, \dots , x_0][y_0, \dots , y_{\ell-1}])$ when $z_i = x_{i+1}$ for $i = -\ell+1, \dots , -1$ and $w_{i-1} = y_i$ for $i = 1, \dots , \ell-1$. We proceed as for the one-sided shifts but now we show $\Sigma_{A^{[\ell]}}$ can be obtained from Σ_C by a sequence of column amalgamations and $\Sigma_{B^{[\ell]}}$ can be obtained from Σ_C by a sequence of row amalgamations.

For $k = 0, \dots , \ell - 1$ define Σ_{A_k} be the subshift of finite type with alphabet

$$L_{A_k} = \{([x_{-\ell+1}, \dots , x_0], [y_0, \dots , y_k]) : \varphi(x) = y\}$$

The transitions are defined by saying the symbol $([z_{-\ell+1}, \dots , z_0], [w_0, \dots , w_k])$ can follow $([x_{-\ell+1}, \dots , x_0], [y_0, \dots , y_k])$ when $z_i = x_{i+1}$ for $i = -\ell+1, \dots , -1$ and $w_{i-1} = y_i$ for $i = 1, \dots , k$. Then $\Sigma_{A^{[\ell]}}$ and Σ_{A_0} are the same since $[x_{-\ell+1}, \dots , x_0]$ determines y_0. Observe that it is possible to go from Σ_{A_k} to $\Sigma_{A_{k+1}}$ by state splittings by followers. This is seen by observing that when two symbols from $L_{A_{k+1}}$ have the same image in L_{A_k} they have the same predecessors and disjoint followers. Since $\Sigma_{A_{\ell-1}}$ is Σ_C we see we can go from $\Sigma_{A^{[\ell]}}$ to Σ_C by a sequence of state splittings by followers and consequently from Σ_A to Σ_C by a sequence of state splittings by followers. This means that the matrices A and C have the same total column amalgamations.

On the other side of Figure 2.1.7 we work up from $\Sigma_{B^{[\ell]}}$ using state splittings by predecessors.

Let $L_{B_k} = \{([x_{-k}, \dots , x_0], [y_0, \dots , y_{\ell-1}]) : \varphi(x) = y\}$ be the alphabet of Σ_{B_k}. The transitions are defined as before. In this case $\Sigma_{B_{k+1}}$ can be obtained from Σ_{B_k} by a sequence of state splittings by predecessors. This shows B and C have the same total row amalgamation. □

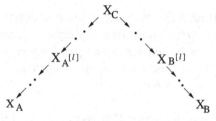

Figure 2.1.7

Given two transition matrices Theorem 2.1.14 gives a necessary and sufficient condition for their subshifts of finite type to be topologically conjugate. It is not known how to determine when this condition is met. It is not even known if this is a decidable problem. This is in contrast to the one-sided case where the condition is much harder to satisfy and can be easily computed. Example 2.1.15 displays transition matrices that define conjugate two-sided shifts but whose one-sided shifts are not conjugate.

Example 2.1.15 The graph of B is obtained from the graph of A in Figure 2.1.8 by a state splitting by predecessors so they describe two topologically conjugate two-sided shifts, but each is its own total column amalgamation so they describe one-sided subshifts of finite type which are not conjugate.

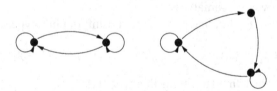

Figure 2.1.8

We have seen that a state splitting or an amalgamation, whether column or row, can be represented by a matrix equation, $A = RS$ and $SR = B$, where R and S are nonnegative integer matrices and either R or S has a special form. This special form for R or S is not necessary.

Lemma 2.1.16 *Suppose A and B are square nonnegative integer matrices and there exist nonnegative integer matrices R and S such that $A = RS$ and $SR = B$. Then the two-sided subshifts of finite type Σ_A and Σ_B defined by A and B are topologically conjugate.*

Proof. Suppose we have the equations as stated $A = RS$ and $SR = B$. Let R define a set of edges from the vertices of G_A to the vertices of G_B where there are R_{ik} edges from vertex i in G_A to vertex k in G_B. Let S define a set of edges from the vertices in G_B to the vertices in G_A in a similar manner. Since $A_{ij} = \sum_k R_{ik} S_{kj}$ we can match uniquely to each edge from vertex i to vertex j in G_A a pair of edges, (r, s) where r goes from vertex i in G_A to a vertex k in G_B and s goes from the vertex k in G_B to vertex j in G_A. Do this for all edges in G_A making sure no two edges in G_A are matched with the same pair (r, s). Similarly, use the equation $SR = B$ to match uniquely to each edge in G_B a pair of edges, (s, r), which goes from G_B to G_A and back to G_B in a consistent manner. This matching defines the conjugacy between Σ_A and Σ_B. Take a point x in Σ_A

$$x = \ldots x_{-1}.x_0 x_1 \ldots = \ldots (r_{-1} s_{-1}).(r_0 s_0)(r_1 s_1) \ldots$$

where the (r_i, s_i) are the pair of edges matched to the edge x_i in G_A. Note that the sequence of r_i's and s_j's without parenthesis

$$\ldots r_{-1} s_{-1}.r_0 s_0 r_1 s_1 \ldots$$

describes a biinfinite path of edges which zig zags back and forth between the vertices of G_A and the vertices of G_B. To describe the conjugacy shift the parentheses.

$$\ldots r_{-1}).(s_{-1} r_0)(s_0 r_1) s_1 \ldots = \ldots .y_0 y_1 \ldots = y$$

where $(s_{i-1} r_i)$ is the pair of edges matched to the edge y_i in G_B. The transitions between the y_i are allowed because the sequence of s_i's and r_j's describes a biinfinite path which zig zags back and forth between G_B and G_A. Shifting the parentheses one way or the other is like squeezing the zig zags closed in different ways. This defines a topological conjugacy.   □

The matrix relation we have defined is not transitive so we make the following definition.

If A and B are square nonnegative integer matrices and there exist nonnegative integer matrices $R_1, \ldots, R_\ell$ and $S_1, \ldots, S_\ell$ with

$$A = R_1 S_1, S_1 R_1 = R_2 S_2, \ldots, S_\ell R_\ell = B,$$

we say that the matrices A and B are *strong shift equivalent*. Putting together Theorem 2.1.14 and Lemma 2.1.16 we have proved the next theorem.

Theorem 2.1.17 *If A and B are square irreducible nonnegative integer matrices then the two-sided subshifts of finite type they define, Σ_A and Σ_B, are topologically conjugate if and only if the matrices are strong shift equivalent.*

Exercises

1. Let
$$A = \begin{bmatrix} 1 & 1 & 1 \\ 1 & 0 & 0 \\ 0 & 1 & 1 \end{bmatrix}.$$

   Compute the transition matrix for the subshift of finite type obtained by splitting the state 1 using the partition of its successors, $f_1 = \{1,2\}$ and $f_2 = \{3\}$.

2. Show that the map defined in Example 2.1.1 is a topological conjugacy.

3. Show that the ℓ-block presentation of a subshift of finite type can be obtained from the original subshift of finite type by a finite sequence of state splittings by followers.

4. Prove Lemma 2.1.2 when X_{B_1} is obtained from X_A by amalgamating states 1_1 and 1_2 and X_{B_2} is obtained from X_A by amalgamating states 1_2 and 1_3.

5. Compute the total column amalgamation for
$$\begin{bmatrix} 2 & 2 & 0 & 0 \\ 1 & 1 & 2 & 2 \\ 1 & 0 & 0 & 0 \\ 0 & 1 & 2 & 2 \end{bmatrix}.$$

6. Show that if A is a $\{0,1\}$ transition matrix the subshift of finite type defined by G_A with alphabet E_A is the two-block presentation of the subshift of finite type defined by G_A with alphabet V_A.

7. For Example 2.1.6 find the matrices R_1, S_1, R_2 and S_2 with $A = R_1 S_1 = R_2 S_2$, $S_1 R_1 = B_1$ and $S_2 R_2 = B_2$.

8. In the proof of Theorem 2.1.10 verify that $X_{A_{k+1}}$ can be obtained from X_{A_k} by a sequence of state splittings.

9. Show that the ℓ-block presentation of Σ_A can be obtained from Σ_A by a sequence of k state splittings by followers together with a sequence of $\ell - k - 1$ state splittings by predecessors for any $0 \le k \le \ell - 1$. Show that this is consistent with Exercise 7 of Section 1.4.

10. Find a matrix with the matrix
$$\begin{bmatrix} 3 & 2 & 1 \\ 1 & 0 & 1 \\ 1 & 0 & 1 \end{bmatrix}$$

    as a one-step column amalgamation and the matrix
$$\begin{bmatrix} 2 & 2 & 1 \\ 1 & 1 & 2 \\ 1 & 1 & 1 \end{bmatrix}$$

    as a one-step row amalgamation.

11. Check to see that the map defined in the proof of Lemma 2.1.16 is a topological conjugacy

12. Let A be a transition matrix and let A^t denote its transpose. Show that the shift, σ, acting on Σ_{A^t} is topologically conjugate to the inverse of the shift, σ^{-1}, acting on Σ_A.

§ 2.2 Algebraic Consequences of Topological Conjugacy

In this section we will examine some of the algebraic consequences of topological conjugacy for two-sided subshifts of finite type. The conditions in this section follow from the description of conjugacies developed in Section 2.1. From Lemma 1.4.5 we know that the transition matrices of conjugate subshifts of finite type have the same characteristic polynomials away from zero. To this point, the characteristic polynomial away from zero is the only computable algebraic invariant of conjugacy available. Examples will show that this is not sufficient to guarantee conjugacy. The first and most important algebraic condition to be discussed is shift equivalence. The other algebraic invariants we develop are consequences of this condition.

If A and B are transition matrices and there exist nonnegative integer matrices R and S and a positive integer ℓ with

$$A^\ell = RS \qquad SR = B^\ell$$
$$AR = RB \qquad SA = BS$$

the matrices A and B are said to be *shift equivalent*. Sometimes this is called *weak shift equivalence*. The positive integer ℓ is called the *lag* of the shift equivalence.

Lemma 2.2.1 *If A, B and C are transition matrices with A shift equivalent to B and B shift equivalent to C then A is shift equivalent to C. If there is a shift equivalence between A and B with lag L then there is a shift equivalence between A and B with lag ℓ for all $\ell \geq L$.*

Proof. Let R, S and ℓ describe a shift equivalence between A and B. Let U, V and k describe a shift equivalence between B and C. A computation shows RU, VS and $\ell + k$ describe a shift equivalence between A and C.

Let R, S and ℓ describe a shift equivalence between A and B. Then AR, S and $\ell + 1$ describe another shift equivalence between A and B.   □

Part of the relationship between strong shift equivalence and weak shift equivalence is exhibited by the next observation.

Observation 2.2.2 *If A and B are strong shift equivalent transition matrices then A and B are shift equivalent.*

Proof. Suppose the matrices $R_1, \dots, R_\ell$ and $S_1, \dots, S_\ell$ give a strong shift equivalence between A and B. Then $R = R_1 R_2 \cdots R_\ell$, $S = S_\ell \cdots S_2 S_1$ and the same ℓ describe a shift equivalence between A and B. □

This leads us to the best known problem of symbolic dynamics. It is usually called *Williams' conjecture* or *the shift equivalence problem*.

Problem 2.2.3 If A and B are shift equivalent, irreducible transition matrices are they strong shift equivalent?

This has been shown to be false if A and B are not irreducible, see the notes. See the notes concerning recent progress on the irreducible case.

Example 2.2.4 These matrices have the same characteristic polynomials away from zero but are not shift equivalent. The two-sided subshifts of finite type they describe are not topologically conjugate.

$$A = \begin{pmatrix} 1 & 2 \\ 2 & 1 \end{pmatrix} \qquad B = \begin{pmatrix} 1 & 4 \\ 1 & 1 \end{pmatrix}$$

Compute

$$c_A^*(x) = c_A(x) = x^2 - 2x - 3 = c_B(x) = c_B^*(x).$$

Suppose R, S and ℓ describe a shift equivalence between A and B. Let $R = \begin{pmatrix} a & b \\ c & d \end{pmatrix}$. Using $AR = RB$ we compute that $b = 2c$ and $d = 2a$. This means that $\det R = 2(a^2 - c^2)$. But $\det A = -3$ so it is impossible to have R and S integral while satisfying $A^\ell = RS$.

Let A be an $n \times n$ transition matrix. Denote by J_A its Jordan form and by J_A^* the principal submatrix of the Jordan form with nonzero entries on the diagonal. It is obtained from J_A by deleting the rows and columns with a zero on the diagonal. We say that J_A^* is the *Jordan form of A away from zero*. Note that $c_A^*(x)$ is the characteristic polynomial of J_A^*. The matrix A acts on the vector space $\mathbb{R}^n$. Let $I_A^R = \cap_{r=0}^{\infty} A^r(\mathbb{R}^n)$. It is the *eventual image of A in $\mathbb{R}^n$* and is the subspace of $\mathbb{R}^n$ corresponding to the block of J_A with nonzero entries on the diagonal. The matrix A maps I_A^R to itself. Restricted to I_A^R, A is an automorphism and can be represented by a matrix whose Jordan form is J_A^*. Let $K_A^R = \{x \in \mathbb{R}^n : A^r x = 0 \text{ for some } r \in \mathbb{N}\}$. This is the *eventual kernel of A*. It is the subspace of $\mathbb{R}^n$ corresponding to the block of J_A with zeros on the diagonal. The matrix A maps K_A^R into itself and for all sufficiently large r (n will do), $A^r(K_A^R) = 0$. Then $\mathbb{R}^n \simeq I_A^R \oplus K_A^R$.

Observation 2.2.5 *If A and B are shift equivalent transition matrices, $J_A^* = J_B^*$.*

Proof. Let A be an $n \times n$ matrix and B an $m \times m$ matrix. By Lemma 2.2.1 we can assume that ℓ is larger than n and m. The shift equivalence equations give the commuting diagram in Figure 2.2.1.

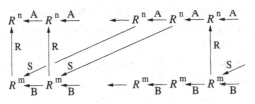

Figure 2.2.1

We have $A^{2\ell} = RB^\ell S$ which means RS defines an automorphism of I_A^R. We also have $B^{2\ell} = SA^\ell R$ so SR defines an automorphism I_B^R. It follows that S defines an isomorphism from I_A^R to I_B^R and that R defines an isomorphism from I_B^R to I_A^R. The shift equivalence equation $AR = RB$ shows A restricted to I_A^R and B restricted to I_B^R are conjugate and consequently $J_A^* = J_B^*$.   □

The condition $J_A^* = J_B^*$ is in some cases strictly stronger than the condition $c_A^*(x) = c_B^*(x)$. But, it is not enough to guarantee topological conjugacy. The two matrices in Example 2.2.4 have the same Jordan form away from zero but are not topologically conjugate. A particularly intriguing case of shift equivalence is when a matrix has a characteristic polynomial of the form $x^\ell(x - k)$.

Lemma 2.2.6 *If A is a transition matrix with characteristic polynomial $c_A(x) = x^\ell(x - k)$, for some $\ell, k \in \mathbb{N}$, then A is shift equivalent to the matrix $[k]$.*

Proof. Suppose the transition matrix A has characteristic polynomial $c_A(x) = x^\ell(x-k)$. The characteristic polynomial shows that A has a nonzero trace which makes it is aperiodic. By the Perron-Frobenius Theorem, A has unique (up to constant multiple), strictly positive left and right Perron eigenvectors. Since A is an integer matrix the equation $(A - kI)x = 0$ has integral solutions. Let r be the unique positive integral right eigenvector whose entries have greatest common divisor 1. We know $A^\ell \geq 0$ and has rank one so there is a nonnegative column vector x and a nonnegative row vector y with $A^\ell = xy$. The product yx is equal to the trace of A^ℓ which is k^ℓ. The equation $A^\ell x = x(yx) = k^\ell x$ shows that x is the right Perron eigenvector for A^ℓ and so a multiple of r. Choose a new rational row vector s so that $A^\ell = rs$ and the same argument will show that s is the left Perron eigenvector for A. We want to show that s is an integer vector. Suppose it is not. Let q be the smallest positive integer such that $qs = \bar{s}$ is integer. Then $A^\ell = (1/q)r\bar{s}$. There is a j such that q doesn't divide $\bar{s}_j$. For this j, let $\bar{q} = q/gcd(\bar{s}_j, q) \neq 1$. The jth column of A^ℓ is $(1/q\bar{s}_j)r$. This means that $\bar{q}$ divides every r_i which is a contradiction. So s is an integer vector. Since $r[k]^\ell = k^\ell r = A^\ell r = r(sr)$ we see that $sr = [k]^\ell$. Also, $Ar = r[k]$ and $sA = [k]s$. The matrices A and $[k]$ are shift equivalent.   □

This is the simplest case. The question of topological conjugacy for full shifts reduces to asking if a matrix with characteristic polynomial $x^\ell(x - k)$ is strong shift equivalent to $[k]$.

An interesting point is that there are two by two matrices which are not shift equivalent to their transposes.

Example 2.2.7 This is due to J. Ashley, the graph in Figure 2.2.2 describes a transition matrix with characteristic polynomial $x^7(x - 2)$. Lemma 2.2.6 shows that the transition matrix is shift equivalent to the matrix [2]. It is not known whether or not the two-sided subshift of finite type it describes is topologically conjugate to the 2-shift.

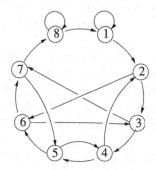

Figure 2.2.2

Lemma 2.2.8 *If A and B are irreducible and aperiodic transition matrices and there exist integer matrices U, V and a $k \in \mathbb{N}$ so that $A^k = UV$, $VU = B^k$, $AU = UB$ and $VA = BV$ then A and B are shift equivalent.*

Proof. The issue is the nonnegativity of the matrices defining the shift equivalence. This is where the aperiodicity of A and B comes into play. Let λ be the Perron value of both A and B, and let ℓ_A and r_A be the left and right Perron eigenvectors for A normalized so that $\ell_A r_A = 1$. They are strictly positive.

From $AU = UB$ and $VA = BV$ we see that $\ell_A U = \ell_B$ is a left Perron eigenvector for B and $Vr_A = r_B$ is a right Perron eigenvector for B. Each must be strictly positive or strictly negative. Since $\ell_B r_B = (\ell_A U)(Vr_A) = \ell_A A^k r_A = \lambda^k$, either both vectors are strictly positive or both are strictly negative. We can assume that they are strictly positive, multiplying U and V by negative one if necessary. The Perron-Frobenius Theorem tells us that $(A/\lambda)^n \to r_A \ell_A$. Then $\frac{1}{\lambda^n}(A^n U) \to r_A(\ell_A U) = r_A \ell_B > 0$ and $\frac{1}{\lambda^n}(VA^n) \to (Vr_A)\ell_A = r_B \ell_A > 0$, which are strictly positive. For sufficiently large n, $A^n U$ and VA^n are strictly positive. We take $R = A^n U$ and $S = VA^n$ and see that R, S and $\ell = 2n + k$ describe a shift equivalence for A and B. $\qquad\square$

The crucial point here is that $(A/\lambda)^n$ converges to a strictly positive matrix. If the transition matrix is not aperiodic this does not happen.

Now we define a family of abelian groups. Let $\mathcal{R}$ be a commutative ring. We are primarily interested in the cases where $\mathcal{R}$ is either $\mathbb{Z}$ or $\mathbb{Z}[t]$ but other rings are sometimes useful. The additive structure of the ring allows us to think of $\mathcal{R}$ as a $\mathbb{Z}$ module. If $a \in \mathcal{R}$ then for $0 \in \mathbb{Z}$, $0a = 0$ is the additive identity in $\mathcal{R}$, for $-1 \in \mathcal{R}$, $(-1)a$ is the additive inverse of a and for $k \in \mathbb{Z}$, $k > 0$, ka is the sum of a with itself k times. We can also think of $\mathcal{R}^n$ as a $\mathbb{Z}$ module using its additive structure. An $n \times n$ integer matrix, A, defines a homomorphism from $\mathcal{R}^n$ to itself. For $\gamma \in \mathcal{R}^n$ define $A\gamma$ by $(A\gamma)_i = \sum_j A_{ij}\gamma_j$. Let $p(x) = a_0 + a_1 x + \cdots + a_d x^d$ be an element of $\mathcal{R}[x]$. Define $p(A)$ to be $a_0 I + a_1 A + \cdots + a_k A^k$. It also defines a homomorphism of $\mathcal{R}^n$ to itself by $p(A)\gamma = a_0 I\gamma + a_1 A\gamma + \cdots + a_k A^k \gamma$.

For the construction let A be a transition matrix and choose $p(x) \in \mathcal{R}[x]$ with a_0 multiplicatively invertible in $\mathcal{R}$. Here we are primarily interested in the cases where $p(x)$ is $1 - x$ in $\mathbb{Z}[x]$ or $1 - tx$ in $(\mathbb{Z}[t])[x]$. We form the additive quotient group

$$\frac{\mathcal{R}^n}{p(A)\mathcal{R}^n}.$$

These groups are the *Bowen-Franks groups*.

Observation 2.2.9 *The matrix A induces an automorphism A_* on the group* $\frac{\mathcal{R}^n}{p(A)\mathcal{R}^n}$.

Proof. Define a group homomorphism

$$\alpha : \mathcal{R}^n \to \frac{\mathcal{R}^n}{p(A)\mathcal{R}^n}$$

by $\alpha(x) = Ax + p(A)\mathcal{R}^n$. We will show that the kernel of α is $p(A)\mathcal{R}^n$. First observe that $\alpha(p(A)x) = p(A)(Ax) + p(A)\mathcal{R}^n = p(A)\mathcal{R}^n$ so $p(A)\mathcal{R}^n \subseteq \ker \alpha$. Next suppose $x \in \ker \alpha$. This means that $Ax = p(A)y$ for some y. Let $z =$

$a_0^{-1}(x - a_1 y - a_2 A y - \cdots - a_d A^{d-1} y)$. Then compute to see $p(A)z = x$. So $\ker \alpha \subseteq p(A)\mathcal{R}$. Consequently A induces an automorphism A_* on the group.
□

Observation 2.2.10 *If A and B are shift equivalent transition matrices then the groups $\frac{\mathcal{R}^n}{p(A)\mathcal{R}^n}$ and $\frac{\mathcal{R}^m}{p(B)\mathcal{R}^m}$ are isomorphic and the induced automorphisms A_* and B_* are conjugate.*

Proof. Suppose R, S and ℓ describe a shift equivalence. Define a group homomorphism

$$S_* : \frac{\mathcal{R}^n}{p(A)\mathcal{R}^n} \to \frac{\mathcal{R}^m}{p(B)\mathcal{R}^m}$$

by $S_*(x + p(A)\mathcal{R}^n) = Sx + p(B)\mathcal{R}^m$. It is well defined since $S(p(A)y) = p(B)(Sy)$. Similarly, using R we define a group homomorphism

$$R_* : \frac{\mathcal{R}^m}{p(B)\mathcal{R}^m} \to \frac{\mathcal{R}^n}{p(A)\mathcal{R}^n}.$$

The equation $A^\ell = RS$ translates to $A_*^\ell = R_* S_*$ on $\frac{\mathcal{R}^n}{p(A)\mathcal{R}^n}$. Since A_*^ℓ is an automorphism, S_* is one-to-one and R_* is onto. Using B^ℓ we see that R_* is one-to-one and S_* is onto. Both R_* and S_* are isomorphisms between the groups. The equation $AR = RA$ becomes $A_* R_* = R_* B_*$ so A_* and B_* are conjugate.
□

A very useful case is when $\mathcal{R} = \mathbb{Z}$ and $p(x) = 1 - x$. The automorphism A_* in this case is the identity since $x - Ax$ is in $(I - A)\mathbb{Z}$ for all x. The group itself contains all the available information. In this case the group $\frac{\mathbb{Z}^n}{(I-A)\mathbb{Z}^n}$ can be easily computed. It is a finitely generated abelian group and so is isomorphic to a direct sum of $\mathbb{Z}/q\mathbb{Z}$'s and $\mathbb{Z}$'s. The matrix $I - A$ is integral and can be diagonalized by a sequence of elementary row and column operations. There are integer matrices with determinant plus or minus one that execute these elementary operations. If U is the matrix that executes the row operations and V is the matrix that executes the column operations we have $U(I - A)V = D$, a diagonal matrix. If $\{d_1, \ldots, d_n\}$ are the resulting diagonal elements, called the *elementary divisors of* $I - A$, then $\frac{\mathbb{Z}^n}{(I-A)\mathbb{Z}}$ is isomorphic to $\oplus \mathbb{Z}/d_i\mathbb{Z}$.

Example 2.2.11 Let A and B be as in Example 2.2.4. Compute to see $\frac{\mathbb{Z}^2}{(I-A)\mathbb{Z}^2}$ is isomorphic to $\mathbb{Z}/2\mathbb{Z} \oplus \mathbb{Z}/2\mathbb{Z}$ and $\frac{\mathbb{Z}^2}{(I-B)\mathbb{Z}^2}$ is isomorphic to $\mathbb{Z}/4\mathbb{Z}$. These are not isomorphic groups so we have another proof that A and B are not shift equivalent.

Next we will see how the shift equivalence equations arise naturally in the setting of inverse limit spaces.

Let $\mathbb{T}^n$ denote the n-dimensional torus and $\pi : \mathbb{R}^n \to \mathbb{T}^n$ be the standard covering map. Then $\mathbb{T}^n$ is isomorphic to $\mathbb{R}^n/\mathbb{Z}^n$ and is a group with addition inherited from $\mathbb{R}^n$ as the operation. Let A be an $m \times n$ matrix with integer entries. It maps $\mathbb{R}^n$ into $\mathbb{R}^m$ and takes $\mathbb{Z}^n$ into $\mathbb{Z}^m$. This condition means that A induces a group homomorphism from $\mathbb{T}^n$ into $\mathbb{T}^m$. The image of of $\mathbb{T}^n$ is a subtorus in $\mathbb{T}^m$. We use A to denote the matrix and the induced map as well. Conversely, any homomorphism from $\mathbb{T}^n$ into $\mathbb{T}^m$ is induced by an $m \times n$ matrix with integer entries. If A is an $n \times n$ integer matrix it induces a homomorphism from $\mathbb{T}^n$ to itself. When A is $n \times n$, let $I_A^T = \cap_{r=0}^{\infty} A(\mathbb{T}^n)$. It is the *eventual image of A in $\mathbb{T}^n$* and is equal to $\pi(I_A^R)$ where I_A^R is the eventual image of A in $\mathbb{R}^n$. If d is the dimension of I_A^R then I_A^T is a d-dimensional subtorus of $\mathbb{T}^n$ and A restricted to it is an onto, δ-to-one homomorphism, where δ is the absolute value of the determinant of J_A^*. Let $K_A^T = \{z \in \mathbb{T}^n : A^r z = 0 \text{ for some } r \in \mathbb{N}\}$. This is the *eventual kernel of A in $\mathbb{T}^n$*. It is an $(n - d)$-dimensional subtorus, $\pi(K_A^R) = K_A^T$ and $\mathbb{T}^n \simeq I_A^T \oplus K_A^T$.

Next we examine some product and inverse limit spaces. As with a finite set, define the sequence space $(\mathbb{T}^n)^{\mathbb{N}}$ and endow it with the product topology. A group operation is defined as coordinate by coordinate addition. This makes it a compact group and the shift transformation, $\sigma : (\mathbb{T}^n)^{\mathbb{N}} \to (\mathbb{T}^n)^{\mathbb{N}}$ is a continuous group homomorphism.

Suppose $\varphi : (\mathbb{T}^n)^{\mathbb{N}} \to (\mathbb{T}^m)^{\mathbb{N}}$ is a continuous group homomorphism which commutes with the shifts. Define a continuous homomorphism $\varphi_0 : (\mathbb{T}^n)^{\mathbb{N}} \to \mathbb{T}^m$ by $\varphi_0(x) = (\varphi(x))_0$. Since φ_0 is a group homomorphism it can be written as a (seemingly infinite) sum $\varphi_0(x) = \sum_j \varphi^j(x_j)$ where each $\varphi^j : \mathbb{T}^n \to \mathbb{T}^m$ is a homomorphism. Each φ^j can be defined by an $m \times n$ integer matrix Φ_j. Then the sum $\varphi_0(x) = \sum_j \Phi_j x_j$ converges for each $x \in \mathbb{T}^n$ and defines a uniformly continuous function. Consequently, only a finite number of the Φ_j are nonzero. Denote these matrices by $\Phi_0, \ldots, \Phi_k$. Since φ commutes with the shifts $(\varphi(x))_i = \sum_j \Phi_j x_{i+j}$, for all i. The homomorphism is defined by these $k + 1$ integer matrices. We can think of φ as block map.

Let A be an $n \times n$ integer matrix and consider it acting on $\mathbb{T}^n$. Define a compact, shift invariant subgroup of $(\mathbb{T}^n)^{\mathbb{N}}$ by

$$\varprojlim_{\mathbb{N}}(\mathbb{T}^n, A) = \varprojlim(\mathbb{T}^n, A) = \{x \in (\mathbb{T}^n)^{\mathbb{N}} : x_i = A x_{i+1}, \forall i \in \mathbb{N}\}.$$

This is the *inverse limit* over $\mathbb{N}$ of A acting on $\mathbb{T}^n$. In ergodic theory it is often called the *natural extension* of A acting on $\mathbb{T}^n$. For $x \in \varprojlim(\mathbb{T}^n, A)$, each $x_i \in I_A^T$ since $x_i = A^n x_{i+n}$. The rule $A x_{i+1} = x_i$ can be thought of as a transition rule which makes the inverse limit space similar to a subshift of finite type but with alphabet I_A^T. The homomorphism A of $\mathbb{T}^n$ induces an automorphism A^* of $\varprojlim(\mathbb{T}^n, A)$ by $A^*(x_0, x_1, \ldots) = (A x_0, x_0, x_1, \ldots)$. Notice that $(A^*)^{-1}$ is the shift restricted to $\varprojlim(\mathbb{T}^n, A)$.

If $\varprojlim(\mathbb{T}^n, A)$ and $\varprojlim(\mathbb{T}^m, B)$ are two inverse limit spaces and $\varphi : \varprojlim(\mathbb{T}^n, A)$ $\rightarrow \varprojlim(\mathbb{T}^m, B)$ is a continuous group homomorphism which commutes with the induced automorphisms, or equivalently the shifts, then φ is again a block map. There are $(k+1)$ $(m \times n)$ integer matrices, $\Phi_0, \dots, \Phi_k$, which define φ. For $x \in \varprojlim(\mathbb{T}^n, A)$, $(\varphi(x))_i = \sum_j \Phi_j x_{i+j}$, for each i. For x in an inverse limit, this expression simplifies. Observe

$$(\varphi(x))_i = \sum_{j=0}^k \Phi_j x_{i+j} = (\sum_{j=0}^k \Phi_j A^{k-j}) x_k = \Phi x_k,$$

where $\Phi = \sum_{j=0}^k \Phi_j A^{k-j}$. The homomorphism φ is defined by a single $m \times n$ integer matrix.

Given two square integer matrices we say their inverse limit spaces are *conjugate* if there is a continuous group isomorphism $\varphi : \varprojlim(\mathbb{T}^n, A) \rightarrow \varprojlim(\mathbb{T}^m, B)$ with $\varphi \circ A^* = B^* \circ \varphi$. Next we see how the shift equivalence equations are related to these groups.

Lemma 2.2.12 *Given A and B integer matrices there exist integer matrices U and V and an $\ell > 0$ such that*

$$A^\ell = UV \qquad AU = UB \qquad VA = BV \qquad VU = B^\ell$$

if and only if their inverse limit spaces, $\varprojlim(\mathbb{T}^n, A)$ and $\varprojlim(\mathbb{T}^m, A)$, are conjugate.

Proof. Given U, V and ℓ as stated, consider the diagram in Figure 2.2.3.

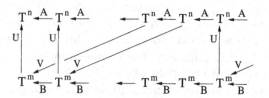

Figure 2.2.3

Define a map $V^* : \varprojlim(\mathbb{T}^n, A) \rightarrow \varprojlim(\mathbb{T}^m, B)$ by $V^*(x_0, x_1, \dots) = (Vx_0, Vx_1, \dots)$. We see that $V^* A^* = B^* V^*$. Do the same for U defining $U^* : \varprojlim(\mathbb{T}^m, B) \rightarrow \varprojlim(\mathbb{T}^n, A)$ with $U^* B^* = A^* U^*$. The matrix equations also show $(A^*)^\ell = U^* V^*$ and $V^* U^* = (B^*)^\ell$. Since A^* and B^* are automorphisms U^* and V^* are isomorphisms.

Conversely, suppose φ is an isomorphism as stated. By the discussion preceeding this lemma, φ and φ^{-1} are $k+1$ block maps for some k and there is an $m \times n$ integer matrix, Φ, which defines φ and an $n \times m$ integer matrix,

Θ, which defines φ^{-1}. For $x \in I_A^T$, $\Phi Ax = B\Phi x$ and for $y \in I_B^T$, $\Theta By = A\Theta y$. When $x \in \varprojlim(\mathbb{T}^n, A)$ with $\varphi(x) = y$ we see $A^{2k}x_{2k} = x_0 = \Theta y_k = \Theta \Phi x_{2k}$. So for $x \in I_A^T$, $A^{2k}x = \Theta \Phi x$ and similarly for $y \in I_B^T$ we have $B^{2k}y = \Phi \Theta y$. For all $x \in \mathbb{T}^n$, $A^n x \in I_A^T$ and for all $y \in \mathbb{T}^m$, $B^m y \in I_B^T$. Let $V = \Phi A^n$, $U = \Theta B^m$ and $\ell = n + m + 2k$. These produce the desired matrix equations. □

Theorem 2.2.13 *Let A and B be two irreducible and aperiodic transition matrices. Then A and B are shift equivalent if and only if the inverse limit spaces, $\varprojlim(\mathbb{T}^n, A)$ and $\varprojlim(\mathbb{T}^m, B)$, are conjugate.*

Proof. This follows from Lemmas 2.2.8 and 2.2.12. □

There is a "dual" version of the relationship of the shift equivalence equations to the inverse limit spaces. References concerning duality theory are contained in the notes at the end of chapter. For the next discussion we use row vectors and let matrices act on the right, $xA = y$. So, an $n \times m$ integer matrix maps $\mathbb{Z}^n$ into $\mathbb{Z}^m$. The *direct sum* over $\mathbb{N}$ of $\mathbb{Z}^n$ is

$$\{x \in (\mathbb{Z}^n)^{\mathbb{N}} : x_i \neq 0 \text{ for finitely many } i\}.$$

It is denoted by $\oplus_{\mathbb{N}}\mathbb{Z}^n$ or $\oplus\mathbb{Z}^n$. The direct sum is a group where the operation is coordinate by coordinate addition in $\mathbb{Z}^n$.

Let A be an $n \times n$ integer matrix. Define the the *direct limit* over $\mathbb{N}$ of A acting on $\mathbb{Z}^n$ to be the quotient of $\oplus\mathbb{Z}^n$ where the kernel is generated by

$$\{\alpha \in \oplus_{\mathbb{N}}\mathbb{Z}^n : \exists r \in \mathbb{N} \text{ with } \alpha_i = 0, \forall i \neq r, r+1 \text{ and } \alpha_{r+1} = -\alpha_r A\}.$$

The direct limit is denoted by $\varinjlim_{\mathbb{N}}(\mathbb{Z}^n, A)$ or $\varinjlim(\mathbb{Z}^n, A)$. It is a group with the operation inherited from the group operation on $\oplus\mathbb{Z}^n$. The matrix A induces an automorphism, A_*, of $\varinjlim(\mathbb{Z}^n, A)$ by $A_*(\alpha_0, \alpha_1, \dots) = (\alpha_0 A, \alpha_1 A, \dots)$.

Abstractly, this can be quickly summarized by duality theory for locally compact topological groups with countable bases for their topologies. Let G be a compact abelian topological group with its operation denoted by $+$. A group character for G is a continuous group homomorphism $\chi : G \to \mathbb{S}^1$. The set of all group characters for G forms a group where the operation is pointwise addition. For group characters χ and χ' define $\chi + \chi'$ by $(\chi + \chi')(x) = \chi(x) + \chi'(x)$. The group of all group characters for G is called the *dual group of G*. It is denoted by $G^{\wedge}$ and the operation is denoted by $+$. The dual group is given the compact-open topology. The dual group of a compact group is discrete, so locally compact. An important point is that when $x \neq y \in G$ there is a $\chi \in G^{\wedge}$ with $\chi(x) \neq \chi(y)$. The dual group of $G^{\wedge}$ is naturally isomorphic to G. A homomorphism τ of G induces a homomorphism $\tau^{\wedge}$ of $G^{\wedge}$ by $(\tau^{\wedge}(\chi))(x) = \chi(\tau(x))$. If the homomorphism is an automorphism then the induced map is also an automorphism. In this discussion we use three types of compact groups with homomorphisms and their dual groups with homomorphisms.

The first type of compact group is the n-dimensional torus, $\mathbb{T}^n$, and the homomorphism is defined by an $n \times n$ integer matrix, A, acting on column vectors by left multiplication. The dual group of $\mathbb{T}^n$ is the n-dimensional integer lattice $\mathbb{Z}^n$. The homomorphism induced by A on $\mathbb{Z}^n$ is also defined by A but now it acts on row vectors by right multiplication. An element $\eta \in \mathbb{Z}^n$ defines a character $\chi_\eta \in (\mathbb{T}^n)^\wedge$ by $\chi_\eta(x) = <\eta, x> = \eta x$ modulo 1, where η is a row vector, x a column vector and the product a real number.

The second type of compact group is the infinite product group, $(\mathbb{T}^n)^\mathbb{N}$. The group homomorphism is the usual shift, σ. The character group of $(\mathbb{T}^n)^\mathbb{N}$ is the direct sum of $\mathbb{Z}^n$ over $\mathbb{N}$, $\oplus_\mathbb{N} \mathbb{Z}^n$. The dual homomorphism, $\sigma^\wedge$, acts on $\oplus_\mathbb{N} \mathbb{Z}^n$ by $(\sigma^\wedge(\alpha))_i = \alpha_{i-1}$ and $(\sigma^\wedge(\alpha))_0 = 0$. An element $\alpha \in \oplus \mathbb{Z}^n$ defines a character, χ_α, with $\chi_\alpha(x) = \sum_i < \alpha_i, x_i >$. The operation is well defined since there are only finitely many nonzero α_i.

The third type of compact group is an inverse limit group, $\underleftarrow{\lim}(\mathbb{T}^n, A)$. The matrix A induces an automorphism, A^*, defined by $A^*(x_0, x_1, \dots) = (Ax_0, x_0, x_1, \dots)$. As previously noted, $(A^*)^{-1}$ is the shift restricted to $\underleftarrow{\lim}(\mathbb{T}^n, A)$. The character group of $\underleftarrow{\lim}(\mathbb{T}^n, A)$ is the *direct limit* over $\mathbb{N}$ of A acting on $\mathbb{Z}^n$, $\underrightarrow{\lim}(\mathbb{Z}^n, A)$. The matrix A induces an automorphism A_* of $\underrightarrow{\lim}(\mathbb{Z}^n, A)$ by $A_*(\alpha_0, \alpha_1, \dots) = (\alpha_0 A, \alpha_1 A, \dots)$. An element $\bar{\alpha} \in \underrightarrow{\lim}(\mathbb{Z}^n, A)$ defines a character by choosing an $\alpha \in \oplus \mathbb{Z}^n$ which represents $\bar{\alpha}$ and letting $\chi_{\bar{\alpha}}(x) = \sum < \alpha_i, x_i >$. The definition does not depend on the choice of α because of the way the two limit spaces are defined.

Given an $n \times n$ transition matrix we define the *dimension group of* Σ_A to be the direct limit over $\mathbb{N}$ of A acting on $\mathbb{Z}^n$, $\underrightarrow{\lim}(\mathbb{Z}^n, A)$. We denote the dimension group by $Dim(A)$. We have seen that it is an additive group and on it A induces an automorphism A_*. We say that the dimension groups of two transition matrices, A and B, are *conjugate* if there is a continuous group isomorphism $\varphi : Dim(A) \to Dim(B)$ with $\varphi \circ A_* = B_* \circ \varphi$. This is exactly the way we did it for the inverse limit spaces of A and B acting on the torus.

Theorem 2.2.14 *Let A and B be two irreducible and aperiodic transition matrices. Then A and B are shift equivalent if and only if the dimension groups, $Dim(A)$ and $Dim(B)$, are conjugate.*

Proof. This is the "dual" statement of Theorem 2.2.13. The dimension group of A is the character group of $\underleftarrow{\lim}(\mathbb{T}^n, A)$ and A_* is the automorphism on the dimension group induced by A^*. Duality theory shows that the dimension groups are conjugate if and only if the inverse limit spaces are conjugate. $\square$

§2.2 Algebraic Consequences of Topological Conjugacy   57

There is another more geometric description of the dimension group. Let A be an $n \times n$ integer matrix. Then A acts on the rational vector space $\mathbb{Q}^n$. Let $I_A^Q = \cap_{r=0}^{\infty}(\mathbb{Q}^n)A^r$. It is the *eventual image of A in $\mathbb{Q}^n$*. The matrix A is an automorphism when restricted to I_A^Q. Define

$$\mathcal{D}_A = \{x \in I_A^Q : xA^j \in \mathbb{Z}^n \text{ for some } j \in \mathbb{N}\}.$$

This is an A invariant additive subgroup of $\mathbb{Q}^n$. Let A_* be the restriction of A to $\mathcal{D}_A$.

Observation 2.2.15 *If A is an $n \times n$ integer matrix then $\varinjlim(\mathbb{Z}^n, A)$ is isomorphic to $\mathcal{D}_A$ and the automorphisms induced by A on each are conjugate.*

Proof. Define a homomorphism, ψ, from $\mathcal{D}_A$ to $\mathcal{D}im(A)$ by taking $x \in \mathcal{D}_A$, choosing a j so that $xA^j \in \mathbb{Z}^n$ and saying $\psi(x) \in \mathcal{D}im(A)$ is 0 except in the jth entry where it is xA^j. Check to see that this doesn't depend on the choice of j. If we take j or $j+1$ to define $\psi(x)$ the difference is one of the generators for the kernel of the quotient map from $\oplus\mathbb{Z}^n$ which defines the direct limit. It is easy to see that the map is onto and has kernel zero. The automorphisms induced by A are conjugate. $\qquad\square$

Example 2.2.16 For the full two-shift the matrix is [2]. It maps $\mathbb{Q}$ onto $\mathbb{Q}$ so that $I_A^Q = \mathbb{Q}$ and $\mathcal{D}_A = \mathbb{Z}[1/2]$, the diadic rationals.

There is a third description of the dimension group. Because of the duality between the inverse and direct limits we are presently using row vectors and letting matrices act by right multiplication. To fit in with this we define another class of Bowen-Franks groups using row vectors and right multiplication. Let $\mathcal{R}$ be a commutative ring and A an $n \times n$ integer matrix. Define the Bowen-Franks groups as before but with

$$\frac{\mathcal{R}^n}{\mathcal{R}^n p(A)}.$$

The automorphism, A_*, is also defined using right multiplication.

Observation 2.2.17 *If A is an $n \times n$ integer matrix then $\varinjlim(\mathbb{Z}^n, A)$ is isomorphic to the Bowen-Franks group $\frac{(\mathbb{Z}[t])^n}{(\mathbb{Z}[t])^n(I-tA)}$ and the automorphisms induced by A on each are conjugate.*

Proof. Define a homomorphism $\psi : \oplus\mathbb{Z}^n \to \frac{(\mathbb{Z}[t])^n}{(\mathbb{Z}[t])^n(I-tA)}$ by $\psi(\alpha) = \sum \alpha_i t^i$. Let

$$K = \{\alpha \in \oplus\mathbb{Z} : \exists r \in \mathbb{N} \text{ with } \alpha_i = 0, \ \forall i \neq r, r+1 \text{ and } \alpha_{r+1} = -\alpha_r A\}.$$

We know that K generates the kernal of the quotient map which defines the direct limit. We will see that it also generates the kernel of ψ. For $\alpha \in K$, $\psi(\alpha) = \alpha_r t^r + (-\alpha_r A)t^{r+1} = \alpha_r t^r (I - tA)$. So K is contained in the kernel of ψ. If $\psi(\alpha) = y(I - tA)$ with $y = \sum_{r=0}^{k} y_r t^r$ and $y_r \in \mathbb{Z}^n$ then $y(I - At) = \sum_{r=0}^{k} (y_r - y_r tA)t^r$. For each r let $\beta^r \in \oplus\mathbb{Z}$ have $\beta_i^r = 0$ when $i \neq r, r+1$ and $\beta_r^r = y_r$, $\beta_{r+1}^r = -y_r A$. Then each $\beta^r \in K$ and $\alpha = \sum \beta^r$. By construction $\psi \circ A_* = A_* \circ \psi$.   □

Lemma 2.2.18 *If A is an $n \times n$ transition matrix then $Dim(A)$ is isomorphic to to the Bowen-Franks group $\frac{(\mathbb{Z}[t])^n}{(\mathbb{Z}[t])^n(I-tA)}$ and the automorphisms induced by A on each are conjugate.*

Exercises

1. Compute the Jordan form away from zero for
    $$\begin{bmatrix} 1 & 2 & 2 \\ 1 & 1 & 1 \\ 1 & 0 & 0 \end{bmatrix}.$$

2. Let $GL(n, \mathbb{Z})$ denote the group of all $n \times n$ integer matrices with determinant plus or minus one. Two $n \times n$ matrices, A and B, are *similar over $GL(n, \mathbb{Z})$* if there is a $Q \in GL(n, \mathbb{Z})$ with $AQ = QB$. Let A and B be two $n \times n$ aperiodic transition matrices. Prove that A and B are shift equivalent if they are similar over $GL(n, \mathbb{Z})$.

3. Let $A, B \in GL(n, \mathbb{Z})$ be two aperiodic transition matrices. Prove that A and B are shift equivalent if and only if A and B are similar over $GL(n, \mathbb{Z})$.

4. Let $A = \begin{bmatrix} 0 & 1 \\ n & m \end{bmatrix}$ with n and m positive integers. Show that A is shift equivalent to its transpose.

5. Compute the Bowen-Franks group $\frac{\mathbb{Z}^3}{(I-A)\mathbb{Z}^3}$ for the matrix
    $$A = \begin{bmatrix} 0 & 0 & 1 \\ 1 & 1 & 0 \\ 2 & 2 & 2 \end{bmatrix}.$$

    Hint: Use elementary row and column operations to diagonalize A.

6. Describe the set of sequences which make up the inverse limit spaces $\varprojlim(\mathbb{T}^1, [2])$ and $\varprojlim(\mathbb{T}^2, \begin{bmatrix} 1 & 1 \\ 1 & 1 \end{bmatrix})$. Describe the induced automorphism on each. Show that they are conjugate. Note: $\varprojlim(\mathbb{S}^1, [2])$ is the 2-adic solenoid.

7. Using the geometric description of the dimension group, $\mathcal{D}_A$, compute the dimension groups of $\begin{bmatrix} 1 & 1 \\ 1 & 1 \end{bmatrix}$ and $\begin{bmatrix} 1 & 1 \\ 1 & 0 \end{bmatrix}$. Use these to conclude (geometrically) that the two matrices are not shift equivalent.

Notes

Section 2.1

The results in this section are due to R.F. Williams in 1970 and are from his remarkable paper *Classification of Subshifts of Finite Type* [Wi2]. The approach used here stresses graphs and symbols. The proofs are done by altering graphs through a series of splittings and amalgamations. Williams stressed partitions of the underlying space and worked through sequences of splittings and amalgamations of generating partitions when examining conjugacies. This makes the arguments appear to be quite different, but at heart they are the same. The results are all contained in Williams' paper. The statement of Theorem 2.1.14 is due to B. Kitchens in the mid-1980's and the proof of Lemma 2.1.16 presented here is from W. Parry and S. Tuncel's book [PT].

Theorems 2.1.14 and 2.1.17 give algebraic conditions on transition matrices for the subshifts of finite type they define to be topologically conjugate. It is not known whether or not these conditions are computable. Even in the simplest cases it is unknown. This difficulty is illustrated by Lemma 2.2.6 and Example 2.2.7. The problem for two by two transition matrices has been examined by J. Cuntz and W. Krieger [CK2] and by K. Baker [Ba].

In the late 1980's J. Wagoner [Wa1] initiated a study of strong shift equivalence using techniques from Algebraic K-theory and considering how elementary conjugacies change Markov partitions.

Section 2.2

Shift equivalence and strong shift equivalence for maps was formulated by R.F. Williams while classifying one-dimensional attractors in 1970 [Wi1]. He was investigating the structure of conjugacies between inverse limit spaces where the state space is a one-dimensional branched manifold (train track) and there is a single bonding map. Soon afterwards he worked on the topological conjugacy problem for subshifts of finite type. In his 1973 paper [Wi2] he defined strong shift equivalence for transition matrices and proved it equivalent to conjugacy of the two-sided subshifts of finite type they define. In the same paper he formulated shift equivalence for transition matrices and "proved " it equivalent to strong shift equivalence. W. Parry found an error in the proof that shift equivalence of transition matrices implies strong shift equivalence of the matrices, see the Errata to [Wi2]. It has never been determined whether or not shift equivalence for irreducible transition matrices implies strong shift equivalence. It (Problem 2.2.3) is the best known problem in symbolic dynamics. In 1992 K.H. Kim and F. Roush published an example [KR2] of two **reducible** transition matrices which are shift equivalent but not strong shift equivalent. Their example makes essential use of an example in the paper [KRW1] by K.H. Kim, F. Roush and J. Wagoner. K. H. Kim and F. Roush have just produced two irreducible transition matrices that are shift equivalent but not strong shift equivalent.

R. Williams proved Lemma 2.2.6 in [Wi2] and J. Ashley discovered Example 2.2.7 in late 1989. Ashley's example has received considerable attention and it is still not known whether or not the subshift of finite type it describes is conjugate to the the two-shift. An example of a two by two transition matrix not shift equivalent to its transpose is $\begin{pmatrix} 19 & 5 \\ 4 & 1 \end{pmatrix}$ which was discovered by Köllmer. There is a proof of this for Köllmer's matrix in [PT] and a discussion of the connection with continued fractions in [CK2].

As pointed out in Section 2.1, it is not known if it is possible to decide whether or not two transition matrices are strong shift equivalent. In contrast, K.H. Kim and F. Roush proved, in 1986, shift equivalence to be decidable [KR1].

The Bowen-Franks groups were defined by R. Bowen and J. Franks in 1977 [BF] as an invariant for flow equivalence of subshifts of finite type. Their work followed the earlier work of W. Parry and D. Sullivan [PS] who showed that the determinant of $(I - A)$ is an invariant of flow equivalence. In 1984 J. Franks [Fk] showed that the group $\mathbb{Z}^n / (I - A)\mathbb{Z}^n$ and the determinant of $(I - A)$ determine the flow equivalence class of the subshift of finite type defined by A.

The inverse limit groups $\varprojlim(T^n, A)$ and their relationship to the shift equivalence equations were used by Williams in his papers [Wi1] and [Wi2]. Today the formulation of these ideas in terms of dimension groups is commonly used.

An excellent reference for duality theory is [HR].

The dimension group was introduced into dynamics by W. Krieger in the late 1970's [Kg1]. Dimension and dimension groups had been used in the study of C^*-algebras and Krieger adapted some of the ideas and applied them to the study of subshifts of finite type [Kg2], [Kg3]. The original definition of dimension group is more geometric than the one given here. It uses stable sets and certain transformation groups. In [Kg3] Krieger proved Theorem 2.2.14 and noted the connection to the Bowen-Franks groups. In [CK1] and [CK2] J.

Cuntz and W. Krieger further studied the dimension group and its relationship to C^*-algebras, the Bowen-Franks groups and subshifts of finite type. Theorems 2.2.13 and 2.2.14 are stated for aperiodic transition matrices. If the transition matrices are periodic there is a positivity condition needed on the conjugacy between the inverse limit groups or the dimension groups. Krieger addressed this problem in [Kg3], including in the definition of the dimension group the notion of a positive cone.

There has been far more work on the relationship between the dimension group, topological conjugacies and the automorphism group (see Chapter 3) than has been mentioned here. Additional references are supplied in Chapter 3.

References

[Ba]   K. Baker, *Strong Shift Equivalence of* 2×2 *matrices of Non-negative Integers,* Ergodic Theory and Dynamical Systems **3** (1983), 501–508.

[BF]   R. Bowen and J. Franks, *Homology for Zero-dimensional Nonwandering Sets,* Annals of Mathematics **106** (1977), 73–92.

[CK1]   J. Cuntz and W Krieger, *A Class of C^*-Algebras and Topological Markov Chains,* Inventiones Mathematicae **56** (1980), 251–268.

[CK2]   J. Cuntz and W. Krieger, *Topological Markov Chains with Dicyclic Dimension Groups,* Journal für die reine und angewandte Mathematik **320** (1980), 44–50.

[Fk]   J. Franks, *Flow Equivalence of Subshifts of Finite Type,* Ergodic Theory and Dynamical Systems **4** (1984), 53–66.

[HR]   E. Hewitt and K. Ross, *Abstract Harmonic Analysis,* Academic Press and Springer-Verlag, 1963.

[KR1]   K.H. Kim and F. Roush, *Decidability of Shift Equivalence,* Dynamical Systems: Proceedings, University of Maryland 1886–87 (J.C. Alexander, ed.), Springer-Verlag, 1988, pp. 374–424.

[KR2]   K.H. Kim and F. Roush, *William's Conjecture is False for Reducible Subshifts,* Journal of the American Mathematical Society **5** (1992), 213–215.

[KRW]   K.H. Kim, F. Roush and J. Wagoner, *Automorphisms of the Dimension Group and Gyration Numbers,* Journal of the American Mathematical Society **5** (1992), 191–212.

[Kg1]   W. Krieger, *On Topological Markov Chains,* Astérisque **50** (1977), 193–196.

[Kg2]   W. Krieger, *On Dimension for a Class of Homeomorphism Groups,* Mathematische Annalen **252** (1980), 87–95.

[Kg3]   W. Krieger, *On Dimension Functions and Topological Markov Chains,* Inventiones Mathematicae **56** (1980), 239–250.

[PS]   W. Parry and D. Sullivan, *A Topological Invariant for Flows on One-dimensional Spaces,* Topology **14** (1975), 297–299.

[PT]   W. Parry and S. Tuncel, *Classification Problems in Ergodic Theory* London Mathematical Society Lecture Series, 67, Cambridge University Press, 1982.

[Wa1]   J. Wagoner, *Markov Partitions and K_2,* Publications Mathematétiques IHES no. 65 (1987), 91–129.

[Wi1] R.F. Williams, *Classification of One-dimensional Attractors,* in Global Analysis, Proceedings of Symposia in Pure and Applied Math (S-S. Chern and S. Smale, eds.), vol. XIV, American Mathematical Society, 1970, pp. 341–361.

[Wi2] R.F. Williams, *Classification of Subshifts of Finite Type,* Annals of Mathematics **98** (1973), 120–153; *Errata* **99** (1974), 380–381.

Chapter 3. Automorphisms

An *automorphism* of a subshift of finite type is a homeomorphism of the subshift of finite type to itself that commutes with the shift. The automorphisms of a subshift of finite type form a group under composition. In this chapter we study the structure of an automorphism group considered as an abstract group and examine how it acts in the space.

The first section begins with a striking result about the automorphism groups of the one and two-sided two-shifts. It says that the automorphism group of the one-sided two-shift is isomorphic to $\mathbb{Z}/2\mathbb{Z}$ while the automorphism group of the two-sided two-shift contains subgroups isomorphic to every finite group. Next there are some observations about arbitrary automorphism groups. In the second section we use the results from Chapter 2 concerning conjugacies and consider an automorphism as a conjugacy from a subshift of finite type to itself. This allows us to describe a set of generators, each of finite order, for the automorphism group of any one-sided subshift of finite type. The third section contains results about subgroups of automorphism groups. Some examples of groups that cannot be embedded in any automorphism group and some examples of groups that can be embedded in automorphism groups are given. Then the finite subgroups of automorphism groups are described. The one-sided and two-sided cases are very different. In section four there is a brief discussion of how the automorphism group as a whole acts on the space. We mention the gyration function and the induced action of an automorphism on the dimension group. In the last section there is a short summary of the main results about the automorphism groups.

§3.1 Automorphisms

Denote the automorphism group of the one-sided subshift of finite type X_A by $Aut(X_A)$ and the automorphism group of the two-sided subshift of finite type Σ_A by $Aut(\Sigma_A)$.

Theorem 3.1.1 *Let $X_{[2]}$ be the one-sided two-shift and $\Sigma_{[2]}$ be the two-sided two-shift then*

(i)   *$Aut(X_{[2]})$ is isomorphic to $\mathbb{Z}/2\mathbb{Z}$ and*
(ii)   *$Aut(\Sigma_{[2]})$ contains every finite group.*

Proof. of (i). The alphabet of the shift is $L_{[2]} = \{0,1\}$. For $a \in \{0,1\}$ let $\bar{a}$ be 1 if a is 0 and 0 if a is 1. Suppose φ is an automorphism of $X_{[2]}$ and it is a k-block map for minimal k. For every $[a_1, \dots, a_k] \in \mathcal{W}([2], k)$, if $a = \varphi([a_1, \dots, a_k])$ then $\bar{a} = \varphi([\bar{a}_1, \dots, a_k])$. If this were false φ would not be one-to-one. This property is called *left permutive*. Assume the minimal coding length k is greater than one. Then there exists a block $[w_1, \dots, w_{k-1}] \in \mathcal{W}([2], k-1)$ with $\varphi([w_1, \dots, w_{k-1}, 0]) \neq \varphi([w_1, \dots, w_{k-1}, 1])$. Since φ is left permutive $\varphi([w_1, w_2, \dots, w_{k-1}, 0]) = \varphi([\bar{w}_1, w_2, \dots, w_{k-1}, 1])$. For each integer $p \geq k$ check to see if there are two blocks in $\mathcal{W}([2], p)$, $A_p = [0, a_2, \dots, a_p]$ and $B_p = [1, b_2, \dots, b_p]$ with the image blocks $\varphi(A_p)$ and $\varphi(B_p)$ of length $p - k + 1$ equal. Let P be the set of p where such a pair of blocks exists. We have seen $k \in P$. We will show that $P = \{p : p \geq k\}$. Suppose $p \in P$ and $k < q < p$. Then $q \in P$ by truncating A_p and B_p. Next suppose P has a largest element m. Let $A_m = 0C$ and $B_m = 1D$ for $C, D \in \mathcal{W}([2], m-1)$. Then $\varphi(a0C) = \varphi(a1D)$ for $a = 0, 1$ or m is not maximal by left permutivity. Continuing we see that for any word $w \in \mathcal{W}([2])$, $\varphi(w0C) = \varphi(w1D)$. But this is a contradiction for $w = [w_1, \dots, w_{k-1}]$ as chosen above. This shows that $P = \{p : p \geq k\}$. Next we see that this contradicts the assumption that φ is one-to-one. For each pair of blocks (A_p, B_p) consider the pair of ℓ-blocks obtained by truncating A_p and B_p. There are infinitely many p so there must be a pair of the resulting ℓ-blocks which is repeated infinitely often. Keep the pairs (A_p, B_p) these came from and discard the rest. Next consider the $(\ell + 1)$-blocks obtained by truncating the remaining pairs (A_p, B_p). Once again there must be a pair of resulting $(\ell + 1)$-blocks which is repeated infinitely often. Keep the pairs (A_p, B_p) these came from and discard the rest. Continue. In $X_{[2]}$ the successive pairs of blocks define nested pairs of disjoint, time zero, cylinder sets with the same images. The infinite intersections are nonempty by compactness. This produces at least one point in the intersection of blocks coming from the $A_p \subseteq [0]_0$ and at least one point in the intersection of the blocks coming from the $B_p \subseteq [1]_0$, that map to the same point. The contradiction means that when φ is an automorphism, it is a one-block map. There are only two such automorphisms: the identity and the flip. So $Aut(X_{[2]})$ is isomorphic to $\mathbb{Z}/2\mathbb{Z}$.   □

Proof. of (ii) We want to show that every finite group is a subgroup of $Aut(\Sigma_{[2]})$. If G is a finite group then G is isomorphic to a subgroup of $S_{|G|}$ where $|G|$ is the cardinality of G and $S_{|G|}$ is the permutation group on $|G|$ elements. It is sufficient to show that S_n can be embedded in $Aut(\Sigma_{[2]})$ for all n. The construction is extremely flexible. Fix n and find r so that there are at least n blocks in $\mathcal{W}([2], r)$ that begin and end with 1. Choose n of these blocks and label them 1 through n. A permutation $\pi \in S_n$ permutes these blocks. For $\pi \in S_n$ define an automorphism of $\Sigma_{[2]}$, φ_π, as follows: Look at Figure 3.1.1.

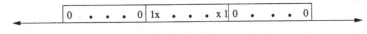

Figure 3.1.1

The automorphism φ_π scans a point x and does nothing unless it sees two blocks of 0's with length r separated by one of the numbered blocks. When this occurs φ_π permutes the block between the zeros by π. The important point is that the string of zero's, called *markers*, keeps the permuted blocks separated. There is no interference and there are no contradictions among the permuted blocks. This embeds each element of S_n into $Aut(\Sigma_{[2]})$.   □

Now we make some general observations about automorphism groups which hold for both one and two-sided subshifts of finite type.

Observation 3.1.2 *Let X be a one or two-sided irreducible subshift of finite type and let $Aut(X)$ denote its automorphism group. Then*

(i)   *$Aut(X)$ is finite or countably infinite,*
(ii)   *$Aut(X)$ is discrete in the compact-open mapping topology and*
(iii) *$Aut(X)$ is residually finite.*

Proof. For statement (*i*) observe that there are only a countable number of block maps from finite alphabets to finite alphabets.

For (*ii*). A subbasis for the compact-open topology on $Aut(X)$ consists of the sets

$$\mathcal{S}(C, U) = \{\varphi \in Aut(X) : \varphi(C) \subseteq U \quad \text{for } C \text{ compact and } U \text{ open}\}.$$

The topology on $Aut(X)$ is made up of all unions of finite intersections of these sets. The group $Aut(X)$ is discrete if every point is open. Each $\varphi \in Aut(X)$ is a block map with a memory m and an anticipation a. Consider the finite collection of subbasis sets $S([i_{-m}, \ldots, i_a], [j])$ where $\varphi([i_{-m}, \ldots, i_a]) = [j]$. The intersection of these sets consists of $\{\varphi\}$.

For (*iii*). The group $Aut(X)$ is residually finite if for every φ not equal to the identity there is a finite group H and a homomorphism $\alpha : Aut(X) \to H$ where $\alpha(\varphi)$ is not the identity in H. Let $Per(X,n)$ be the σ-periodic points in X with least period n. An automorphism permutes these points. So each automorphism defines an element of $S_{Per(X,n)}$, the group of permutations of $Per(X,n)$. The map $\alpha : Aut(X) \to S_{Per(X,n)}$ which sends each automorphism to the permutation it defines is a homomorphism. The periodic points are dense in X so only the identity fixes all periodic points.   □

Exercises

1. Why does a construction similar to the one used to prove Theorem 3.1.1(ii) not work for the one-sided two-shift?
2. Describe automorphisms of the one-sided three-shift of order two and three.
3. Give an example of an automorphism of infinite order of the two-sided two-shift.
4. Give an example of a group that is countably infinite but not residually finite.

§ 3.2 Automorphisms as Conjugacies

In Chapter 2 we examined topological conjugacies between subshifts of finite type. An automorphism of a subshift of finite type is a self-conjugacy so many of the ideas in Chapter 2 are of use when examining automorphisms. In Chapter 2 we saw that conjugacies between one-sided subshifts of finite type are easier to describe and more rigid than conjugacies between two-sided subshifts of finite type. This difference is mirrored in the size and complexity of the automorphism groups. The difference has already been vividly illustrated by Theorem 3.1.1.

Let φ be an automorphism of x, an irreducible subshift of finite type. Consider φ as a conjugacy from X to itself. In Section 2.1 we saw how to construct a subshift of finite type X_C "over" X when X is a one-sided subshift of finite type. The subshift of finite type X_C exhibits the conjugacy. The map φ is obtained by going up one side of the diagram described in the proof of Theorem 2.1.10 and down the other side. It is exhibited as a sequence of state splittings by followers followed by a sequence of amalgamations by followers all compatible with matrix column amalgamations. Furthermore, X is a column amalgamation of X_C. In Section 2.2 we examined the two-sided situation. We composed φ with a power of the shift and then constructed a subshift of finite type Σ_C "over" X which exhibited φ. In this case φ (composed with a power of the shift) is decomposed into a sequence of state splittings by predecessors followed by a sequence of amalgamations by followers all compatible with matrix operations. Here we have X a row amalgamation of Σ_C on the one hand and a column amalgamation of Σ_C on the other hand.

Next we use the description of one-sided conjugacies to describe a set of generators for the automorphism group of a one-sided shift. Each of the generators will have finite order. The main tools for this were developed in Section 2.1 when we examined topological conjugacies between one-sided subshifts of finite type. Given an $n \times n$ transition matrix A recall the definition of an elementary column amalgamation. An elementary amalgamation of A produces a matrix B which is an $(n-1) \times (n-1)$ matrix and where we have the equations $A = RS$ and $SR = B$ with S a zero-one matrix having no all zero rows and a single 1 in each column. A matrix such as S is called a *subdivision* matrix. A *one-step amalgamation* of A produces a matrix B that is defined by finding an $n \times (n - k)$ matrix R, $0 \leq k < n$, and an $(n - k) \times n$ subdivision matrix S with $A = RS$ and taking $SR = B$. Notice B can be obtained from A by a sequence of k elementary amalgamations. We define the *total one-step amalgamation*, A_1, of A as follows. Suppose there are $n - k$, $0 \leq k < n$, distinct columns in A. Let R be an $n \times (n - k)$ matrix made up of the $(n - k)$ distinct columns of A. The matrix R is unique up to right multiplication by a permutation matrix. Any other is RP for some $(n - k)$ permutation matrix P; i.e. we are allowed to rearrange the columns. Once R is fixed there is a unique $(n - k) \times n$ subdivision matrix S with $A = RS$. Then $A_1 = SR$. If we change R by rearranging the columns, taking RP instead, then we must rearrange the rows of S appropriately, taking $P^{-1}S$. So $A_1 = P^{-1}SRP$. We could have proceeded the other way around, by first choosing an $(n - k) \times n$ subdivision matrix, S, whose i^{th} and j^{th} columns agree if and only if the ith and j^{th} columns of A agree. We are free up to a rearrangement of the rows of S. Any other one is PS for a permutation matrix P. Once S is fixed R is determined, and $A_1 = SR$. We have proved the following lemma.

Lemma 3.2.1 *Given a square nonnegative integer matrix A the total one-step amalgamation of A is uniquely determined up to conjugation by a permutation matrix.*

It now makes sense to speak of *the* total one-step amalgamation of a matrix. It is also clear that if we relabel the rows and columns of A we still have the same total one-step amalgamations, $(PAP^{-1}), = A$, when P is a permutation matrix. This leads to the following observation.

Lemma 3.2.2 *Given A, R, S, with $A_1 = SR$ the total one-step amalgamation of A and R_1, S_1, with S_1, R_1, a one-step amalgamation of A, there is a unique subdivision matrix S_2 with $S_2 S_1 = S$. And then*

$$A = RS = R_1 S_1, \qquad S_1 R_1 = R_2 S_2, \qquad S_2 R_2 = SR = A_1$$

where R_2 is a uniquely determined matrix containing columns of $S_1 R_1$. See Figure 3.2.1.

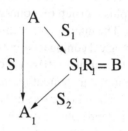

Figure 3.2.1

Proof. Notice both S_1 and S have full rank so both have right inverses. Also, if $\{1, \ldots, n\}$ indexes the vertices of A, then S_1 and S both define partitions of $\{1, \ldots, n\}$ and the partition defined by S_1 is a refinement of the one defined by S. Let S_2 be a matrix with rows indexed by the elements of the partition defined by S, columns indexed by the elements of the partition defined by S_1, and put a one in the ij^{th} entry if the jth element of S_1's partition is contained in the ith element of the one defined by S. Clearly, S_2 is a subdivision matrix and $S = S_2 S_1$. Now because S_1 has a right inverse we see that $A = RS = RS_2 S_1 = R_1 S_1$ implies that $R_1 = RS_2$. Let $R_2 = S_1 R$ so that $S_1 R_1 = S_1 RS_2 = R_2 S_2$ and $S_2 R_2 = S_2 S_1 R = SR = A_1$. $\qquad \square$

Given the matrix equations, $A = RS$ and $SR = B$, where S is a subdivision matrix, we must still define a graph homomorphism from G_A to G_B to define the one-step conjugacy from X_A to X_B. The matrix S determines which vertices in G_A map to which vertices in G_B but it may not determine which edges to map to which edges. There may still be choices. To define a map on the edges first divide the vertices into equivalence classes determined by their image in G_B. Two vertices are related if and only if S maps them to the same vertex in G_B. For a vertex i in G_A denote its equivalence class of vertices in G_A by $[i]_A$. Denote by $[i]_B$ its image vertex in G_B. For each pair of vertices i and j in G_A number the edges in $E_A(i, j)$ from 1 to $|E_A(i, j)|$. For each pair of vertices $[i]_B$ and $[j]_B$ in G_B number each edge in $E_B([i]_B, [j]_B)$ from 1 to $|E_B([i]_B, [j]_B)|$. These two numberings determine the map on edges. An edge e in $E_A(i, j)$ is mapped to the edge in $E_B([i]_B, [j]_B)$ with the same number. The matrix equations $A = RS$ and $SR = B$ show that this is well defined. The map determines an equivalence relation on the edges of G_A. As for vertices, denote the equivalence class of an edge in G_A by $[e]_A$ and its image in G_B by $[e]_B$. Once this is done we have defined a graph homomorphism from G_A to G_B and an elementary conjugacy from X_A to X_B. We say that the conjugacy is *compatible* with S. The two choices for edge numberings are reflected in Lemmas 3.2.3 and 3.2.4.

Lemma 3.2.3 *Suppose $A = RS$, $SR = B$ is a one-step amalgamation of A, with φ and $\varphi' : X_A \to X_B$ two one-step amalgamations compatible with S. Then $\varphi = \kappa \circ \varphi' \circ \tau$ where $\kappa : X_B \to X_B$ and $\tau : X_A \to X_A$ are automorphisms defined by graph automorphisms of G_B and G_A, respectively, that fix the vertices.*

Proof. There are two types of choices available in defining a graph homomorphism compatible with S. The first is in the numbering of the edges in G_A. This determines the equivalence classes of edges. There is one numbering for φ and one for φ'. Define τ to be the graph automorphism of G_A that fixes the vertices and takes the numbering for φ to the one for φ'. The second choice is the numbering of the edges in G_B. Define κ to be the graph automorphism of G_B that fixes the vertices and changes the numbering for φ' to the numbering for φ □

Lemma 3.2.4 *Let A, R, and S be matrices with $A_1 = SR$ the total one-step amalgamation of A. Suppose there are matrices R_1, S_1, R_2, and S_2 with $A = R_1 S_1$, and $B = S_1 R_1 = R_2 S_2$ a one-step amalgamation of A, $S_2 R_2 = A_1$ with $S_2 S_1 = S$ and $\varphi_1 : X_A \to X_B$ compatible with S_1. There is a $\varphi_2 : X_B \to X_{A_1}$ compatible with S_2 so that $\varphi_2 \circ \varphi_1$ is compatible with S. Moreover, if we also have φ compatible with S we may choose φ_2 so that $\varphi = \varphi_2 \circ \varphi_1 \circ \tau$ where $\tau : X_A \to X_A$ is defined by a graph automorphism of G_A that fixes the vertices.*

Proof. We have the diagram in Figure 3.2.1 and φ_1 compatible with S_1. The matrices S, S_1 and S_2 define the equivalence relations on V_A, V_A and V_B respectively, which we denote by $[\cdot]$, $[\cdot]_1$ and $[\cdot]_2$. They also define correspondences between the equivalence classes and the vertices in V_{A_1}, V_B and V_{A_1}. We see $[i] = \cup [i']_1$, where the union is over all $[i']_1$ in $[[i]_1]_2$. The map φ_1 comes from an equivalence relation $[\cdot]_1$ on $E_A(i, [j]_1)$ defined by a numbering of each edge of $E_A(i, j')$, and a correspondence between the equivalence classes of $E_A(i, [j]_1)$ and the edges in $E_B([i]_1, [j]_1)$. Label an edge $[e]_1 \in E_B$ by (n, i) where n is the number of each $e' \in [e]_1$ in G_A and i is the beginning vertex in G_A of each edge $e' \in [e]_1$. To define φ_2 first define an equivalence relation, $[\cdot]_2$, on edges in G_B. Say $[e]_1$ is related to $[e']_1$ if $[e]_1, [e']_1$ are in the same $E_B([i]_1, [j]_1]_2)$ and they have the same (n, i') label. Now make a one-to-one correspondence between

$[\cdot]_2$ classes in $E_B([[i]_1]_2, [[j]_1]_2)$ and edges in

$$E_{A_1}([[i]_1]_2, [[j]_1]_2) = E_{A_1}([i], [j]).$$

The map φ_2 is now well-defined and compatible with S_2. The map $\varphi_2 \circ \varphi_1$ is defined by the numbering of the edges in G_A which defines φ_1 and is compatible with S.

The second assertion follows from Lemma 3.2.3. If φ is compatible with S, $\varphi = \kappa \circ \varphi_2 \circ \varphi_1 \circ \tau$. But $\kappa \circ \varphi_2$ is just another one-step amalgamation compatible with S_2.   □

We will single out two special types of automorphisms of subshifts of finite type and then show any automorphism can be decomposed into a composition of these special automorphisms. If τ is a graph automorphism of G_A then for any pair of vertices i, j there is a smallest positive integer $r(i, j)$ with $\tau^{r(i,j)}(i) = i$ and $\tau^{r(i,j)}(j) = j$. Define the *first return map of $E_A(i,j)$* to be the map $\tau^{r(i,j)}$: $E_A(i,j) \rightarrow E_A(i,j)$. It is a permutation of $E_A(i,j)$. If G is a group of graph automorphisms of G_A and (i,j) is a pair of vertices in G_A the *return maps of G on (i,j)* are the maps $\gamma : E_A(i,j) \rightarrow E_A(i,j)$ for all $\gamma \in G$ where $\gamma(i) = i$ and $\gamma(j) = j$. The *return maps of G* are the return maps for all pairs of vertices. We single out two special types of graph automorphisms. A graph automorphism is a *vertex automorphism* if the first return map on $E_A(i,j)$ is the identity for every pair of vertices i and j. A graph automorphism is a *simple automorphism* if it fixes the vertices. Any graph automorphism can be written as a vertex automorphism composed with a simple automorphism (in either order). An automorphism of X_A is *defined by a vertex automorphism* if it is defined by a vertex automorphism of G_A. An automorphism φ is a *simple automorphism* if it is conjugate to an automorphism φ' of $X_{A'}$ where φ' is defined by a simple graph automorphism of $G_{A'}$, for some A'. Simple automorphisms are simple no matter how the subshift of finite type is presented.

Theorem 3.2.5 *The automorphism group of a one-sided subshift of finite type is generated by simple automorphisms and automorphisms defined by vertex automorphisms of the total amalgamation.*

Proof. Let X_A be the one-sided subshift of finite type, where A is totally amalgamated, and let φ be an automorphism of X_A. Define $C = A_0$ as in the proof of Theorem 2.1.10. Complete the diagram in Figure 3.2.2 as was done in Figure 2.1.3.

We need to be slightly careful. Examine a single one-step diamond as in Figure 3.2.3.

We require D^δ to be the total one-step amalgamation of D^α, $D^\alpha = RS$, $SR = D^\delta$ and $S_2 S_1 = \bar{S}_2 \bar{S}_1 = S$. It is possible by Lemma 3.2.4. We require

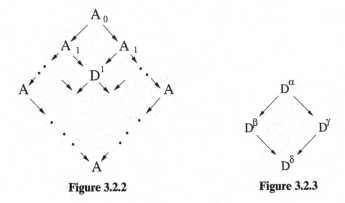

Figure 3.2.2   Figure 3.2.3

this of each one-step diamond in the diagram of Figure 3.2.2. The matrices down the lower left and right sides of Figure 3.2.2 are all A since they are amalgamations of A which is already a total amalgamation. We have one-step conjugacies $\varphi_i : X_{A_i} \to X_{A_{i+1}}$ and $\bar{\varphi}_i : X_{\bar{A}_i} \to X_{\bar{A}_{i+1}}$ with the appropriate S's that are supplied by the original φ with

$$\varphi = \bar{\varphi}_r \circ \cdots \circ \bar{\varphi}_0 \circ \varphi_0^{-1} \circ \cdots \circ \varphi_r^{-1}.$$

That is, the map is obtained by going up the upper left hand side and down the upper right hand side is φ. Starting at the top of the diagram and working down we apply Lemma 3.2.4 in each one-step diamond to choose compatible one-step conjugacies. This is shown in Figure 3.2.4.

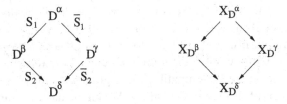

Figure 3.2.4

At each stage we have φ_1 and $\bar{\varphi}_1$ defined and compatible with S_1 and $\bar{S}_1$, but φ_2 and $\bar{\varphi}_2$ are not yet defined. By Lemma 3.2.4 we can choose φ_2 so that $\varphi_2 \circ \varphi_1$ is compatible with $S_2 S_1 = \bar{S}_2 \bar{S}_1$. Then we can choose $\bar{\varphi}_2$ so that $\bar{\varphi}_2 \circ \bar{\varphi}_1$ is compatible with $S_2 S_1$ and so that $\varphi_2 \circ \varphi_1 \circ \tau = \bar{\varphi}_2 \circ \bar{\varphi}_1$ where τ is an automorphism of X_{D^α} that comes from a simple graph automorphism of G_{D^α}.

This gives the diagram in Figure 3.2.5.

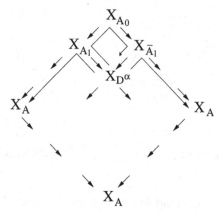

Figure 3.2.5

Now

$$\bar{\varphi}_2 \circ \bar{\varphi}_1 \circ \varphi_1^{-1} \circ \varphi_2^{-1} = \varphi_2 \circ \varphi_1 \circ \tau \circ \varphi_1^{-1} \circ \varphi_2^{-1}.$$

Let ψ be the conjugacy that goes from X_{D^1} to X_{A^1} and down the left side to X_A. Let $\bar{\psi}$ be the conjugacy that goes from X_{D^1} to $X_{\bar{A}^1}$ and down the right side to X_A. Then

$$\varphi = \bar{\psi} \circ \bar{\varphi}_2 \circ \bar{\varphi}_1 \circ \varphi_1^{-1} \circ \varphi_2^{-1} \circ \psi^{-1}$$
$$= \bar{\psi} \circ \psi^{-1} \circ ((\psi \circ \varphi_2 \circ \varphi_1) \circ \tau \circ (\psi \circ \varphi_2 \circ \varphi_1)^{-1}).$$

So $\bar{\psi} \circ \psi^{-1}$ is an automorphism of X_A, and φ is equal to $\bar{\psi} \circ \psi^{-1}$ preceded by a simple automorphisms of X_A. We continue working down the diagram in this way until we get φ equal to an automorphism ξ preceded by a sequence of simple automorphisms, and ξ is the automorphism that is obtained by going down the lower left side of the diagram in Figure 3.2.5 and up the lower right hand side. At each stage ξ is compatible with a graph automorphism of G_A, so ξ is defined by a graph automorphism of G_A. We know that any graph automorphism can be decomposed into a vertex automorphism preceded by a a simple graph automorphism. This means that φ is equal to a vertex automorphism of X_A preceded by a sequence of simple automorphisms.   □

Corollary 3.2.6 *The automorphism group of a one-sided subshift of finite type is generated by elements of finite order.*

Lemma 3.2.7 *Let A be a transition matrix and B be an amalgamation of A. Then the largest entry of A is less than or equal to the largest entry of B*

Proof. This follows immediately from the matrix equations for a one-step amalgamation.   □

Corollary 3.2.8 *Let X_A be a one-sided subshift of finite type with A a totally amalgamated zero-one transition matrix. Then the automorphism group of X_A is isomorphic to the group of graph automorphisms of G_A.*

Proof. This follows from Lemma 3.2.7 because all the matrices that occur in Figure 3.2.5 are zero-one matrices and there are no simple automorphisms.   □

Example 3.2.9 The one-sided Golden mean subshift of finite type defined by the transition matrix

$$\begin{pmatrix} 1 & 1 \\ 1 & 0 \end{pmatrix}$$

has a trivial automorphism group.

Lemma 3.2.10 *Let $A \neq [n]$ be an irreducible matrix whose total amalgamation is $[n]$, then every entry of A is strictly less than n.*

Proof. Any matrix arrived at by an elementary state splitting of $[n]$ has this property. Then apply Lemma 3.2.7.   □

We now have another proof of Theorem 3.1.1(i).

Corollary 3.2.11 *$Aut(X_{[2]})$ is isomorphic to $\mathbb{Z}/2\mathbb{Z}$.*

Proof. From Lemma 3.2.10 we see that any matrix with [2] as its total amalgamation is either a zero-one matrix or [2]. The only simple automorphism is defined by the one nontrivial automorphism $G_{[2]}$.   □

Suppose Σ_A is a two-sided subshift of finite type and φ is an automorphism. There is an integer k so that both φ and φ^{-1} can be defined by block maps with anticipation and memory k. In the proof of Theorem 2.1.14 the map φ was replaced by the map $\psi = \sigma^{-k} \circ \varphi$. Then ψ is defined by a block map with memory $2k$ and anticipation 0 while ψ^{-1} is defined by a block map with memory 0 and anticipation $2k$. The maps ψ and ψ^{-1} were used to build Σ_C. When ψ is an automorphism Σ_A is both a row and column amalgamation of Σ_C. But the resulting diagram cannot be analyzed as easily as in the one-sided case. The diagram exhibiting ψ does produce a strong shift equivalence from Σ_A to itself which is compatible with ψ. The difference in the diagrams between the one-sided and the two-sided cases reflects the fact that the shift is not an automorphism of a one-sided subshift of finite type but it is an automorphism of a two-sided one. In light of this and Theorem 3.2.5 a natural question arises.

Problem 3.2.12 Is the automorphism group of the two-sided two-shift generated by elements of finite order and the shift?

Very recently H. Kim, F. Roush and J. Wagoner have produced a two-sided irreducible and aperiodic subshift of finite type whose automorphism group is not generated by elements of finite order. See the notes.

Exercises

1. Define $\varphi \in Aut(X_{[3]})$ by saying φ scans a point and does nothing unless it sees 012 in which case it changes 0 to 1, or it sees 112 in which case it changes the first 1 to 0, or it sees 02 in which case it changes 0 to 1, or it sees 12 in which case it changes 1 to 0. What is the order of φ? Write down the transition matrix for X_C as in the proof of Theorem 3.2.5. Show that φ can be decomposed into a composition of two simple automorphisms each of order two.

2. Why does an argument similar to the one used to prove Theorem 3.2.5 not work for two-sided subshifts of finite type? Compare this situation to the one for conjugacy of one-sided subshifts of finite type (Theorem 2.1.10) verses the question of the decidability of conjugacy for two-sided subshifts of finite type.

§ 3.3 Subgroups of the Automorphism Group

In this section we will examine the problem of describing the subgroups of an automorphism group. We say a group G is *contained* in an automorphism group if G is isomorphic to a subgroup of the automorphism group. We begin with some examples of groups that cannot be contained in any automorphism group.

Observation 3.3.1 *Let X be a one or two-sided irreducible subshift of finite type and let $Aut(X)$ denote its automorphism group. None of following groups can be contained in $Aut(X)$.*

(i)   *An infinite simple group.*

(ii) *S_∞, the group of permutations of a countable set which fix all but a finite number of elements.*

(iii) *A nontrivial divisible group.*

Proof. An infinite simple group is not residually finite since there are no nontrivial homomorphisms from it to any finite group. By Observation 3.1.2 (*iii*) an infinite simple group cannot be a subgroup of $Aut(X)$.

The group A_∞ is the subgroup of S_∞ consisting of all even permutations. It is infinite and simple.

A group is divisible if each element contains roots of all orders. Let G be a nontrivial divisible group and φ an element of G not equal to the identity. Let H be a finite group and $\alpha : G \to H$ a homomorphism which doesn't send φ to the identity. Let K be the kernal of α and n be the cardinality of G/K. Suppose there is a ψ with $\psi^n = \varphi$. The nth power of every coset of K is K, so $K = (\psi K)^n = \psi^n K = \varphi K$. Then φ is in K which is a contradiction. $\qquad \square$

There is another type of group which cannot be contained in any automorphism group but for very different reason. To understand these groups we need some combinatorial group theory. Let $S = \{a_1, a_1^{-1}, a_2, a_2^{-1}, \dots\}$ be a finite or countably infinite set of pairs, a_i and a_i^{-1}, of symbols. Let $\mathcal{W}(S)$ be the set of all finite words on the symbols in S. Use a_i^k to denote the word of k a_i's, a_i^{-k} to denote the word of k a_i^{-1}'s and ϕ to denote the empty word. Define an equivalence relation on $\mathcal{W}(S)$ by the following rule. Two words are related if it is possible to transform one word into the other by a finite sequence of the following two operations;

(i)   inserting $a_i a_i^{-1}$ or $a_i^{-1} a_i$ between two successive symbols of a word or
(ii)   deleting $a_i a_i^{-1}$ or $a_i^{-1} a_i$ from a word.

The elements of the free group based on S are these equivalence classes. Multiplication of two equivalence classes is defined by concatenating an element from each class. A word $a_{j_1}^{\varepsilon_1} \cdots a_{j_\ell}^{\varepsilon_\ell}$ is reduced if $\varepsilon_t \in \mathbb{Z}, \varepsilon_t \neq 0$ and $a_{i_t} \neq a_{i_{t+1}}$ for each t. Each equivalence class contains exactly one reduced word and we use 1 to denote the class containing the empty word. The cardinality of S determines the free group up to isomorphism. If there are k pairs of symbols in S we say that the group is the *free group on k symbols* and denote it by F_k or $< a_1, \dots, a_k >$. If there are countably many pairs of symbols in S we say that the group is the *free group on countably many symbols* and denote it by F_∞ or $< a_1, a_2, \dots >$. Curiously, every free group on two or more symbols is isomorphic to a subgroup of every other free group on two or more symbols (see Exercise 2).

Consider the free group F based on the set S. Let $r_1, r_2, \dots \in \mathcal{W}(S)$. Let K denote the normal subgroup of F generated by the equivalence classes of $r_1, r_2, \dots$. We say that $< a_1, a_2, \dots ; r_1, r_2 \dots >$ is a *presentation* of the quotient group F/K. If there are finitely many a_j and r_i we say that the group is *finitely presented*. The r_i are the *relators* of the presentation.

Given a finitely presented group G based on the set S and a word $w \in \mathcal{W}(S)$ we can ask whether or not w is equivalent to the empty word. Given a finitely presented group G we say G has a *solvable word problem* if there is a Turing machine that can answer this question for every word. If there is no such

Turing machine we say G has an *unsolvable word problem*. There are finitely presented groups with unsolvable word problems.

A free group has a solvable word problem since it is possible to determine the unique reduced word in any word's equivalence class. A finite free product of cyclic groups also has a solvable word problem. These groups are defined as follows. A cyclic group has a presentation $< a; a^q >$ for some $q > 0$. It is isomorphic to $\mathbb{Z}/q\mathbb{Z}$. A finite free product of cyclic groups is defined by a finite collection of positive integers $q_1, \ldots, q_s$. A presentation for the group is $< a_1, \ldots, a_s; a_1^{q_1}, \ldots, a_s^{q_s} >$. A word $a_{j_1}^{\varepsilon_1} \cdots a_{j_\ell}^{\varepsilon_\ell}$ in such a group is *reduced* if $0 < \varepsilon_t < q_{j_t}$ and $a_{j_t} \neq a_{j_{t+1}}$ for each t. As for a free group each equivalence class contains exactly one reduced word and it is possible to determine the unique reduced word in the equivalence class of any word.

Now we apply these observations about finitely presented groups to our study of automorphism groups.

Observation 3.3.2 *Suppose G is a finitely presented group with an undecidable word problem. Then G cannot be contained in the automorphism group of any one or two-sided subshift of finite type.*

Proof. Suppose G is such a group and it is contained in $Aut(X)$ for some subshift of finite type X. Let $\{a_1, \ldots, a_k; r_1, \ldots, r_\ell\}$ be a presentation for G. Each a_i corresponds to an automorphism φ_i of X. Each φ_i has an explicit description as a block map. Given any finite word in the generators, $a_{i_1} \cdots a_{i_m}$, $\varphi_{i_1} \circ \cdots \circ \varphi_{i_m}$ is a explicit block map. Given an explicit block map it is easy to determine whether or not it defines the identity map on the shift space. $\quad\square$

Some finitely presented groups are contained in automorphism groups.

Observation 3.3.3 *If G is a free product of finitely many cyclic groups then it is isomorphic to a subgroup of the automorphism group of the one or two-sided n-shift for all sufficiently large n.*

Proof. Let G be a group with presentation $< a_1, \ldots, a_k; a_1^{q_1}, \cdots, a_k^{q_k} >$.

Let $N = q_1 + \cdots + q_k$. We prove the statement for the one-sided N-shift. The same proof works for both one and two-sided n-shifts when $n \geq N$. Let $\{j_t : 1 \leq t \leq k, 0 \leq j < q_t\}$ be the alphabet for $X_{[N]}$. For $t = 1, \ldots, k$ define $\tau_t \in Aut(X_{[N]})$ of order q_t by saying τ_t scans a point and does nothing unless it sees a 0_s, $s \neq t$. Then it applies the permutation to the symbol preceeding 0_s which sends the symbol j_t to $(j + 1 \mod q_t)_t$. It leaves all other symbols unchanged. The symbols 0_s for $s \neq t$ are the markers for τ_t.

Fix an element $g \in G$ and let $a_{j_0}^{\varepsilon_0} a_{j_1}^{\varepsilon_1} \cdots a_{j_\ell}^{\varepsilon_\ell}$ be the unique reduced word equivalent to g. Define a map $\psi : G \to Aut(X_{[N]})$ by sending g to $\psi_g =$

$\tau_{j_0}^{\varepsilon_0} \circ \cdots \circ \tau_{j_\ell}^{\varepsilon_\ell}$. This is a homomorphism. We prove that the kernel of ψ is the identity in G. Let $a_{j_0}^{\varepsilon_0} \cdots a_{j_\ell}^{\varepsilon_\ell}$ be a reduced word and $\tau_{j_0}^{\varepsilon_0} \circ \cdots \circ \tau_{j_\ell}^{\varepsilon_\ell}$ the automorphism it defines. Let $x \in X_{[N]}$ be a point with

$$x_i = (q_{j_i} - \varepsilon_i)_{j_i} \quad \text{for} \quad 0 \le i \le \ell,$$
$$x_i = 0_s \quad \text{for} \quad s \ne j_\ell \quad \text{and} \quad i > \ell.$$

Observe that
$$(\tau_{j_\ell}^{\varepsilon_\ell}(x))_i \quad \text{is unchanged for} \quad i \ne \ell$$
$$(\tau_{j_\ell}^{\varepsilon_\ell}(x))_\ell = 0_{j_\ell}.$$

Similarly
$$(\tau_{j_t}^{\varepsilon_t} \circ \cdots \circ \tau_{j_\ell}^{\varepsilon_\ell}(x))_i \quad \text{is unchanged for} \quad i < t \quad \text{and} \quad i > \ell$$
$$(\tau_{j_t}^{\varepsilon_t} \circ \cdots \circ \tau_{j_\ell}^{\varepsilon_\ell}(x))_t = 0_{j_t}$$

while the other coordinates depend on j_i and ε_i for $t < i \le \ell$. This depends on the fact that $a_{j_t}^{\varepsilon_t} \cdots a_{j_\ell}^{\varepsilon_\ell}$ is reduced. Working forward through the individual maps we see that while $x_0 \ne 0_s$ for any s at any time $(\tau_{j_0}^{\varepsilon_0} \circ \cdots \circ \tau_{j_\ell}^{\varepsilon_\ell}(x))_0 = 0_{j_0}$. Therefore $a_{j_0}^{\varepsilon_0} \cdots a_{j_\ell}^{\varepsilon_\ell}$ is not in the kernel of ψ.   □

A simple example illustrates the idea behind the proof.

Example 3.3.4 Let the group G have the presentation $< a_1, a_2, a_3; a_1^3, a_2^2, a_3^2 >$ so $N = 7$. The alphabet for $X_{[7]}$ is $\{0_1, 1_1, 2_1, 0_2, 1_2, 0_3, 1_3, \}$. Take for the reduced word $a_3 a_1^2 a_2 a_1$ and let x be the point $1_3 1_1 1_2 2_1 0_2 0_2 \cdots$.

Corollary 3.3.5 *For $n \ge 6$ the automorphism group of the one- or two-sided n-shift contains the free product of three copies of the cyclic group with two elements.*

Corollary 3.3.6 *For $n \ge 6$ the automorphism groups of the one- or two-sided n-shifts contain the free group on a finite or countably infinite set of generators.*

Proof. A presentation for the free product of three copies of the cyclic group with two elements is $< a_1, a_2, a_3; a_1^2, a_2^2, a_3^2 >$. The two elements $a_1 a_2$ and $a_1 a_3$ in this group generate a subgroup isomorphic to the free group on two generators. By our earlier discussion and Exercise 2 the free group on two generators has subgroups isomorphic to all the desired free groups.   □

These observations lead to the natural question.

Problem 3.3.7 Characterize the finitely presented groups that are contained in the automorphism group of a subshift of finite type.

In what follows we examine finite subgroups of automorphism groups.

Lemma 3.3.8 *Suppose X is an irreducible one or two-sided subshift of finite type and G is a finite subgroup of $Aut(X)$. Then X is conjugate to a subshift of finite type Y where each element of G is defined by a graph automorphism of the transition graph defining Y.*

Proof. Let $\mathcal{P}_A$ be the time zero partition of X, define $\mathcal{P}' = \vee g(\mathcal{P}_A)$, over all $g \in G$. It is a finite, open-closed partition of X. If $P_i \in \mathcal{P}'$, then $g(P_i) = P_j$ for some j. Given $x \in X$, associate x with its $(\mathcal{P}', \sigma)$ name. That is, $x' \in (\mathcal{P}')^{\mathbb{N}}$ or $(\mathcal{P}')^{\mathbb{Z}}$ where $(x')_n$ is the element of $\mathcal{P}'$ that contains $\sigma^n(x)$. Let X' be the collection of all $\mathcal{P}'$ names that arise in this way. Each element of G acts on X' as a one-block map. There is a conjugacy between (X', σ) and (X, σ). Now go to a higher block presentation of X' to get Y, a one-step subshift of finite type. Each element of G induces an automorphism of Y defined by a graph automorphism of the transition graph describing Y.   □

Recall from Section 3.2 the definition of the return maps for a group of graph automorphisms. The next lemma is stated for one-sided subshifts of finite type but also applies to two-sided ones and for two-sided ones we can use either a row or column amalgamation.

Lemma 3.3.9 *Suppose G is a group of graph automorphisms of G_A, and every return map is the identity. Then there is an isomorphism $\psi : G \to G'$, where G' is a group of graph automorphisms of the total one-step column amalgamation of A, and a graph homomorphism $\varphi : G_A \to G_{A_1}$, compatible with the amalgamation. Furthermore, the induced map, $\varphi : X_A \to X_{A_1}$, conjugates the G and G' actions.*

Proof. As in the construction of a one-step conjugacy compatible with a fixed subdivision matrix, number the edges in $E_A(i,j)$ from 1 to A_{ij} for each pair of vertices. Do this so that the group G acting on G_A preserves the numbering of the edges. Observe that this is possible if and only if every return map is the identity. That is the hypothesis. Now proceed exactly as in the construction of a one-step conjugacy and define a graph homomorphism using this numbering. The equivalence relation on the vertices is clearly preserved by G, and we define G to act on V_{A_1} by $g([i]) = [g(i)]$.

Consider the edges. Suppose $e, e' \in [e]$. Then $e \in E_A(i,j)$, $e' \in E_A(i,j')$ for some i and $j' \in [j]$. Also the two are labelled with the same number. For

$$g \in G, \quad g(e) \in E_A(g(i), g(j)), \quad g(e') \in E_A(g(i), g(j')), \quad g(j') \in [g(j)],$$

and the numbers are unchanged. This means that G preserves the equivalence relation on E_A. We define the action of G on the edges and so on G_{A_1}.   □

Next we classify the finite subgroups of the one-sided full shifts.

Example 3.3.10 Consider the one-sided three-shift $X_{[3]}$ with alphabet $\{0,1,2\}$. We see that S_3, the group of permutations of a three element set, is contained in $Aut(X_{[3]})$ since any permutations of the symbols defines an automorphism. Next define τ_2 to be an automorphism of order two by saying that τ_2 scans a point and does nothing unless it sees a 2 preceded by a 0 or a 1. Then it adds 1 modulo 2. The symbol 2 is the marker. We can define τ_0 and τ_1 similarly. The automorphism τ_1 scans a point and does nothing unless it sees a 1 preceded by a 0 or a 2 in which case it permutes the 1 and the 0 or 2 which procedes it?. The last one automorphism τ_0 is defined as expected. Each of these three automorphisms has order two but the composition of two has infinite order. To see this examine $\tau_1 \circ \tau_2$ in Figure 3.3.1 and observe the marker 2 moves to the left.

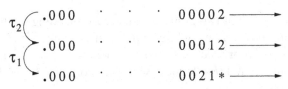

Figure 3.3.1

Since the length of the string of 0's is arbitrary the order of the automorphism is infinite. We now have automorphisms of order two, three and infinity in $Aut(X_{[3]})$. Fix $\ell \in \mathbb{N}$. Define an automorphism $\tau_{(2,\ell)}$ by letting it scan a point looking for the marker 2 as before. When it sees a 2 preceded by a 0 or 1 it adds 1 modulo 2 and carries to the left. It keeps adding and carrying until it runs into a 2 or it has carried to the ℓth place preceeding the marker. In either case it stops. This automorphism will have order 2^ℓ since a block of ℓ 0's and 1's preceeding a 2 will cycle through all blocks of 0's and 1's of length ℓ. We have automorphisms of order three, 2^ℓ and infinity in $Aut(X_{[3]})$. Is there an automorphism of order six?

Before proceeding we need a few definitions and results about finite groups. Let G and H be finite groups and π an onto homomorphism from G to H. Letting K be the kernel of π we have the short exact sequence

$$\{1\} \longrightarrow K \longrightarrow G \longrightarrow H \longrightarrow \{1\}.$$

The subgroup K is normal in G, H is isomorphic to the quotient group G/K and we say G is an *extension* of K by H. Given a finite group G a *normal series* is a chain

$$G = G_0 \supseteq G_1 \supseteq \cdots G_{\ell-1} \supseteq G_\ell = \{1\},$$

where G_{i+1} is a normal subgroup of G_i. The *factor groups* of a normal series are the groups

$$G_0/G_1, G_1/G_2, \ldots, G_{\ell-1}/G_\ell.$$

The group G_i is an extension of G_{i+1} by the factor group G_i/G_{i+1}. A *composition series* is a normal series where each G_{i+1} is a maximal normal subgroup in G_i. A normal series is a composition series if and only if each factor group is simple. The Jordan-Hölder Theorem states that for a finite group there is a one-to-one correspondence between the factor groups of any two composition series with the corresponding groups isomorphic. The factor groups of a composition series for G are called the *composition factors* of G. The original group G is built from the composition factors by a sequence of group extensions. If G is an extension of K by H then the composition factors of G are the same as the union of the composition factors of K and the composition factors of H.

The *direct product* of two groups is the group consisting of all pairs (k, h) such that $k \in K$ and $h \in H$ with the group operation defined as $(k', h')(k, h) = (k'k, h'h)$. The direct product of K and H is denoted by $K \times H$ and contains isomorphic copies $\bar{K}$ and $\bar{H}$ of K and H with $\bar{K}, \bar{H}$ normal in $K \times H$, $\bar{K} \cap \bar{H} = \{1\}$ and $\bar{K}\bar{H} = K \times H$.

A *semi-direct product* of two groups K and H is defined by a homomorphism $\Theta : H \to Aut(K)$. Letting $\Theta_h \in Aut(K)$ denote the image of h, the semi-direct product consists of all pairs (k, h) such that $k \in K$ and $h \in H$ with the group operation defined as $(k', h')(k, h) = (\Theta_h(k')k, h'h)$. It is denoted by $K \times_\Theta H$ and is the direct product when Θ is the constant homomorphism $\Theta \equiv 1$. The semi-direct product contains isomorphic copies $\bar{K}$ and $\bar{H}$ of K and H with $\bar{K}$ normal in $K \times_\Theta H$, $\bar{K} \cap \bar{H} = \{1\}$ and $\bar{K}\bar{H} = K \times_\Theta H$. The semi-direct product $K \times_\Theta H$ is an extension of K by H and so the composition factors of the semi-direct product are the same as the union of the composition factors of K and H.

For a finite set E let S_E denote the group of permutations of E and let S_n denote the group of permutations of $\{1, \ldots, n\}$.

For convenience we will construct some special groups. For $\gamma \in S_n^{\{1,\ldots,n\}}$ and $j \in \{1, \ldots, n\}$, $\gamma_j \in S_n$. Define a semi-direct product $S_n^{\{1,\ldots,n\}} \times_\Theta S_n$ with $\Theta : S_n \to Aut(S_n^{\{1,\ldots,n\}})$ defined by $(\Theta_h(\gamma))_j = \gamma_{h(j)}$. This particular semi-direct product is usually called the *wreath product*. The group $S_n^{\{1,\ldots,n\}} \times_\Theta S_n$ acts on $\{1, \ldots, n\}^2$ by $(\gamma, h)(i, j) = (\gamma_j(i), h(i))$. Every composition factor of $S_n^{\{1,\ldots,n\}} \times_\Theta S_n$ is isomorphic to a subgroup of S_n. In addition, if K and H are isomorphic to subgroups of S_n then $S_n^{\{1,\ldots,n\}} \times_\Theta S_n$ contains an isomorphic copy of every extension of K by H. Define by induction $Z_n^1 = S_n$, $Z_n^2 = S_n^{\{1,\ldots,n\}} \times_\Theta S_n$ and $Z_n^k = S_n^{\{1,\ldots,n\}^{k-1}} \times_\Theta Z_n^{k-1}$ with $\Theta : Z_n^{k-1} \to Aut(S_n^{\{1,\ldots,n\}^{k-1}})$ defined by $(\Theta_h(\gamma))_{(j_1,\dots,j_{k-1})} = \gamma_{h(j_1,\dots,j_{k-1})}$. The group Z_n^k acts on $\{1,\dots,n\}^k$ by $(\gamma,h)(i_1,\dots,i_k) = (\gamma_{(i_2,\dots,i_k)}(i_1),h(i_2,\dots,i_k))$. Every composition factor of Z_n^k is isomorphic to a subgroup of S_n. In addition, if G is a group and every composition factor of G is isomorphic to a subgroup of S_n then G is isomorphic to a subgroup of Z_n^k for some k.

Now we apply these ideas about finite groups to our study of automorphism groups.

Observation 3.3.11 *If G is a finite subgroup of $Aut(X_{[n]})$ then either:*

(i) *the composition factors of G are isomorphic to subgroups of S_{n-1} or;*
(ii) *G is isomorphic to a subgroup G' of S_n.*

Proof. Use Lemma 3.3.8 to get an X_A conjugate to $X_{[n]}$ with G acting as a group of graph automorphisms on G_A. Let

$$P_A = \{(i,j) \in V_A \times V_A : A_{ij} > 0\}.$$

Define a homomorphism $\nu_1 : G \to S_{P_A}$ by $\nu_1(g)(i,j) = (g(i),g(j))$. Let $G_1 = \nu_1(G)$ and $g_1 = \nu_1(g) \in G_1$. Define the wreath product $S_{n-1}^{P_A} \times_\Theta G_1$ as discussed with $\Theta : G_1 \to Aut(S_{n-1}^{P_A})$ and $(\Theta_h(\gamma))_{(i,j)} = \gamma_{h(i,j)}$. It acts on $\{1,\dots,n-1\} \times P_A$ by $(\gamma,h)(r,(i,j)) = (\gamma_{(i,j)}(r),h(i,j))$. For each $(i,j) \in P_A$ number the edges in $E_A(i,j)$ from 1 to A_{ij}. By Lemma 3.2.10 either all $A_{ij} < n$ or we are done. Identify E_A with the corresponding subset of $\{1,\dots,n-1\} \times P_A$. Define an embedding, ε_1, from G into $S_{n-1}^{P_A} \times_\Theta G_1$ by G's action on $E_A \subseteq \{1,\dots,n-1\} \times P_A$, $g(r,(i,j)) = (\gamma,g_1)(r,(i,j)) = (\gamma_{(i,j)}(r),g_1(i,j))$. This forces the definition of $\Theta : G_1 \to Aut(S_{n-1}^{P_A})$ since

$$g'g(r,(i,j)) = g'(\gamma_{(i,j)}(r),g_1(i,j))$$
$$= (\gamma'_{g_1(i,j)}(\gamma_{(i,j)}(r)),g_1'g_1(i,j))$$
$$= (\Theta_{g_1}(\gamma')\gamma)_{(i,j)}(r),g_1'g_1(i,j)).$$

Let K_1 be the intersection of $\varepsilon_1(G)$ and the kernel of the quotient map $S_{n-1}^{P_A} \times_\Theta G_1 \to G_1$. Since K_1 is isomorphic to a subgroup of $S_{n-1}^{P_A}$ all its composition factors are isomorphic to subgroups of S_{n-1}. We have

$$\{1\} \longrightarrow K_1 \longrightarrow G \longrightarrow G_1 \longrightarrow \{1\}.$$

The group G_1 acts on G_A by taking its action on P_A and preserving the edge numbering of G_A. By Lemma 3.3.9, G_1 induces a conjugate G_1 action on the total one-step amalgamation of X_A. The action is defined by a G_1 action on G_{A_1}. Repeat this process reducing until reaching a matrix B whose total one-step amalgamation is $[n]$. It is acted on by G_ℓ with identity return maps. We have

$$G \xrightarrow{\nu_1} G_1 \xrightarrow{\nu_2} \cdots G_{\ell-1} \xrightarrow{\nu_{\ell-1}} G_\ell,$$

where each ν_i is onto and each K_i has all of its composition factors isomorphic to subgroups of S_{n-1}.
There are two cases:

a) B is $s \times s$ for some $s < n$;
b) B is the $n \times n$ matrix of all 1's.

In case (a), G_ℓ is determined by its action on V_B and is isomorphic to a subgroup of S_{n-1}. Then G has all its composition factors isomorphic to subgroups of S_{n-1}.

In case (b), all the matrices from A down to B are zero-one matrices. This means that $K_i \simeq \{1\}$, for all i and $G \simeq G_\ell$. But, G_ℓ is determined by its action on V_B and so is isomorphic to a subgroup G' of S_n.   □

Example 3.3.12 Consider the automorphism $\tau_{(2,2)}$ of Example 3.3.10. It generates a finite subgroup of $Aut(X_{[3]})$ which is isomorphic to $\mathbb{Z}/4\mathbb{Z}$. First apply Lemma 3.3.8 to obtain the graphs in Figure 3.3.2. In the left hand graph the edges are labelled to describe the conjugacy with $X_{[3]}$. In the right hand graph the edges from vertex i to vertex j are labelled from 1 to A_{ij} as in the proof of Observation 3.3.11.

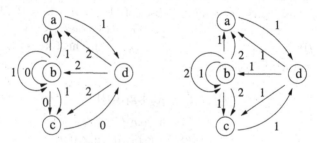

Figure 3.3.2

The automorphism $\tau_{(2,2)}$ as described in terms of the right hand graph interchanges the vertices a and c and interchanges the labels on the edges from b to a when it sends them to edges from b to c. That is all it does. In terms of the proof of Observation 3.3.11

$$P_A = \{(a,d),\ (b,a),\ (b,b),\ (b,c),\ (c,d),\ (d,a),\ (d,b),\ (d,c),\ (d,d)\},$$

so S_{P_A} is isomorphic to S_9. The map $\nu_1 : \mathbb{Z}/4\mathbb{Z} \to S_{P_A}$ sends 1 to the pair of two cycles $\{((b,a),(b,c)),\ ((d,a),(d,c))\}$ in S_{P_A}. So G_1 is isomorphic to $\mathbb{Z}/2\mathbb{Z}$.

The map $\varepsilon_1 : \mathbb{Z}/4\mathbb{Z} \to S_2^{P_A} \times_\Theta G_1$ is

$$\varepsilon_1(0) = (id, id),$$
$$\varepsilon_1(1) = (\gamma^1, \nu_1(1)),$$
$$\varepsilon_1(2) = (\gamma^2, id),$$
$$\varepsilon_1(3) = (\gamma^3, \nu_1(1)),$$

where

$$\gamma^1_{(i,j)} = (12) \quad \text{if} \quad (i,j) = (b,a)$$
$$id \quad \text{otherwise}$$
$$\gamma^2_{(i,j)} = (12) \quad \text{if} \quad (i,j) = (b,a),\ (b,c)$$
$$id \quad \text{otherwise}$$
$$\gamma^3_{(i,j)} = (12) \quad \text{if} \quad (i,j) = (b,c)$$
$$id \quad \text{otherwise}$$

Then $\varepsilon_1(\mathbb{Z}/4\mathbb{Z})$ acts by $(\gamma^t, g_1)(r, (i,j)) = (\gamma^t_{(i,j)}(r), g_1(i,j))$. This results in the exact sequence

$$\{0\} \longrightarrow \mathbb{Z}/2\mathbb{Z} \longrightarrow \mathbb{Z}/4\mathbb{Z} \longrightarrow \mathbb{Z}/2\mathbb{Z} \longrightarrow \{0\}.$$

The total one-step amalgamation of A is given by the graph in Figure 3.3.3.

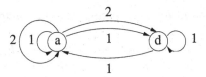

Figure 3.3.3

The group $G_1 \simeq \mathbb{Z}/2\mathbb{Z}$ pushes down to an action on this defined by permuting the two edges between a and d.

We will show that if a finite group has all its composition factors isomorphic to a subgroup of S_{n-1} then it is isomorphic to a subgroup of $Aut(X_{[n]})$. The construction is a generalization of the construction of the map $\tau_{(2,\ell)}$ in Example 3.3.10. We use a marker to signal the automorphism when to act and then carry to the left. We use the groups Z_n^k defined in the discussion about finite groups and the fact that if G is a group with all its composition factors isomorphic to subgroups of S_{n-1} then it is isomorphic to a subgroup of Z_{n-1}^k for some k. We show that every Z_{n-1}^k can be embedded in $Aut(X_{[n]})$.

Construction 3.3.13 Begin by numbering the symbols of $X_{[n]}$ 0 through $n - 1$. We will define the action of $(\gamma, g) \in Z_{n-1}^2$ on $X_{[n]}$ using 0 as a marker and "carrying" once to the left. Define the automorphism by

$$((\gamma, g)(x))_i = \begin{cases} g(x_i) & \text{if } x_i \neq 0 \text{ and } x_{i+1} = 0 \\ \gamma_{x_{i+1}}(x_i) & \text{if } x_i, x_{i+1} \neq 0 \text{ and } x_{i+2} = 0 \\ x_i & \text{otherwise.} \end{cases}$$

This embeds Z_{n-1}^2 into $Aut(X_{[n]})$. 0 is the marker and we "carry" once to the left. Similarly, we embed Z_{n-1}^3 into $Aut(X_{[n]})$. We again use 0 for a marker but now carry twice to ths left. For $(\delta, (\gamma, g)) \in Z_{n-1}^3$ define

$$((\delta, (\gamma, g))(x))_i = \begin{cases} (\gamma, g)(x_i) & \text{if } x_{i+1} \text{ or } x_{i+2} = 0 \\ (\Theta_{(\gamma, g)}(\delta))_{(x_{i+1}, x_{i+2})}(x_{i+3}) & \text{if } x_{i+1}, x_{i+2} \neq 0 \text{ and } x_{i+3} = 0 \\ x_i & \text{otherwise.} \end{cases}$$

We embed all Z_{n-1}^k in this way.

This proves the next theorem.

Theorem 3.3.14 *A finite group G is isomorphic to a subgroup of $Aut(X_{[n]})$ if and only if either:*

(i)   *it is isomorphic to a subgroup of S_n; or*
(ii) *all its composition factors are isomorphic to subgroups of S_{n-1}.*

Corollary 3.3.15 *If φ is an automorphism of the full n-shift with finite order, and n is not prime then φ has order $p_1^{e_1} \cdots p_t^{e_t}$ for primes $p_i < n$ and $e_i \in \mathbb{Z}^+$. If n is prime then φ has order $p^{e_1} \cdots p_t^{e_t}$ for primes $p_i < p$ and $e_i \in \mathbb{Z}^+$ or it has order n. Moreover, all of these orders occur.*

Example 3.3.16 Consider $X_{[3]}$ as in Example 3.3.10. Corollary 3.3.15 shows that there is no automorphism of $X_{[3]}$ with order six.

Theorem 3.3.14 completely characterizes the finite subgroups of the automorphism groups of one-sided full shifts. Theorem 3.1.1 shows that the automorphism group of the two-sided two-shift contains every finite group. Next we will determine some further properties of the automorphism groups of two-sided subshifts of finite type.

In all of our constructions of automorphisms we have used markers to signal the automorphism when to act. The next lemma guarantees the existence of the markers we will need in the constructions that follow.

Lemma 3.3.17 *Let Σ_A be an irreducible, aperiodic, two-sided subshift of finite type with positive entropy. For all sufficiently high block presentations of Σ_A we can find:*

(i) *A marking symbol $m \in L_{A[k]}$ so that the subshift of finite type obtained by deleting the symbol m from the alphabet of $\Sigma_{A[k]}$ is irreducible, aperiodic and has entropy at least $h(\Sigma_A, \sigma) - \varepsilon$, or*

(ii) *A marking block $M \in \Sigma_{A[k]}$ of arbitrarily long length that cannot overlap itself.*

We leave the proof of this lemma as Exercise 10 and use it to prove the next theorem.

Theorem 3.3.18 *The automorphism group of any irreducible, aperiodic, two-sided subshift of finite type with positive entropy contains the automorphism group of any two-sided full shift.*

Proof. Let Σ_A denote the subshift of finite type. By Lemma 3.3.17 we may assume that the alphabet of Σ_A contains the marking symbol m. Fix n and let $\{1, \dots, n\}$ be the alphabet for $\Sigma_{[n]}$. For sufficiently large k we can choose n^2 distinct blocks in $\mathcal{W}(A, k)$ which do not contain the symbol m but which can be preceded and followed by m. Let $w(i,j)$ for $i, j = 1, \dots, n$ denote these blocks.

Let $\varphi \in Aut(\Sigma_{[n]})$ and take ℓ so that φ can be expressed as a block map with memory and anticipation ℓ. Define $\bar{\varphi} \in Aut(\Sigma_A)$ as follows. Define the image of a block

$$mw(i_{-\ell}, j_{-\ell}) \cdots mw(i_0, j_0) \cdots mw(i_\ell, j_\ell)m$$

to be the block $w(i', j')$ where

$$i' = \varphi([i_{-\ell}, \dots, i_\ell]), \qquad j' = \varphi([j_\ell, j_{\ell-1}, \dots, j_{-\ell}]).$$

Note that the order of the j's is reversed. This defines $\bar{\varphi}$ on all points made up of alternating $w(i,j)$ and m's. It must be extended to all of Σ_A.

For $0 \le t < \ell$ and x a block not of the form $mw(i,j)$ define the image of a block

$$xmw(i_{-t}, j_{-t}) \cdots mw(i_0, j_0) \cdots mw(i_\ell, j_\ell)m$$

to be the block $w(i', j')$ where

$$i' = \varphi([j_{t'}, j_{t'-1}, \dots, j_{-t}, i_{-t}, i_{-t+1}, \dots, i_\ell])$$

and

$$j' = \varphi([j_\ell, j_{\ell-1}, \dots, j_{-t}, i_{-t}, i-t+1, \dots, i_{t'}]),$$

where $t' = \ell - 2t - 1$. This can be explained for i' by saying $\bar\varphi$ scans $[i_0, \dots, i_\ell]$ and then begins scanning to the left of i_0. When it scans from i_{-t} to look for i_{-t-1} it hits a wall and bounces back scanning the j's. The same is true for j'.

For $0 \le t < \ell$ and y a block not of the form $w(i,j)m$ define the image of a block

$$mw(i_{-\ell}, j_{-\ell}) \cdots mw(i_0, j_0) \cdots mw(i_t, j_t)my$$

as for the block above but with the ends reversed.

For $0 \le s, t < \ell$, x a block not of the form $mw(i,j)$ and y a block not of the form $w(i,j)m$ define the image of a block

$$xmw(i_{-s}, j_{-s}) \cdots mw(i_0, j_0) \cdots mw(i_t, j_t)my$$

to be

$$xmw(i'_{-s}, j'_{-s}) \cdots mw(i'_0, j'_0) \cdots mw(i'_t, j'_t)my,$$

where φ sends the periodic point defined by the block

$$[i_{-s}, \dots, i_t, j_t, j_{t-1}, \dots, j_{-s}]$$

to the periodic point block defined by the block

$$[i'_{-s}, \dots, i'_t, j'_t, j'_{t-1}, \dots, j'_{-s}].$$

This is consistent with the definition of $\bar\varphi$ when there is no y to the right of $w_{(i_0,j_0)}$ or x to the the left. On all other blocks $\bar\varphi$ is the identity. The map $\bar\varphi$ doesn't change the marker structure of a point. It changes only the blocks $w(i,j)$ which are always changed to blocks of the same form. The map is well-defined and preserves composition of maps, $\overline{(\varphi \circ \psi)} = \bar\varphi \circ \bar\psi$. The construction defines an embedding.   □

Corollary 3.3.19 *The automorphism group of any irreducible, aperiodic, two-sided subshift of finite type with positive entropy contains every finite group.*

Corollary 3.3.20 *For any $n \ge 2$ the automorphism group of the two-sided two-shift and the automorphism group of the two-sided n-shift contain the same subgroups.*

This leads to the question.

Problem 3.3.21 Are the automorphism groups of the two-sided two-shift and the two-sided three-shift isomorphic?

Another indication of the size and complexity of the automorphism groups of two-sided subshifts of finite type comes from the next theorem.

In Section 3.1 we let $Per(\Sigma_A, n)$ denote the points in Σ_A with least σ-period n and noted that any automorphism permutes these points. Moreover, the automorphism must preserve their σ-obits and the order on these orbits. If we let $Orb(\Sigma_A, n)$ denote the σ-orbits of length n then any automorphism permutes these orbits.

The subgroup of a group composed of all elements which commute with every element in the group is called the *center* of the group.

Theorem 3.3.22 *Let Σ_A be an irreducible and aperiodic subshift of finite type with positive entropy. The center of $Aut(\Sigma_A)$ consists of the powers of the shift.*

Proof. By Lemma 3.3.17 we may assume Σ_A contains marking blocks of arbitrarily long length which cannot overlap themselves. Fix a marking block M of some length k.

For $n \geq 0$ let $\mathcal{B}(M, n) \subseteq \mathcal{W}(\Sigma_A, n)$ consist of all blocks which do not contain M but can be preceded and followed by M. Let $S_{\mathcal{B}(M,n)}$ denote the group of permutations of $\mathcal{B}(M, n)$. For $\pi \in S_{\mathcal{B}(M,n)}$ define $\psi_\pi \in Aut(\Sigma_A)$ by saying ψ_π scans a point and does nothing unless it sees a block of the form MuM for $u \in \mathcal{B}(M, n)$ in which case it changes the block to $M\psi_\pi(u)M$. This is a standard marker automorphism. We want to show that only powers of the shift can commute with all such automorphisms.

Let $\mathcal{P}(M, n)$ be the set of σ-periodic points of period $k + n$ defined by the blocks MuM for $u \in \mathcal{B}(M, n)$ where the successive occurrences of M are overlapped. These points have period $k + n$. Let $\mathcal{O}(M, n)$ denote the orbits of these points. The group of automorphisms $\{\psi_\pi : \pi \in S_{\mathcal{B}(M,n)}\}$ induces all possible permutations of $\mathcal{O}(M, n)$. Also, each ψ_π fixes all points of σ-period $k + n$ not in $\mathcal{P}(M, n)$.

Let $\varphi \in Aut(\Sigma_A)$ and suppose φ is in the center of $Aut(\Sigma_A)$. If $\mathcal{O}(M, n)$ consists of more than one orbit, φ defines a permutation of $\mathcal{O}(M, n)$. Otherwise, φ maps one of the orbits in $\mathcal{O}(M, n)$ to an orbit not in $\mathcal{O}(M, n)$ and since it commutes with all the ψ_π it maps every orbit in $\mathcal{O}(M, n)$ to the same orbit outside of $\mathcal{O}(M, n)$. The only permutation of a set with two or more elements which commutes with every permutation of the set is the identity. Consequently, φ defines the identity permutation on $\mathcal{O}(M, n)$.

Let $\gamma \in \mathcal{O}(M, n)$. Then there is a j so that for $x \in \gamma$, $\varphi(x) = \sigma^j(x)$. Let γ' be another orbit in $\mathcal{O}(M, n)$. Let ψ_π be one of our automorphisms which takes γ' to γ. For $y \in \gamma'$, $\varphi(y) = \psi_\pi^{-1} \circ \varphi \circ \psi_\pi(y) = \sigma^j(y)$. This says φ is σ^j on each point in $\mathcal{P}(M, n)$.

Suppose φ has memory m and anticipation a. It is possible to find a marker block M and an n so that every block of length $m + a + 1$ occurs in $\mathcal{B}(M, n)$. We know $\varphi = \sigma^j$ for some j on $\mathcal{P}(M, n)$. It follows that $-m \leq j \leq a$ and φ is σ^j. $\qquad\square$

Exercises

1. Let the group G have presentation $< a_1, a_2; a_1^2, a_2^4 >$. Describe embeddings of G into the automorphism groups of the one and two-sided six-shifts.

2. Let F_k denote the free group on k symbols. Show that F_3 can be embedded in F_2. Hint: let $< 0, 1 >$ be a presentation for F_2 and consider the words 01 and 001. Generalize the embedding to F_k for any k and to F_∞.

3. Let τ_2 be the automorphism of $X_{[3]}$ with order two described in Example 3.3.10. Let τ be the automorphism of $X_{[3]}$ with order three defined by the permutation of symbols (012). What is the order of $\tau \circ \tau_2$?

4. For the one-sided four-shift find automorphisms with order two, three, four, six, nine and one with infinite order.

5. Show that $\mathbb{Z}/2\mathbb{Z} \oplus \mathbb{Z}/2\mathbb{Z}$ and $\mathbb{Z}/4\mathbb{Z}$ can be embedded in $Aut(X_{[3]})$ but $\mathbb{Z}/6\mathbb{Z}$ cannot.

6. Let A be the transition matrix $\begin{bmatrix} 1 & 1 \\ 2 & 0 \end{bmatrix}$. Find an automorphism of order four of X_A.

7. Describe the automorphism found in Exercise 6 as defined by a graph automorphism as described in Lemma 3.3.9.

8. Let A be the transition matrix

    $$\begin{bmatrix} 1 & 1 & 1 \\ 1 & 0 & 1 \\ 1 & 1 & 0 \end{bmatrix}.$$

    Show that $Aut(X_A)$ is isomorphic to $\mathbb{Z}/2\mathbb{Z}$ and $Aut(\Sigma_A)$ contains every finite group.

9. Use the Perron-Frobenius Theorem to show that there is no automorphism of the two-sided two-shift whose square is the shift. Show that there is an automorphism of the two-sided four-shift whose square is the shift. Use Theorem 3.3.22 to conclude that the automorphism groups of the two-shift and the four-shift are not isomorphic.

10. Prove Lemma 3.3.17.

§ 3.4 Actions of Automorphisms

In this section we will restrict our attention to two-sided subshifts of finite type and examine how automorphisms act on the space. In Section 3.1 and 3.3 we let $Per(\Sigma_A, n)$ denote the points with least σ-period n. Let $Orb(\Sigma_A, n)$ denote the σ-orbits of length n. As noted there, an automorphism of Σ_A defines a permutation of each of these sets. There is a homomorphism from $Aut(\Sigma_A)$ to

$S_{Per(\Sigma_A,n)}$ and to $S_{Orb(\Sigma_A,n)}$ for each n. An old and interesting question is the following.

Problem 3.4.1 Is there an irreducible subshift of finite type with two fixed points which cannot be interchanged by any automorphism?

The gyration function is used to keep track of some of this periodic point information. Fix a $\varphi \in Aut(\Sigma_A)$ and an $n \geq 0$. For each $\gamma \in Orb(\Sigma_A, n)$ choose a base point $x_\gamma \in \gamma$. There is a well-defined $m(\varphi, x_\gamma) \in \mathbb{Z}/n\mathbb{Z}$ such that $\varphi(x_\gamma) = \sigma^{m(\varphi, x_\gamma)}(x_{\varphi(\gamma)})$. Define the nth *gyration number* to be $g_n(\varphi) = \sum m(\varphi, x_\gamma)$ where the sum is over all $\gamma \in Orb(\Sigma_A, n)$ and let $g_n(\varphi) = 0$ if $Orb(\Sigma_A, n)$ is empty. Define the *gyration function* $g : Aut(\Sigma_A) \to \prod \mathbb{Z}/n\mathbb{Z}$, where the product is over all $n \in \mathbb{N}$, by $g(\varphi) = (g_1(\varphi), g_2(\varphi), \dots)$.

Proposition 3.4.2 *The gyration function* $g : Aut(\Sigma_A) \to \prod \mathbb{Z}/n\mathbb{Z}$ *is a well-defined homomorphism.*

Proof. We need only show that each g_n is a well-defined homomorphism. Let $\varphi \in Aut(\Sigma_A)$. Suppose $\gamma \in Orb(\Sigma_A, n)$ and we replace the base point x_γ by $\sigma(x_\gamma)$.

There are two possibilities, either $\varphi(\gamma) = \gamma$ or not. If $\varphi(\gamma) = \gamma$ then $m(\varphi, \sigma(x_\gamma)) = m(\varphi, x_\gamma)$ and no other $m(\varphi, x_{\gamma'})$ is affected. If $\varphi(\gamma) \neq \gamma$ then $m(\varphi, \sigma(x_\gamma)) = m(\varphi, x_\gamma) + 1$, the new $m(\varphi, x_{\varphi^{-1}(\gamma)})$ is equal to the old one minus 1 and no other $m(\varphi, x_{\gamma'})$ is affected. So, the sum for the gyration number is unchanged and the function is well-defined.

To see g_n is a homomorphism suppose ψ is also in $Aut(\Sigma_A)$ and $\gamma \in Orb(\Sigma_A, n)$. Then $m(\varphi \circ \psi, x_\gamma) = m(\phi, x_{\psi(\gamma)}) + m(\psi, x_\gamma)$ so $g_n(\varphi \circ \psi) = g_n(\varphi) + g_n(\psi)$. □

Finite collections of periodic points can form compact sets which are invariant under the action of $Aut(\Sigma_A)$. The next theorem shows that these are the only nontrivial compact invariant sets.

Theorem 3.4.3 *If* $X \subseteq \Sigma_A$ *is a compact set which is invariant under the action of* $Aut(\Sigma_A)$ *then it is either finite or all of* Σ_A.

The proof of the theorem will follow from the next lemma by noting that any infinite compact σ-invariant set contains a point with an infinite σ-orbit. The lemma is stated without proof and a reference is included in the notes. The proof is similar to the proof of Theorem 3.3.22 but more delicate. It involves constructing marker automorphisms that carry the point's orbit into any cylinder set.

Lemma 3.4.4 *Suppose Σ_A is an irreducible and aperiodic subshift of finite type. Let $x \in \Sigma_A$ and suppose it is not a periodic point. Then the orbit of x under the action of $Aut(\Sigma_A)$ is dense in Σ_A.*

Let Σ_A be an irreducible subshift of finite type. Recall from Section 2.2 the inverse limit group $\varprojlim(\mathbb{T}^n, A)$ with its associated automorphism A^*. Let $\varphi \in Aut(\Sigma_A)$. We can consider φ as a conjugacy from Σ_A to itself as we did in Section 3.2 and apply Theorems 2.1.14 and 2.1.17. This decomposes φ into a sequence of elementary conjugacies which are described by a sequence of elementary matrix equations. Each elementary matrix equation defines an automorphism of the group $\varprojlim(\mathbb{T}^n, A)$ which commutes with A^*. So φ induces an automorphism of $\varprojlim(\mathbb{T}^n, A)$ which commutes with A^*. This leads to the next observations.

Observation 3.4.5 *Let Σ_A be an irreducible subshift of finite type. There is a nontrivial homomorphism from $Aut(\Sigma_A)$ to the group of automorphisms of $\varprojlim(\mathbb{T}^n, A)$ and the image of each element commutes with A^*.*

Observation 3.4.6 *Let Σ_A be an irreducible subshift of finite type. There is a nontrivial homomorphism from $Aut(\Sigma_A)$ to the group of automorphisms of the dimension group of Σ_A and the image of each element commutes with A_*.*

Exercises

1. Let $\tau_{(2,2)}$ be the automorphism of Example 3.3.10. Compute the gyration numbers $g_1(\tau_{(2,2)})$ through $g_5(\tau_{(2,2)})$.
2. Let Σ_A be an irreducible subshift of finite type and $\varphi \in Aut(\Sigma_A)$ have finite order. Show that the gyration numbers, $g_p(\varphi)$, are zero for all sufficiently large primes p.
3. Show that an infinite, compact, σ-invariant set contains a point that is not periodic.
4. Recall the geometric definition of the dimension group preceeding Observation 2.2.15. Determine the dimension group for the three-shift as we did for the two-shift in Example 2.2.16. Compute the automorphisms induced on the dimension group by the shift, by a permutation of the symbols and by the map τ_2 from Example 3.3.10.

§ 3.5 Summary

Here, we quickly summarize some of the main results on automorphism groups.

Summary 3.5.1 *Remarks on the automorphism groups of one-sided subshifts of finite type.*

(i) $Aut(X_{[2]})$ *is isomorphic to* $\mathbb{Z}/2\mathbb{Z}$ *and* $Aut(X_{[n]})$ *is infinite for all* $n \geq 3$.
(ii) $Aut(X_A)$ *is finite or countably infinite and residually finite.*
(iii) $Aut(X_{[n]})$ *is generated by elements of finite order.*
(iv) *A finitely presented group with an undecidable word problem cannot be contained in* $Aut(X_A)$.
(v)   *Any countably generated free group is contained in* $Aut(X_{[n]})$ *for* $n \geq 6$.
(vi) *Any finite free product of cyclic groups is contained in* $Aut(X_{[n]})$ *for all sufficiently large n.*
(vii) *The finite subgroups of* $Aut(X_{[n]})$ *are completely characterized by their composition factors.*

Summary 3.5.2 *Remarks on automorphism groups of two-sided subshifts of finite type. Let* Σ_A *be an irreducible and aperiodic subshift of finite type with positive entropy.*

(i) $Aut(\Sigma_A)$ *is countably infinite and residually finite.*
(ii) *A finitely presented group with an undecidable word problem cannot be contained in* $Aut(\Sigma_A)$.
(iii) *Any countably generated free group is contained in* $Aut(\Sigma_A)$.
(iv) *Any finite free product of cyclic groups is contained in* $Aut(\Sigma_A)$.
(v)   *Every finite group is contained in* $Aut(\Sigma_A)$.
(vi) *The center of* $Aut(\Sigma_A)$ *consists of the powers of the shift.*
(vii) *The only compact sets invariant under the action of* $Aut(\Sigma_A)$ *are finite or all of* Σ_A.

Notes

Section 3.1

The investigation of the automorphism groups was begun for the two-sided two-shift in the 1960's by Hedlund and his coworkers. Their results can be found in [H1] and [H2]. Theorem 3.1.1(i) was proved by E. Coven in about 1965 using polynomials to describe block maps. The proof used here is from G.A. Hedlund's 1969 paper [H2] where he proved a left permutive map of the two-sided two-shift is defined by a one-block map and a power of the shift. The one-sided result is not stated explicitly. Hedlund proved Theorem 3.1.1(ii) in his paper and attributed it to M. Curtis, G.A. Hedlund and R. Lyndon. The proof included here is theirs. Observation 3.1.2 is a 'folk' theorem and can be found stated explicitly in [BLR].

Section 3.2

The results concerning generators for the automorphism groups of one-sided subshifts of finite type (3.2.1 through 3.2.10) are due to M. Boyle, J. Franks and B. Kitchens [BFK] in 1990. The methods used are an outgrowth of the methods developed by R. Williams when examining conjugacies in [Wi2]. The work was motivated by the question of whether or not there is an automorphism of the one-sided three-shift with order six. References for the group theory used are [Ha] and [Rn]. Shortly after that work J. Ashley identified a particularly useful set of generators for the automorphism group of any one-sided full-shift [Ay1]. Ashley's work was motivated by the work of P. Blanchard, R. Devaney and L. Keen [BDK]. They were investigating the topology of the parameter space of third degree complex polynomials with respect to the dynamics of the polynomials as maps of the complex plane to itself. They used Ashley's generators to examine the fundamental group of a dynamically defined set of polynomials. J. Wagoner in the late 1980's developed a more general approach than the one used in [BFK]. His approach was geared toward examining automorphisms of two-sided subshifts of finite type [Wa1], [Wa2] and [Wa3]. Problem 3.2.12 is probably the second best known problem in symbolic dynamics and goes back to Hedlund in the 1960's. As stated at the end of the section H. Kim, F. Roush and J. Wagoner [KRW2] have produced an example of an irreducible and aperiodic subshift of finite type whose automorphism group is not generated by elements of finite order. Their arguments use gyration numbers and the action induced by an automorphism on the dimension group. These were very briefly described in Section 3.4.

Section 3.3

The investigation of the isomorphism class and the subgroups of the automorphism group has been an area of great interest. Observation 3.3.1 is due to M. Boyle, D. Lind and D. Rudolph in [BLR]. The result concerning S_∞ (3.3.1 (ii)) was also discovered by H. Kim and F. Roush. An excellent discussion of finitely presented groups can be found in [MKS]. Observation 3.3.2 is due to B. Kitchens in the mid-1980's and 3.3.3 to 3.3.6 can be found in [BLR]. The classification of the finite subgroups of one-sided subshifts of finite type, Theorem 3.3.14 and its corollaries, is due to M. Boyle, J. Franks and B. Kitchens in

[BFK]. The use of markers to produce automorphisms goes back to Hedlund [H2] as was seen in Theorem 3.1.1. One of the problems in investigating the automorphism groups is that the only constructions known either use symmetries of the transition graph or some version of a marker automorphism. Theorem 3.3.18 which says the automorphism group of any irreducible and aperiodic subshift of finite type with positive entropy contains the automorphism group of every full shift was proved by H. Kim and F. Roush [KR3] in 1990. Theorem 3.3.22 was proved by J.P. Ryan [Ry1], [Ry2] in 1975.

Section 3.4

Question 3.4.1 about interchanging fixed points by an automorphism was asked by R. Williams when he was trying to find an another invariant for topological conjugacy. The idea was to find an invariant that is different from shift equivalence. The gyration function was defined by M. Boyle and W. Krieger [BK1] in 1987 to investigate the actions of automorphisms on the periodic points. It is a very useful tool. Theorem 3.4.3 which concerns compact sets invariant under the action of the automorphism group was proved by M. Boyle, D. Lind and D. Rudolph [BLR]. The automorphism induced on the dimension group and the relationship to the gyration function has received a great deal of attention in the last few years. There has been a great deal of work on this and its relationship to topological conjugacy. A number of references are included.

References

[Ay1] J. Ashley, *Marker Automorphisms of the One-sided d-shift*, Ergodic Theory and Dynamical Systems **10** (1990), 247–262.

[BF] M. Boyle and U.-R. Fiebig, *The Action of Inert Finite Order Automorphisms on Finite Subsystems of the Shift*, Ergodic Theory and Dynamical Systems **11** (1991), 413–425.

[BFK] M. Boyle, J. Franks and B. Kitchens, *Automorphisms of One-sided Subshifts of Finite Type*, Ergodic Theory and Dynamical Systems **10** (1990), 421–449.

[BDK] P. Blanchard, R. Devaney and L. Keen, *The Dynamics of Complex Polynomials*, Inventiones Mathematicae **104** (1991), 545–580.

[BK1] M. Boyle and W. Krieger, *Periodic Points and Automorphisms of the Shift*, Transactions of the American Mathematical Society **302** (1987), 125–149.

[BK2] M. Boyle and W. Krieger, *Automorphisms and Subsystems of the Shift*, Journal für die reine und angewandte Mathematik **437** (1993), 13–28.

[BLR] M. Boyle, D. Lind and B. Rudolph, *The Automorphism Group of a Shift of Finite Type*, Transactions of the American Mathematical Society **306** (1988), 71–114.

[Ha] M. Hall, Jr., *The Theory of Groups*, Chelsea Publishing Co., 1976.

[H1]   G.A. Hedlund, *Transformations Commuting with the Shift,* Topological Dynamics (J. Auslander and W. Gottschalk, eds.), W.A. Benjamin, 1968.

[H2]   G.A. Hedlund, *Endomorphisms and Automorphisms of the Shift Dynamical System,* Mathematical Systems Theory **3** no. 4 (1969), 320–375.

[KR3]   K.H. Kim and F. Roush, *On the Automorphism Groups of Subshifts,* PU.M.A Series B **1** (1990), 203–230.

[KR4]   K.H. Kim and F. Roush, *On the Structure of Inert Automorphisms of Subshifts,* PU.M.A Series B **2** (1991), 3–22.

[KRW2] K.H. Kim, F. Roush and J. Wagoner, *Inert Actions on Periodic Points,* preprint.

[MKS]   W. Magnus, A. Karrass and D. Solitar. *Combinatorial Group Theory,* Dover Publications, 1976.

[Rn]   J. Rotman, *An Introduction to the Theory of Groups,* 3rd. ed., Allyn and Bacon, Inc., 1984.

[Ry1]   J.P. Ryan, *The Shift and Commutivity,* Mathematical Systems Theory **6** (1973), 82–85.

[Ry2]   J.P. Ryan, *The Shift and Commutivity II,* Mathematical Systems Theory **8** (1975), 249–250.

[Wa1]   J. Wagoner, *Markov Partitions and K_2,* Publications Mathematétiques IHES no. 65 (1987), 91–129.

[Wa2]   J. Wagoner, *Triangle Identities and Symmetries of a Subshift of Finite Type,* Pacific Journal of Mathematics **144** (1990), 181–205.

[Wa3]   J. Wagoner, *Eventual Finite Order Generation for the Kernel of the Dimension Group Representation,* Transactions of the American Mathematical Society **317** (1990), 331–350.

[Wi2]   R.F. Williams, *Classification of Subshifts of Finite Type,* Annals of Mathematics **98** (1973), 120–153; *Errata* **99** (1974), 380–381.

Chapter 4. Embeddings and Factor Maps

In this chapter we will examine some questions about embeddings and factor maps. An embedding is a continuous, invertible, shift commuting map from one subshift of finite type into another. A factor map is a continuous, shift commuting map from one subshift of finite type onto another. We will concentrate on two-sided subshifts of finite type and then see how these results carry over to one-sided subshifts of finite type.

In the first section we begin with some basic observations about factor maps and then see that there are two very different types of factor maps. One type of factor map is uniformly bounded-to-one. The other type of factor map is uncountable-to-one on some points. In the second section we investigate some of the basic properties of the bounded-to-one factor maps and find some necessary conditions for their existence. The third section contains some special constructions involving bounded-to-one factor maps. In the fourth section we produce necessary and sufficient conditions for one subshift of finite type to be properly embedded in another. Surprisingly, the same reasoning leads to necessary and sufficient conditions for the existence of an unbounded-to-one factor map between two subshifts of finite type. The exercises at the end of section 4.4 summarize the situation for factor maps between one-sided subshifts of finite type.

§ 4.1 Factor Maps

Let Σ_A and Σ_B be two subshifts of finite type. A continuous, onto map $\varphi : \Sigma_A \to \Sigma_B$ that commutes with the shifts is called a *factor map*. The subshift of finite type Σ_B is also said to be a *factor* of Σ_A. If the map φ is one-to-one then it is a conjugacy as discussed in Chapter 2. Recall from Theorem 1.4.9 that any factor map is a block map with a memory m and an anticipation a.

Example 4.1.1 Define a one-block map $\varphi : \{0,1,2\}^{\mathbb{Z}} \to \{0,1\}^{\mathbb{Z}}$ by $\varphi(0) = 0$, $\varphi(1) = \varphi(2) = 1$.

Example 4.1.2 As in Example 1.4.7 define a 2-block factor map $\varphi : \{0,1\}^{\mathbb{Z}} \to \{0,1\}^{\mathbb{Z}}$ by $(\varphi(x))_i = x_i + x_{i+1} \mod 2$.

Example 4.1.3 Let Σ_A be defined by the graph in Figure 4.1.1.

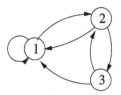

Figure 4.1.1

We will define two different maps, φ and ψ, from Σ_A to the two-shift. Denote the two-shift by to $\{r,b\}^{\mathbb{Z}}$ and label each edge in G_A by either r or b. This is an *edge labelling*. Think of this as a coloring of each edge by either red or blue. An edge in G_A corresponds to a 2-block in Σ_A. The edge labelling then defines a 2-block map from Σ_A into $\{r,b\}^{\mathbb{Z}}$. The map sends the 2-block represented by an edge to its label. Define the map φ by the edge labelling on the left and the map ψ by the edge labelling on the right in Figure 4.1.2.

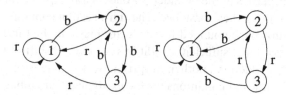

Figure 4.1.2

Observe that both φ and ψ map Σ_A onto $\{r,b\}^{\mathbb{Z}}$. This follows since in G_A every vertex has two outgoing edges and in each labelling one of the outgoing edges is labelled r and the other b. Look at either labelling and pick a finite block of r's and b's. Choose a vertex in G_A and observe the string of r's and b's is like a string of directions. It guides you through a sequence of vertices in G_A and produces a block in Σ_A which is mapped onto your chosen block. Both maps φ and ψ are 2-block factor maps from Σ_A to $\{r,b\}^{\mathbb{Z}}$.

Example 4.1.4 Define Σ_A by the graph on the left in Figure 4.1.3 and the Golden Mean subshift of finite type by the graph on the right.

Define a one-block factor map from Σ_A onto the Golden Mean subshift of finite type by sending each 1_k to 1 and 2 to 2.

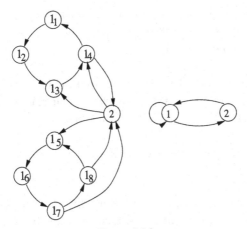

Figure 4.1.3

The factor maps in Examples 4.1.2 and 4.1.3 have an important special property. Let $\varphi : \Sigma_A \to \Sigma_B$ be a one-block factor map. If for each $i \in L_A$, φ maps the follower set of i in a one-to-one way onto the follower set of $\varphi(i)$, we say φ is *right-resolving*. If for each $i \in L_A$, φ maps the predecessor set of i in a one-to-one way onto the predecessor set of $\varphi(i)$, we say φ is *left-resolving*. The factor maps in Examples 4.1.2 and 4.1.3 are right-resolving factor maps when considered as one-block maps from the two-block presentations of the domain shifts. It is easily seen that a right or left-resolving map which takes the alphabet of the domain shift onto the alphabet of the image shift is automatically an onto map and each point in the image shift has a bounded number of preimages in the domain. The number of preimages of a point is bounded by the number of elements in the alphabet of the domain shift.

Lemma 4.1.5 *Suppose $\varphi : \Sigma_A \to \Sigma_B$ is a factor map. There is a higher block presentation of Σ_A so that the factor map induced by φ from this higher block presentation of Σ_A to Σ_B is a one-block factor map.*

Proof. By Theorem 1.4.8 we know that φ is an ℓ-block map for some ℓ. We also know that $\ell = a + m + 1$ where a is the anticipation and m is the memory of the map. Let ψ be the conjugacy from Σ_A to it's ℓ-block presentation defined by $\psi(x)_i = [x_{i-m}, \ldots , x_{i+a}]$. Then the induced map $\varphi \circ \psi^{-1} : \Sigma_{A^{[\ell]}} \to \Sigma_B$ is a one-block map with no memory or anticipation. □

It is often useful to assume a factor map is a one-block map.

Proposition 4.1.6 *Let Σ_A and Σ_B be two irreducible subshifts of finite type. Suppose Σ_B is a factor of Σ_A. Then the topological entropy of Σ_B is less than or equal to the topological entropy of Σ_A.*

Proof. We may use Lemma 4.1.5 to assume that the factor map $\varphi : \Sigma_A \to \Sigma_B$ is a one-block map. This means that for each k, φ maps $W(A,k)$ onto $W(B,k)$. Consequently, $|W(A,k)| \geq |W(B,k)|$ for all k and

$$h(\Sigma_A, \sigma) = \lim_{k\to\infty} \frac{1}{k} \log |W(A,k)| \geq \lim_{k\to\infty} \frac{1}{k} \log |W(B,k)| = h(\Sigma_B, \sigma). \quad \square$$

Next we observe that there are two very different types of factor maps. Suppose $\varphi : \Sigma_A \to \Sigma_B$ is a one-block factor map. If there exist two blocks $[i_1, \ldots, i_k] \neq [j_1, \ldots, j_k] \in W(A)$ with $i_1 = j_1$, $i_k = j_k$ and $\varphi([i_1, \ldots, i_k]) = \varphi([j_1, \ldots, j_k])$ we say the map φ has a *diamond*.

Theorem 4.1.7 *Suppose $\varphi : \Sigma_A \to \Sigma_B$ is a one-block factor map between irreducible subshifts of finite type and Σ_A has positive entropy. Then either*

i.   *φ is uniformly bounded-to-one,*
ii. *φ has no diamonds and*
iii. *$h(\Sigma_A, \sigma) = h(\Sigma_B, \sigma)$.*

or

i.   *φ is uncountable-to-one on some point,*
ii. *φ has a diamond and*
iii. *$h(\Sigma_A, \sigma) > h(\Sigma_B, \sigma)$.*

Proof. First we observe that if φ has no diamonds then any block in $W(B)$ has at most $|L_A|^2$ preimages. By compactness any point in Σ_B has at most $|L_A|^2$ preimages. Conversely, if φ has a diamond we can construct a point in Σ_B with uncountable many preimages. To do this let $[i_1, \ldots, i_k]$, $[j_1, \ldots, j_k] \in W(A)$ be a pair of blocks that form a diamond. Let $x \in \Sigma_A$ be any point where $[i_1, \ldots, i_k]$ occurs infinitely often and $y = \varphi(x)$. Then y has uncountably many preimages because any set of substitutions of the block $[j_1, \ldots, j_k]$ for the block $[i_1, \ldots, i_k]$ in x will result in a new point which also maps to y.

If φ has no diamonds then as we observed, for each k $|W(A,k)| \leq |L_A|^2 |W(B,k)|$ so $h(\Sigma_A, \sigma) = h(\Sigma_B, \sigma)$. Conversely, suppose $[i_1, \ldots, i_k]$, $[j_1, \ldots, j_k] \in W(A)$ is a pair of blocks that form a diamond. Form a new subshift of finite type $\Sigma \subset \Sigma_A$ by excluding all points in Σ_A containing the block $[j_1, \ldots, j_k]$. The map φ restricted to Σ maps Σ onto Σ_B. This amounts to taking the transition matrix for the k-block presentation of A and putting all zeros in the row and column corresponding to the symbol $[j_1, \ldots, j_k]$. Call the new matrix $\hat{A}$. We now have that φ restricts to a factor map from $\Sigma_{\hat{A}}$ to Σ_B so $h(\Sigma_{\hat{A}}, \sigma) \geq h(\Sigma_B, \sigma)$. Finally, the Perron-Frobenius Theorem 1.3.5(e) tells us that $h(\Sigma_A, \sigma) > h(\Sigma_{\hat{A}}, \sigma)$. $\square$

Corollary 4.1.8 *Suppose* $\varphi : \Sigma_A \to \Sigma_B$ *is a factor map between irreducible subshifts of finite type and* Σ_A *has positive entropy. Then either*

i. φ *is uniformly bounded-to-one and*
ii. $h(\Sigma_A, \sigma) = h(\Sigma_B, \sigma)$

or

i. φ *is uncountable-to-one on some point and*
ii. $h(\Sigma_A, \sigma) > h(\Sigma_B, \sigma)$.

The first type of map in Theorem 4.1.7 is said to be a *finite-to-one* factor map and the second type is said to be an *infinite-to-one* factor map.

Example 4.1.9 Example 4.1.1 is an infinite-to-one factor map while Examples 4.1.2-4.1.4 are finite-to-one factor maps.

In the proof of Theorem 4.1.7 we see the advantage of taking the factor map to be a one-block map, then finding a simple combinatorial property, diamonds, and using this property to give a simple proof which illuminates the crucial topological property of the map. This idea will be further exploited in our examination of finite-to-one factor maps.

Next we observe that the situation for finite-to-one factor maps for one-sided subshifts of finite type is the same as for two-sided subshifts of finite type.

Observation 4.1.10 *Let A and B be irreducible transition matrices. There is a factor map from the two-sided subshift of finite type defined by A to the two-sided subshift of finite type defined by B if and only if there is a factor map from the one-sided subshift of finite type defined by A to the one-sided subshift of finite type defined by B.*

Proof. Let A and B be irreducible transition matrices. A factor map $\varphi : X_A \to X_B$ is defined by a block map with zero memory and anticipation a. Such a block map also defines a factor map from Σ_A to Σ_B.

A factor map $\varphi : \Sigma_A \to \Sigma_B$ is defined by a block map with memory m and anticipation a. The same block will also define a factor map from X_A to X_B with zero memory and anticipation $a + m$. $\square$

Next we observe that the distinction between finite-to-one and infinite-to-one factor maps also holds for one-sided subshifts of finite type.

Observation 4.1.11 *Suppose φ is a factor map between irreducible one-sided subshifts of finite type X_A and X_B. Further suppose X_A has positive entropy. Then either*

i.   *φ is uniformly bounded-to-one and*
ii. *$h(X_A, \sigma) = h(X, \sigma)$*

or

i.   *φ is uncountable-to-one on some point and*
ii. *$h(X_A, \sigma) > h(X_B, \sigma)$.*

Proof. The proof is the same as for Theorem 4.1.7 and Corollary 4.1.8. First go to a higher block presentation of X_A so φ is a one-block map. Then observe that the map is finite-to-one if and only if there are no diamonds and proceed as before.   □

Let A be an irreducible transition matrix. Recall from Section 1.3 the definition of the period of A and the definition of the cyclic subsets of the indices of A. If A has period p then we say the one and two-sided subshifts of finite type defined by A have *period p*. Suppose X_A has period p. Then σ^p acting on X_A divides it into p disjoint irreducible and aperiodic subshifts of finite type, $X_{A_0}, \ldots, X_{A_{p-1}}$. The alphabet for X_{A_k} is the cyclic subset of the indices C_k. The same is true for the two-sided subshift of finite type Σ_A. This leads to the next observation about factor maps.

Observation 4.1.12 *Let X and Y be irreducible one or two-sided subshifts of finite type. If there is a factor map from X to Y then the period of Y divides the period of X.*

Proof. Let $\varphi : X \to Y$ be a factor map. Suppose X has period p. We know σ^p acting on X divides it into p irreducible and aperiodic subshifts of finite type, $X_0, \ldots X_{p-1}$. The image under φ of each X_k is irreducible and aperiodic. So the period of Y divides the period of X.   □

Exercises

1.   Show that the map φ in Example 4.1.2 is both left and right-resolving.
2.   Show that the maps φ and ψ in Example 4.1.3. are right-resolving but not left-resolving.
3.   Show that the map φ in Example 4.1.4 is neither left nor right-resolving.
4.   Prove that a right-resolving map which takes the alphabet of the domain subshift of finite type onto the alphabet of the range subshift of finite type is onto.

5. Show that the map φ in Example 4.1.4 does not have a diamond and is uniformly bounded-to-one.

6. Show that the map in Example 4.1.1 has a diamond and find a point in the two-shift with uncountably many preimages.

§4.2 Finite-to-one Factor Maps

In this section we examine finite-to-one factor maps. As it was for topological conjugacies, our goal here is to explain the structure of the maps and to find necessary and sufficient conditions for the existence of a factor map between two subshifts of finite type. In Chapter 2 we were able to show that any conjugacy can be decomposed into a finite composition of elementary conjugacies and each elementary conjugacy has a simple structure. We are unable to do this for finite-to-one factor maps. There are no known necessary and sufficient conditions for the existence of a conjugacy between subshifts of finite type and similarly there are no known necessary and sufficient conditions for the existence of a finite-to-one factor map. In Chapter 2 we found a number of necessary conditions for the existence of a conjugacy and in this section we will formulate several necessary conditions for the existence of a finite-to-one factor map.

Theorem 4.2.1 *Let* $\varphi : \Sigma_A \to \Sigma_B$ *be a finite-to-one factor map between irreducible subshifts of finite type. Then there exists a positive integer d such that:*

i. $|\varphi^{-1}(y)| \geq d$ *for every* $y \in \Sigma_B$ *and*

ii. φ *is exactly d-to-one on the doubly transitive points.*

Proof. For statement (i) suppose $\varphi : \Sigma_A \to \Sigma_B$ is a finite-to-one, one-block factor map between irreducible subshifts of finite type. Let $w = [w_1, \ldots, w_\ell] \in \mathcal{W}(B)$. Define

$$d(w) = \min_{1 \leq t \leq \ell} |\{j \in L_A : j = x_t \text{ for some } [x_1, \ldots, x_\ell] \in \varphi^{-1}(w)\}|.$$

Then define

$$d = \min_{w \in \mathcal{W}(B)} \{d(w)\}.$$

Let $y \in \Sigma_B$. By compactness we can see that for all sufficiently large n

$$|\varphi^{-1}(y)| \geq |\{j \in L_A : j = x_0 \text{ for some } [x_{-n}, \ldots, x_n] \in \varphi^{-1}([y_{-n}, \ldots, y_n])\}|$$
$$\geq d.$$

For statement (ii) let $w = [w_1, \ldots, w_\ell] \in \mathcal{W}(B)$ be such that $d(w) = d$. Let t be a coordinate where

$$d = |\{j \in L_A : j = x_t \text{ for some } [x_1, \dots, x_\ell] \in \varphi^{-1}(w)\}|$$

and

$$\mathcal{M} = \{i_1, \dots, i_d\} = \{j \in L_A : j = x_t \text{ for some } [x_1, \dots, x_\ell] \in \varphi^{-1}(w)\}.$$

Let $v = [v_1, \dots v_k]$ be a block with $wvw \in \mathcal{W}(B)$. Then there exists a permutation π_v of $\{1, \dots, d\}$ and d blocks, $v^1, \dots, v^d \in \mathcal{W}(A, \ell + k + 1)$ so that:

i.   for $m = 1, \dots, d$, v^m begins with the symbol i_m and ends with the symbol $i_{\pi_v(m)}$;

ii.   if $[x_1, \dots, x_{k+2\ell}] \in \varphi^{-1}(wvw)$ then $[x_t, \dots, x_{\ell+k+t}] = v^m$ for some m.

Figure 4.2.1 describes this pictorially.

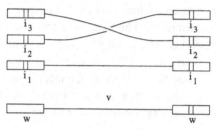

Figure 4.2.1

It says that there are exactly d possible middles lying over two occurrences w. If there were more than d we could string together a long stretch $wvwv \cdots vw$ and force a diamond. By definition there cannot be less than d. Since every doubly transitive point has w occurring infinitely often to the left and to the right of the time zero coordinate a doubly transitive point has exactly d preimages.   □

When $\varphi : \Sigma_A \to \Sigma_B$ is a finite-to-one factor map, the positive integer d of Theorem 4.2.1. is the *degree* of φ. The map φ is also said to be *d-to-one almost everywhere*, which we abbreviate as *d-to-one a.e.*. If $w \in \mathcal{W}(B)$ has $d(w) = d$ then w is said to be a *magic word* or a *resolving block*. A coordinate t where the minimum occurs is said to be a *distinguished or magic coordinate*. In what follows we will carefully exploit the fact that there is a permutation of the symbols lying over the distinguished coordinate in two successive occurrences of a magic word.

Example 4.2.2 The map in Example 4.1.2 is 2-to-1 on every point and every word is a magic word when the map is viewed as a one-block map from the two-block presentation of the 2-shift to the 2-shift. The map φ in Example 4.1.3 is 1-to-1 almost everywhere. It is not a conjugacy and rr is a magic word when viewed as a map from the two-block presentation of Σ_A to $\{r, b\}^{\mathbb{Z}}$. The map ψ in Example 4.1.3 is 2-to-1 almost everywhere. It is 3-to-1 on the point of all r's and bb is a magic word when viewed as a one-block map. The map in Example 4.1.4 is 1-to-1 almost everywhere. It is not a conjugacy and 2 is a magic word.

Observation 4.2.3 *Let* $\varphi : \Sigma_A \rightarrow \Sigma_B$ *be a finite-to-one factor map between irreducible subshifts of finite type. Suppose* Σ_A *has period* p *and* Σ_B *has period* q. *Then the degree of* φ *is* $m(p/q)$ *for some* $m \in \mathbb{N}$.

Proof. By Observation 4.1.12 we know p/q is an integer. We also know σ^p acting on Σ_A divides it into p irreducible and aperiodic subshifts of finite type, $\Sigma_{A_0}, \ldots, \Sigma_{A_{p-1}}$, and σ^p acting on Σ_B divides it into q irreducible and aperiodic subshifts of finite type, $\Sigma_{B_0}, \ldots, \Sigma_{B_{q-1}}$. Suppose $\varphi(\Sigma_{A_0}) = \Sigma_{B_0}$ and has degree m. Then $\varphi(\Sigma_{A_k}) = \Sigma_{B_k} \mod q$ for each k and each has degree m. Exactly p/q of the Σ_{A_k} are mapped onto each Σ_{B_ℓ}. $\square$

Theorem 4.2.4 *Suppose* Σ_A *and* Σ_B *are irreducible and aperiodic subshifts of finite type with the same entropy. If there exists a factor map from* Σ_A *to* Σ_B *then there is a one-to-one almost everywhere factor map from* Σ_A *to* Σ_B.

Proof. Suppose φ is a d-to-one almost everywhere factor map from Σ_A to Σ_B and $d > 1$. We choose a special symbol $a \in L_B$ and three blocks w, u and v in $\mathcal{W}(B)$ satisfying the following conditions:

(i) w is a magic word whose first symbol is a and a occurs nowhere else in w;

(ii) u and v are distinct blocks of the same length not containing a with $wuw, wvw \in \mathcal{W}(B)$;

(iii) $\pi_u \neq \pi_v$.

It may be necessary to go to a higher block presentation of Σ_A and Σ_B to guarantee the existence of the special symbol a. Making sure u and v have the same length and $\pi_u \neq \pi_v$ depends on the aperiodicity of A. Let ℓ be the length of w, t the distinguished coordinate,

$$\mathcal{M} = \{i_1, \ldots, i_d\} = \{j \in L_A : j = x_t \text{ for some } [x_1, \ldots, x_\ell] \in \varphi^{-1}(w)\}$$

and k the length of u and v. We may assume $\pi_u(i_1) \neq \pi_v(i_1)$. Note the three sets

$$\mathcal{B} = \{i \in L_A : i = x_1 \text{ for some } [x_1, \ldots, x_\ell] \in \varphi^{-1}(w)\}$$
$$\{i \in L_A : i = y_1 \text{ for some } [y_1, \ldots, y_{2\ell+k}] \in \varphi^{-1}(wuw)\}$$
$$\{i \in L_A : i = z_1 \text{ for some } [z_1, \ldots, z_{2\ell+k}] \in \varphi^{-1}(wvw)\}$$

are identical. We will define a new factor map $\bar{\varphi}$ from Σ_A to Σ_B with degree less than d. Define $\bar{\varphi}$ to agree with φ except on the blocks in $\varphi^{-1}(wuw)$ and $\varphi^{-1}(vwv)$ having i_1 in the tth coordinate; that is, over the first occurrence of the distinguished coordinate. For a block $[y_1, \ldots, y_{2\ell+k}] \in \varphi^{-1}(wuw)$ with $y_t = i_1$ define its image under $\bar{\varphi}$ to be wvw. For a block $[z_1, \ldots, z_{2\ell+k}] \in \varphi^{-1}(wvw)$ with $z_t = i_1$ define its image under $\bar{\varphi}$ to be wuw. We simply exchange a few images.

This can be done by a $4\ell+2k-1$ block map. The map $\bar{\varphi}$ must look $2\ell+k-1$ coordinates into the past and $2\ell+k-1$ coordinates into the future to determine whether or not it is looking at an entry inside an occurrence of a block in $\varphi^{-1}(wuw)$ or $\varphi^{-1}(wvw)$. Once this is determined the map is well-defined. If $\bar{\varphi}$ is not looking at an entry inside an occurrence of a block in $\varphi^{-1}(wuw)$ or $\varphi^{-1}(wvw)$ then it has the same output as φ. If it is looking at an entry inside an occurrence of a block in $\varphi^{-1}(wuw)$ or $\varphi^{-1}(wvw)$ the output is still well-defined since blocks in $\varphi^{-1}(wuw)$ and $\varphi^{-1}(wvw)$ can overlap themselves or each other only over w. This is because wuw and wvw have the special symbol a occurring only at the beginning of w. The symbol a is acting almost like a "marker" in keeping the middles of the blocks in $\varphi^{-1}(wuw)$ and $\varphi^{-1}(wvw)$ separated.

To see that the new map $\bar{\varphi}$ maps Σ_A onto Σ_B, first observe that when not inside an occurrence of a block from $\bar{\varphi}^{-1}(wuw)$ or $\bar{\varphi}^{-1}(wvw)$, $\bar{\varphi}$ is a one block map. Then observe that the sets

$$\{i \in L_A : i = y_1 \text{ for some } [y_1, \ldots, y_{2\ell+k}] \in \bar{\varphi}^{-1}(wuw)\}$$
$$\{i \in L_A : i = z_1 \text{ for some } [z_1, \ldots, z_{2\ell+k}] \in \bar{\varphi}^{-1}(wvw)\}$$

are still equal to the set

$$\mathcal{B} = \{i \in L_A : i = x_1 \text{ for some } [x_1, \ldots, x_\ell] \in \bar{\varphi}^{-1}(w)\}.$$

This insures that any block in $\mathcal{W}(B)$ which begins and ends with a combination of wuw and wvw has a preimage under $\bar{\varphi}$.

We made sure $\pi_u(1) \neq \pi_v(1)$. This means $\bar{\varphi}^{-1}(wuw)$ will not have $i_{\pi_u(1)}$ over the distinguished coordinate in the second occurrence of w and $\bar{\varphi}^{-1}(wvw)$ will not have $i_{\pi_v(1)}$ over the distinguished coordinate in the second occurrence of w. The degree of φ is less than d since $d(wuw) = d(wvw) = d-1$. Continue until a map of degree one is produced. $\qquad \square$

Next we consider a special case where necessary and sufficient conditions are known for the existence of finite-to-one factor maps.

Theorem 4.2.5 *Let A be an irreducible transition matrix. There exists a finite-to-one factor map from Σ_A onto the full n-shift if and only if $h(\Sigma_A, \sigma) = \log n$.*

Proof. The condition $h(\Sigma_A, \sigma) = \log n$ is necessary by Corollary 4.1.8

Suppose A is a zero-one transition matrix with Perron value n. There is unique right Perron eigenvector with positive integer entries having greatest common divisor one. Let r be this vector. Notice every entry of r is one if and only if A has row sum n. If every row of A has n ones we can define a finite-to-one factor map onto the full n-shift by an edge labelling. At each vertex we label the outgoing edges by $1, \ldots, n$. The labelling defines a two-block factor map onto the n-shift. This is exactly as in Example 4.1.3. Considered as a one-block map the factor map is right-resolving. If r is not all ones we define a sequence of state splitting which produces a new transition matrix B with n ones in each row and Σ_B conjugate to Σ_A. Then define the factor map by an edge labelling.

Suppose r is not all ones and r_1 is a largest entry of r. We will split the symbol 1 into two new symbols, 1_1 and 1_2. The resulting transition matrix will have a right Perron eigenvector, r', which agrees with the vector r in every entry except r'_{1_1} and r'_{1_2}. The entries r'_{1_1} and r'_{1_2} will be positive integers strictly smaller than r_1. To define the splitting use the fact that r is an eigenvector for n so $r_1 = (1/n)\Sigma r_j$, where the sum is over all $j \in f(1)$, the followers of 1. We partition $f(1)$ into two sets f_1 and f_2 so Σr_j is divisible by n when the sum is over either f_1 or f_2. Then use the two sets f_1 and f_2 to define a state splitting as we did in Section 2.1. The vector r' with $r'_j = r_j$ for $j \neq 1$, $r'_{1_1} = (1/n)\sum_{j \in f_1} r_j$ and $r'_{1_2} = (1/n)\sum_{j \in f_2} r_j$ is a Perron eigenvector with integer entries for the new transition matrix.

We need to show that we can partition the followers of 1 as desired. It follows from the pigeon hole principle. Let $f(1) = \{j_1, \ldots, j_k\}$. Then $k \geq n$ since r_1 is a maximal entry of r. Form the n distinct sums, $\sum_{k=1}^{m} r_{j_k}$ for $m = 1, \ldots, n$. Consider the sums modulo n. Suppose the sum $r_{j_1} + \cdots + r_{j_M}$ is congruent to 0 $\mod n$. Then so is the sum $r_{j_{M+1}} + \cdots r_{j_k}$. Take $f_1 = \{j_1, \ldots, j_M\}$ and $f_2 = \{j_{M+1}, \ldots, j_k\}$. If no sum is congruent to zero $\mod n$ then two are equal, say the M_1^{st} and M_2^{nd}. In this case take $f_1 = \{j_{M_1+1}, \ldots, j_{M_2}\}$ and $f_2 = \{j_1, \ldots, j_{M_1}, j_{M_2+1}, \ldots, j_k\}$.

Next divide every entry of r' by the greatest common divisor of the entries. If the resulting vector is all 1's we are done, if not repeat the operation. Each time the operation is repeated one of the largest entries of the eigenvector has been reduced. Eventually this results in a matrix B whose eigenvector is all 1's. □

This leads to an interesting unsolved problem.

Problem 4.2.6 This is known as the *road coloring problem*. Suppose A is an irreducible and aperiodic $\{0, 1\}$ transition matrix with exactly two 1's in each row. Let G_A be its associated directed graph. An *edge coloring* is an edge labelling of G_A by two colors, red and blue, so that at each vertex one of the outgoing edges is labelled red and the other blue. For a fixed edge coloring a word in red and blue is thought of as a set of directions. Someone is at a vertex and traverses a path in the graph by successively taking the red or blue edge as the word dictates. A *magic word* for an edge coloring is a word in reds and blues so that there is a distinguished vertex and no matter where a person begins, at the end of the directions the person is at the distinguished vertex. It is a set of directions that works for everyone. Question: given any such graph is there always an edge coloring with a magic word? It asks if there is a one-to-one almost everywhere right-resolving factor map, defined by an edge labelling, from the subshift of finite type to the full 2-shift. At first glance this sounds impossible but it is easily seen to be true if the graph has a vertex with a self loop.

Next we examine some necessary conditions for the existence of a factor map. The conditions developed all result from the following construction

Lattice Construction 4.2.7 Let $\varphi : \Sigma_A \to \Sigma_B$ be a finite-to-one, one-block factor map between irreducible subshifts of finite type.

We will construct an A invariant lattice, $\mathcal{L} \subseteq \mathbb{Z}^{L_A}$, and an onto linear map $\theta : \mathcal{L} \to \mathbb{Z}^{L_B}$ with $\theta \circ A = B \circ \theta$. The map θ can be described by an $|L_B| \times |L_A|$, $\{0,1\}$ matrix and will extend to a map from the rational span of $\mathcal{L}$, $\mathcal{L}_{\mathbb{Q}} \subseteq \mathbb{Q}^{L_A}$, onto $\mathbb{Q}^{L_B}$.

Suppose φ has degree d, $w = [w_1, \ldots, w_\ell]$ is a magic word, t is the distinguished coordinate and

$$\mathcal{M} = \{i_1, \ldots, i_d\} = \{j \in L_A : j = x_t \text{ for some } [x_1, \ldots, x_\ell] \in \varphi^{-1}(w)\}.$$

We will use the structure of the blocks in $\varphi^{-1}(wvw)$, for blocks $wvw \in \mathcal{W}(B)$. For each $v = [v_1, \ldots, v_k]$ with $vw \in \mathcal{W}(B)$ and for each $i \in \mathcal{M}$ define a vector $\bar{v}^{(v,i)} \in \mathbb{Z}^{L_A}$ by

$$(\bar{v}^{(v,i)})_j = \begin{cases} 1 & \text{if } j = x_1 \text{ for some } [x_1, \ldots, x_{\ell+k}] \in \varphi^{-1}(vw) \text{ with } x_{k+t} = i \\ 0 & \text{otherwise.} \end{cases}$$

There are a finite number of distinct $\bar{v}^{(v,i)}$. Let $\mathcal{L}$ be the sublattice of $\mathbb{Z}^{L_A}$ generated by $\{\bar{v}^{(v,i)}\}$.

We want to show $\mathcal{L}$ is A invariant. Fix a $\bar{v}^{(v,i)}$ and observe $A\bar{v}^{(v,i)} = \sum \bar{v}^{(av,i)}$ where the sum is over all $a \in L_B$ with $avw \in \mathcal{W}(B)$. This follows since φ is an onto one-block map.

Now define a linear map $\theta : \mathcal{L} \to \mathbb{Z}^{L_B}$. For each $a \in L_B$ fix a word $u^a = [u_1^a, \ldots, u_m^a] \in \mathcal{W}(B)$ with $u_m^a = a$ and $wu^a \in \mathcal{W}(B)$. Define a vector $\bar{u}^a \in \mathbb{Z}^{L_A}$ by

$$(\bar{u}^a)_j = \begin{cases} 1 & \text{if } j = x_{\ell+m} \text{ for some } [x_1, \ldots, x_{\ell+m}] \in \varphi^{-1}(wu^a) \\ 0 & \text{otherwise.} \end{cases}$$

Let $\bar{v}^{(v,i)}$ be one of the vectors that generates $\mathcal{L}$ and $x \cdot y$ denote the usual dot product of vectors x and y. Observe

$$\bar{u}^a \cdot \bar{v}^{(v,i)} = \begin{cases} 1 & \text{if } v \text{ begins with } a \\ 0 & \text{otherwise.} \end{cases}$$

This follows from the discussion of the structure of the blocks in $\varphi^{-1}(wuvw)$. There can be only one middle in any preimage that has i over the distinguished coordinate in the second occurrence of w. Define the map θ coordinate by coordinate from $\mathcal{L}$ to $\mathbb{Z}^{L_B}$ by

$$(\theta(v))_a = \bar{u}^a \cdot v.$$

The map θ is linear by construction. We have observed that

$$\theta(\bar{v}^{(v,i)}) = e_a$$

where $v_1 = a$ and e_a is a standard basis vector in $\mathbb{Z}^{L_B}$. So θ maps $\mathcal{L}$ onto $\mathbb{Z}^{L_B}$. The map θ is described by the $|L_B| \times |L_A|$, $\{0,1\}$ matrix Θ which has the transpose of the vector $\bar{u}^a$ in the ath row. The matrix Θ shows θ extends to a linear map of $\mathcal{L}_\mathbb{Q}$ onto $\mathbb{Q}^{L_B}$.

Let A and B be transition matrices. Recall from Section 2.2 the definition of the Jordan form away from zero, J_A^*, for the matrix A. Observation 2.2.5 states that when Σ_A and Σ_B are topologically conjugate $J_A^* = J_B^*$. We will see how the Jordan forms away from zero are related when there is a finite-to-one factor map between two subshifts of finite type.

If A is a square matrix, a principal submatrix of A is obtained by selecting a collection of indices (for the rows and columns) and then discarding the row and column that corresponds to each of the selected indices.

Theorem 4.2.8 *Suppose $\varphi : \Sigma_A \to \Sigma_B'$ is a finite-to-one factor map between irreducible subshifts of finite type. Then J_B^* is a principal submatrix of J_A^*.*

Proof. By going to a higher block presentation of A we may assume $\varphi : \Sigma_A \to \Sigma_B$ is a one-block map.

Apply Construction 4.2.7 to produce $\mathcal{L}_\mathbb{Q}$ and the onto, linear map $\theta : \mathcal{L}_\mathbb{Q} \to \mathbb{Q}_{L_B}$. Standard linear algebra (Exercise 8) arguments show the Jordan form of B is a principal submatrix of the Jordan form of A restricted to $\mathcal{L}_\mathbb{Q}$ and hence a principal submatrix of the Jordan form of A. The reason we can draw a conclusion only about the Jordan form away from zero is that we must go to a higher block presentation of Σ_A so that φ is a one-block map and we can apply Construction 4.2.7. $\qquad\square$

Corollary 4.2.9 *Suppose* $\varphi : \Sigma_A \to \Sigma_B$ *is a finite-to-one factor map between irreducible subshifts of finite type. Then* $c_B^*(x)$ *divides* $c_A^*(x)$ *in* $\mathbb{Z}[x]$.

Corollary 4.2.10 *Suppose* $\varphi : \Sigma_A \to \Sigma_B$ *is a finite-to-one factor map between irreducible subshifts of finite type and the degrees of* $c_A^*(x)$ *and* $c_B^*(x)$ *are the same. Then* $J_A^* = J_B^*$.

Corollary 4.2.11 *A finite-to-one factor of the full n-shift is shift-equivalent to the full n-shift.*

Proof. If a subshift of finite type is a finite-to-one factor of the full n-shift then its characteristic polynomial away from zero is $x - n$. By Lemma 2.2.6 such a transition matrix is shift equivalent to the matrix $[n]$.   □

The next example shows Theorem 4.2.5 does not generalize to arbitrary subshifts of finite type.

Example 4.2.12 This is an example of two subshifts of finite type with the same entropy but no common equal entropy factor. Let Σ_A and Σ_B be defined by the irreducible and aperiodic transition matrices

$$A = \begin{bmatrix} 0 & 0 & 1 \\ 1 & 0 & 1 \\ 0 & 1 & 0 \end{bmatrix} \quad B = \begin{bmatrix} 0 & 0 & 0 & 0 & 1 \\ 1 & 0 & 0 & 0 & 0 \\ 0 & 1 & 0 & 0 & 0 \\ 0 & 0 & 1 & 0 & 0 \\ 0 & 0 & 0 & 1 & 1 \end{bmatrix}.$$

Then $c_A(x) = x^3 - x - 1$ and $c_B(x) = x^5 - x^4 - 1 = (x^3 - x - 1)(x^2 - x - 1)$. The two matrices have the same Perron eigenvalue. Since $c_A(x)$ is irreducible over $\mathbb{Z}$ Corollary 4.2.9 says any finite-to-one factor of Σ_A must have the same zeta function. Any factor of Σ_B must have a fixed point since Σ_B has a fixed point. No subshift of finite type satisfies both of these conditions.

Example 4.2.13 We will describe the generators for the lattice $\mathcal{L}$ and the matrix Θ for the map φ of Example 4.1.4 Let [2] be the magic word w for the construction. The distinct generators that occur are $\bar{v}^{([1],2)}$, $\bar{v}^{([11],2)}$, $\bar{v}^{([111],2)}$, $\bar{v}^{([1111],2)}$ and $v^{([\phi],2)}$, where $[\phi]$ is the empty word. The vectors are:

$$\begin{bmatrix} 0 \\ 0 \\ 0 \\ 1 \\ 0 \\ 0 \\ 1 \\ 1 \\ 0 \end{bmatrix} \begin{bmatrix} 0 \\ 0 \\ 1 \\ 0 \\ 0 \\ 1 \\ 1 \\ 0 \\ 0 \end{bmatrix} \begin{bmatrix} 0 \\ 1 \\ 0 \\ 0 \\ 1 \\ 1 \\ 0 \\ 0 \\ 0 \end{bmatrix} \begin{bmatrix} 1 \\ 0 \\ 0 \\ 0 \\ 1 \\ 0 \\ 0 \\ 1 \\ 0 \end{bmatrix} \begin{bmatrix} 0 \\ 0 \\ 0 \\ 0 \\ 0 \\ 0 \\ 0 \\ 0 \\ 1 \end{bmatrix}.$$

The map θ is described by the matrix

$$\Theta = \begin{bmatrix} 0 & 0 & 1 & 1 & 1 & 0 & 0 & 0 & 0 \\ 0 & 0 & 0 & 0 & 0 & 0 & 0 & 0 & 1 \end{bmatrix},$$

where the first row is the transpose of $\bar{u}^1$ with $u^1 = [1]$ and the second row is the transpose of $\bar{u}^2$ with $u^2 = [\phi]$.

Next we use the subblattice $\mathcal{L}$ defined in Construction 4.2.7 to get some information about the inverse limit spaces, the dimension groups and the Bowen-Franks groups defined in Section 2.2. Let us recall some notation from Chapter 2. For a transition matrix A the subspace $I_A^R = \cap_{r=0}^\infty A^r(\mathbb{R}^{L_A})$ is the eventual image of A in $\mathbb{R}^{L_A}$. The matrix A also acts on the torus $\mathbb{T}^{L_A}$ where there is an eventual image $I_A^T = \cap_{r=0}^\infty A^r(\mathbb{T}^{L_A})$ and an inverse limit space $\varprojlim(\mathbb{T}^{L_A}, A)$. As in Construction 4.2.7 let $\mathcal{L}_\mathbb{Q} \subseteq \mathbb{Q}^{L_A}$ be the rational span of $\mathcal{L}$. Let $\mathbb{T}_\mathcal{L}$ be the subtorus of $\mathbb{T}^{L_A}$ that is the closure of the image of $\mathcal{L}_\mathbb{Q}$ under the projection map $\mathbb{R}^{L_A} \to \mathbb{T}^{L_A}$. The subtorus $\mathbb{T}_\mathcal{L}$ is invariant under A acting on $\mathbb{T}^{L_A}$ and we define the inverse limit $\varprojlim(\mathbb{T}_\mathcal{L}, A)$ and its induced automorphism A^*.

Lemma 4.2.14 *The inverse limit space* $\varprojlim(\mathbb{T}_\mathcal{L}, A)$ *is an* A^* *invariant subgroup of* $\varprojlim(\mathbb{T}^{L_A}, A)$. *If the eventual image of* A *in* $\mathbb{T}^{L_A}$ *is contained in* $\mathbb{T}_\mathcal{L}$ *then* $\varprojlim(\mathbb{T}_\mathcal{L}, A) = \varprojlim(\mathbb{T}^{L_A}, A)$.

Proof. The first inclusion is the natural one defined by the inclusion $\mathbb{T}_\mathcal{L} \subseteq \mathbb{T}^{L_A}$. The inverse limit spaces depend only on the eventual images. When the eventual image of A in $\mathbb{T}^{L_A}$ is contained in $\mathbb{T}_\mathcal{L}$ the two inverse limits are identical. $\square$

Lemma 4.2.15 *Suppose* $\varphi : \Sigma_A \to \Sigma_B$ *is a finite-to-one factor map between irreducible subshifts of finite type. Then there is an onto homomorphism* θ^* : $\varprojlim(\mathbb{T}_\mathcal{L}, A) \to \varprojlim(\mathbb{T}^{L_B}, B)$ *that commutes with the induced automorphisms* A^* *and* B^*.

Proof. Assume φ is a one-block map and apply Construction 4.2.7 to produce the onto map $\theta : \mathcal{L}_\mathbb{Q} \to \mathbb{Q}^{L_B}$. The map θ is described by the $\{0,1\}$ matrix Θ so it defines an onto homomorphism $\theta : \mathbb{T}_\mathcal{L} \to \mathbb{T}^{L_B}$ which commutes with the maps A and B. This defines a coordinate by coordinate homomorphism θ^* : $\varprojlim(\mathbb{T}_\mathcal{L}, A) \to \varprojlim(\mathbb{T}^{L_B}, B)$ which commutes with the induced automorphisms A^* and B^*.

Theorem 4.2.16 *Suppose* $\varphi : \Sigma_A \to \Sigma_B$ *is a finite-to-one factor map between irreducible and aperiodic subshifts of finite type and the degrees of* $c_A^*(x)$ *and* $c_B^*(x)$ *are the same. Then* A *and* B *are shift equivalent.*

Proof. By Theorem 2.2.13 we need only to prove that the inverse limit spaces, $\varprojlim(\mathbb{T}^{L_A}, A)$ and $\varprojlim(\mathbb{T}^{L_B}, B)$ together with their induced automorphisms are conjugate.

Assume φ is a one-block map and apply Construction 4.2.7. When $c_A^*(x)$ and $c_B^*(x)$ have the same degree the eventual image of A in $\mathbb{R}^{L_A}$ is contained in the real span of $\mathcal{L}$. So the eventual image of A in $\mathbb{T}^{L_A}$ is contained in $\mathbb{T}_{\mathcal{L}}$ and by Lemma 4.2.14 $\varprojlim(\mathbb{T}^{L_A}, A)$ and $\varprojlim(\mathbb{T}_{\mathcal{L}}, A)$ are identical.

By Lemma 4.2.15 there is an onto homomorphism $\theta^* : \varprojlim(\mathbb{T}_{\mathcal{L}}, A) \to \varprojlim(\mathbb{T}^{L_B}, B)$ which commutes with A^* and B^*. We need to show that θ^* is invertible.

Since the degree of $c_A^*(x)$ and $c_B^*(x)$ are the same the integer matrix Θ defines an invertible map from $\mathcal{L}_{\mathbb{Q}} \cap I_A^R$ onto $\mathbb{Q}^{L_A} \cap I_B^R$. Then Θ restricts to an invertible map from $\mathbb{Z}^{L_A} \cap \mathcal{L}_{\mathbb{Q}} \cap I_A^R$ onto $\mathbb{Z}^{L_B} \cap I_B^R$. As a result it defines a continuous isomorphism from I_A^T in $\mathbb{T}_{\mathcal{L}}$ to I_B^T. This isomorphism which commutes with A and B defines a coordinate by coordinate isomorphism from $\varprojlim(\mathbb{T}^{L_A}, A)$ to $\varprojlim(\mathbb{T}^{L_B}, B)$ which commutes with A^* and B^*. $\qquad\square$

Two subshifts of finite type are said to be *weakly conjugate* when each is a factor of the other.

Corollary 4.2.17 *If two irreducible and aperiodic subshifts of finite type are weakly conjugate then their transition matrices are shift equivalent.*

The converse to Corollary 4.2.17 is also true and we state the following result without further proof. References are supplied in the notes.

Theorem 4.2.18 *Two irreducible and aperiodic subshifts of finite type are weakly conjugate if and only if they are shift equivalent.*

Another necessary condition for the existence of a finite-to-one factor map involves the Bowen-Franks groups.

Proposition 4.2.19 *Suppose $\varphi : \Sigma_A \to \Sigma_B$ is a finite-to-one factor map between irreducible subshifts of finite type, then there is a quotient map from the Bowen-Franks group $\dfrac{\mathbb{Z}^{L_A}}{(I-A)\mathbb{Z}^{L_A}}$ to the Bowen-Franks group $\dfrac{\mathbb{Z}^{L_B}}{(I-B)\mathbb{Z}^{L_B}}$.*

Proof. Assume φ is a finite-to-one factor map and apply Construction 4.2.7. Let $\bar{\mathcal{L}} = \mathcal{L}_{\mathbb{Q}} \cap \mathbb{Z}^{L_A}$. We will first show there is a quotient map from the Bowen-Franks group $\dfrac{\mathbb{Z}^{L_A}[t]}{(I-tA)\mathbb{Z}^{L_A}[t]}$ to the group $\dfrac{\bar{\mathcal{L}}[t]}{(I-tA)\bar{\mathcal{L}}[t]}$. Then we will show $\dfrac{\mathbb{Z}^{L_B}[t]}{(I-tB)\mathbb{Z}^{L_B}[t]}$ is isomorphic to a subgroup of the group $\dfrac{\bar{\mathcal{L}}[t]}{(I-tA)\bar{\mathcal{L}}[t]}$. Finally we will specialize to the groups of interest and show there is a quotient map from $\frac{\mathbb{Z}^{L_A}}{(I-A)\mathbb{Z}^{L_A}}$ to $\frac{\mathbb{Z}^{L_B}}{(I-B)\mathbb{Z}^{L_B}}$.

Lemma 4.2.14 shows that $\varprojlim(\mathbb{T}_{\mathcal{L}}, A)$ is an A^* invariant subgroup of $\varprojlim(\mathbb{T}^{L_A}, A)$. Duality theory states there is an onto homomorphism from the dual group of $\varprojlim(\mathbb{T}^{L_A}, A)$, which is isomorphic to $\varinjlim(\mathbb{Z}^{L_A}, A)$, to the dual group of $\varprojlim(\mathbb{T}_{\mathcal{L}}, A)$, which is isomorphic to $\varinjlim(\bar{\mathcal{L}}, A)$. Moreover, the homomorphism commutes with the induced automorphism A^*. By Observation 2.2.18 $\varinjlim(\mathbb{Z}^{L_A}, A)$ is isomorphic to the Bowen-Franks group $\frac{\mathbb{Z}^{L_A}[t]}{(I-tA)\mathbb{Z}^{L_A}[t]}$ and $\varinjlim(\bar{\mathcal{L}}, A)$ is isomorphic to $\frac{\bar{\mathcal{L}}[t]}{(I-tA)\bar{\mathcal{L}}[t]}$ with their induced automorphisms.

Lemma 4.2.15 shows there is an onto homomorphism from the group $\varprojlim(\mathbb{T}_{\mathcal{L}}, A)$ to the group $\varprojlim(\mathbb{T}^{L_B}, B)$ which commutes with their induced automorphisms. Duality theory now shows the dual group of $\varprojlim(\mathbb{T}^{L_B}, B)$, which is isomorphic to $\varinjlim(\mathbb{Z}^{L_B}, B)$, is isomorphic to an A^* invariant subgroup of the dual group of $\varprojlim(\mathbb{T}_{\mathcal{L}}, A)$, which is isomorphic $\varinjlim(\bar{\mathcal{L}}, A)$. Here Observation 2.2.18 shows $\varinjlim(\mathbb{Z}_{L_B}, A)$ is isomorphic to the Bowen-Franks group $\frac{\mathbb{Z}^{L_B}[t]}{(I-tB)\mathbb{Z}^{L_B}[t]}$ and its induced automorphism.

Now we specialize to the desired Bowen-Franks groups and use the fact that they are finitely generated abelian groups. Substitute $t = 1$ and observe $\frac{\bar{\mathcal{L}}}{(I-A)\bar{\mathcal{L}}}$ is a quotient group of $\frac{\mathbb{Z}^{L_A}}{(I-A)\mathbb{Z}^{L_A}}$ and $\frac{\mathbb{Z}^{L_B}}{(I-B)\mathbb{Z}^{L_B}}$ is isomorphic to a subgroup of $\frac{\bar{\mathcal{L}}}{(I-A)\bar{\mathcal{L}}}$. The special structure of finitely generated abelian groups shows that if one finitely generated abelian group is isomorphic to a subgroup of another then it is also a quotient group.   □

Example 4.2.20 Consider the two subshifts of finite type defined in Example 2.2.4. In Example 2.2.11 we computed to see that $\frac{\mathbb{Z}^{L_A}}{(I-A)\mathbb{Z}^{L_A}}$ is isomorphic to to $\mathbb{Z}/2\mathbb{Z} \oplus \mathbb{Z}/2\mathbb{Z}$ and that $\frac{\mathbb{Z}^{L_B}}{(I-B)\mathbb{Z}^{L_B}}$ is isomorphic to to $\mathbb{Z}/4\mathbb{Z}$. Proposition 4.2.19 shows that neither subshift of finite type is a factor of the other.

Exercises

1. Let Σ_A be defined by the graph in Figure 4.2.2 and define a factor map φ from Σ_A onto the Golden Mean subshift by dropping the subscripts on the elements in the alphabet of Σ_A. Find the degree of φ and find a magic word for it.

2. Find a one-to-one a.e. right-resolving map from the subshift of finite type defined in Example 2.2.7 onto the full two-shift.

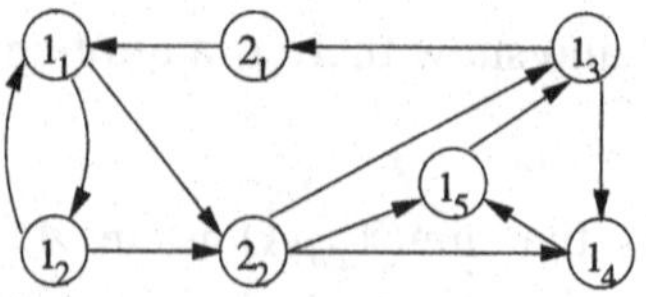

Figure 4.2.2

3. Extend Theorem 4.2.4 to all irreducible subshifts of finite type. Let φ : $\Sigma_A \to \Sigma_B$ be a finite-to-one factor map. Suppose Σ_A has period p and Σ_B has period q. Show there is a degree p/q factor map from Σ_A to Σ_B.

4. Given the transition matrices A and B

$$A = \begin{bmatrix} 1 & 1 & 1 \\ 1 & 0 & 0 \\ 1 & 0 & 0 \end{bmatrix}.B = \begin{bmatrix} 0 & 0 & 1 & 1 \\ 0 & 0 & 1 & 0 \\ 0 & 0 & 0 & 1 \\ 1 & 1 & 0 & 1 \end{bmatrix}.$$

Use the method in the proof of Theorem 4.2.5 to conjugate Σ_A and Σ_B to subshifts of finite type defined by transition matrices with row sum two.

5. For the transition matrices A and B in Problem 4 find one-to-one a.e right-resolving maps from Σ_A and Σ_B onto the full two-shift (Solve the road coloring problem).

6. Suppose $\varphi : \Sigma_A \to \Sigma_B$ is a right-resolving factor map. Define the matrix $R \in \{0,1\}^{L_A \times L_B}$ by

$$R_{ij} = \begin{cases} 1 & \text{if } \varphi(i) = j \\ 0 & \text{otherwise.} \end{cases}$$

Show $AR = RB$. Prove the analogous statement for left-resolving maps.

7. Compute the lattice $\mathcal{L}$ and the matrix Θ from Construction 4.2.7 for the maps φ and ψ in Example 4.1.3

8. For the proof of Theorem 4.2.8 show that if A is an $n \times n$ matrix, B is an $m \times m$ matrix and Θ is an $m \times n$ rank m matrix with $\Theta A = B\Theta$ then the Jordan form of B is a principal submatrix of the Jordan form of A. Then show that if V is an A invariant subspace of $\mathbb{R}^n$, then the Jordan form of A restricted to V is a principal submatrix of the Jordan form of A. Example 4.1.3

§4.3 Special Constructions Involving Factor Maps

In this section we will investigate some special properties and constructions having to do with finite-to-one factor maps.

First we single out a useful construction and apply it in three different settings. In Sections 4.1 and 4.2 we saw the usefulness of passing to a higher block presentation of a subshift of finite type so that we can assume a factor map is a one-block map with no memory or anticipation. This greatly simplifies many of the arguments. We think of this as a *recoding*. To recode means to find a subshift of finite type conjugate to the domain shift and a subshift of finite type conjugate to the image shift, and then take the induced factor map between the new subshifts. The point of a recoding is to exhibit a topological or dynamical property of the factor map as a simple combinatorial property of the recoded factor map.

Recoding Construction 4.3.1 Let $\varphi : \Sigma_A \to \Sigma_B$ be a one-block factor map between irreducible subshifts of finite type. Let $\ell \in \mathbb{N} = \{0, 1, 2, \ldots\}$ and $t \in \{1, \ldots, \ell\}$. Define a new subshift of finite type, $\Sigma_{\hat{A}}$, whose alphabet consists of equivalence classes of words in $\mathcal{W}(A, \ell)$. Two words $u = [u_1, \ldots, u_\ell]$ and $v = [v_1, \ldots, v_\ell] \in \mathcal{W}(A, \ell)$ are related if

$$\varphi(u) = \varphi(v) \qquad \text{and} \qquad u_t = v_t.$$

This defines an equivalence relation on $\mathcal{W}(A, \ell)$. Define transitions between the equivalence classes by saying $[v]$ can follow $[u]$ if and only if there is a $u' \in [u]$ and a $v' \in [v]$ so that v' can follow u' in the ℓ-block presentation of Σ_A. That means $u'_{i+1} = v'_i$ for $i = 1, \ldots, \ell - 1$. There is natural one-block map ψ_A from $\Sigma_{\hat{A}}$ to Σ_A defined by sending an equivalence class to the tth coordinate of the blocks in the equivalence class. Each word in $\mathcal{W}(A, \ell)$ belongs to one equivalence class so ψ_A has an ℓ-block inverse. The subshift of finite type $\Sigma_{\hat{A}}$ is conjugate to Σ_A. Define $\Sigma_{\hat{B}}$ to be the ℓ-block presentation of Σ_B with the natural conjugacy $\psi_B : \Sigma_{\hat{B}} \to \Sigma_B$ defined by $\psi_B([i_1, \ldots, i_\ell]) = i_t$. The one-block map $\hat{\varphi} : \Sigma_{\hat{A}} \to \Sigma_{\hat{B}}$ which sends each equivalence class to its image under φ is the map induced by φ. It is equal to $\psi_B^{-1} \circ \varphi \circ \psi_A$.

In Section 4.2 we saw that any finite-to-one factor map between irreducible subshifts of finite type has a magic word. Magic words are crucial to understanding finite-to-one factor maps. The first application of the recoding construction shows we can recode a factor map so it has a *magic symbol*, that is, a magic word of length one.

Proposition 4.3.2 *Let $\varphi : \Sigma_A \to \Sigma_B$ be a finite-to-one factor map between irreducible subshifts of finite type. Then there is a subshift of finite type $\Sigma_{\hat{A}}$*

conjugate to Σ_A and a subshift of finite type $\Sigma_{\hat{B}}$ conjugate to Σ_B where the factor map, $\hat{\varphi} : \Sigma_{\hat{A}} \to \Sigma_{\hat{B}}$, induced by φ has a magic word of length one.

Proof. Let $\varphi : \Sigma_A \to \Sigma_B$ be a one-block factor map between irreducible subshifts of finite type. Let w be a magic word of length ℓ and t be the distinguished coordinate. Apply Construction 4.3.1 to Σ_A, Σ_B and φ with this ℓ and t. Consider the recoded map $\hat{\varphi}$. The symbol $[w] \in L_{\hat{A}}$ is the only symbol which $\hat{\varphi}$ maps to the symbol $\varphi(w) \in L_{\hat{B}}$. So $\varphi(w)$ is a magic word of length one.   □

Early in Section 4.1 we discussed left and right-resolving maps. These maps have very nice combinatorial properties. Any left or right-resolving map between subshifts of finite type is easily seen to be a finite-to-one factor map. In practice, these are usually the only kind of finite-to-one factor maps constructed. The usefulness of this idea was demonstrated in the proof of Theorem 4.2.5. The definition of resolving is in terms of symbols and predecessor or follower sets. It is not a topological or dynamical definition. There is a dynamical property which captures the essence of the definition. Two points, x and y, in a subshift of finite type with metric d are forwardly asymptotic when $d(\sigma^n(x), \sigma^n(y)) \to 0$ as $n \to +\infty$. Two points are negatively asymptotic when $d(\sigma^n(x), \sigma^n(y)) \to 0$ as $n \to -\infty$. It's easily seen that $d(\sigma^n(x), \sigma^n(y)) \to 0$ as $n \to +\infty$ if and only if there is an $N \in \mathbb{Z}$ and $x_i = y_i$ for all $i \geq N$ and $d(\sigma^n(x), \sigma^n(y)) \to 0$ as $n \to -\infty$ if and only if there is an $N \in \mathbb{Z}$ and $x_i = y_i$ for all $i \leq N$. We say a factor map $\varphi : \Sigma_A \to \Sigma_B$ is *right-closing* if $x = y$ whenever $\varphi(x) = \varphi(y)$ and x and y are negatively asymptotic. Similarly, we say a factor map $\varphi : \Sigma_A \to \Sigma_B$ is *left-closing* if $x = y$ whenever $\varphi(x) = \varphi(y)$ and x and y are forwardly asymptotic. Observe that a right-resolving map is right-closing and a left-resolving map is left-closing. The second application of the recoding construction demonstrates that the closing property captures the crucial dynamical property of resolving maps.

Proposition 4.3.3 *Let $\varphi : \Sigma_A \to \Sigma_B$ be a factor map between irreducible subshifts of finite type. Then φ is right-closing if and only if there is a subshift of finite type $\Sigma_{\hat{A}}$ conjugate to Σ_A and a subshift of finite type $\Sigma_{\hat{B}}$ conjugate to Σ_B so that the factor map, $\hat{\varphi} : \Sigma_{\hat{A}} \to \Sigma_{\hat{B}}$, induced by φ is right-resolving. The same is true for left-closing and left-resolving factor maps.*

Proof. If $\varphi : \Sigma_A \to \Sigma_B$ is right-resolving, then it is right-closing.

Suppose $\varphi : \Sigma_A \to \Sigma_B$ is a right-closing factor map. Assume it is one-block map. There is an $\ell \in \mathbb{N}$ so that if $u = [u_1, \dots, u_\ell], v = [v_1, \dots, v_\ell] \in \mathcal{W}(A, \ell)$ with $u_1 = v_1$ and $\varphi(u) = \varphi(v)$ then $u_2 = v_2$. Before proving such an ℓ exists note that a factor map is right-resolving if and only if such an ℓ exists and is equal to two. To see ℓ exists first note that if φ has a diamond it is not right-closing. Next suppose ℓ does not exist. Then for each $k \in \mathbb{N}$ there are a pair of blocks $u^k, v^k \in \mathcal{W}(A, k)$ with $u_1^k = v_1^k$, $\varphi(u^k) = \varphi(v^k)$ and $u_i^k \neq v_i^k$ for $1 < i \leq k$. We can choose the pairs of blocks so that when $k \leq k'$, $u_i^k = u_i^{k'}$ and $v_i^k = v_i^{k'}$ for $1 \leq i \leq k$. Let $x \in \Sigma_A$ be a point with $x_0 u_1^k = x_0 v_1^k \in \mathcal{W}(A, 2)$. For each k, $\hat{u}^k = (\cdots x_{-1} x_0 u^k]$ and $\hat{v}^k = (\cdots x_{-1} x_0 v^k]$ are nonempty, disjoint, closed sets. We have $\hat{u}^k \supseteq \hat{u}^{k+1}$ and $\hat{v}^k \supseteq \hat{v}^{k+1}$. So the nested intersections $\cap \hat{u}^k$ and $\cap \hat{v}^k$ are nonempty. If $y \in \cap \hat{u}^k$ and $z \in \cap \hat{v}^k$ then $y \neq z$, y and z are negatively asymptotic and $\varphi(y) = \varphi(z)$. This contradicts the assumption that φ is right-closing. Consequently, there is such an ℓ.

Now apply Construction 4.3.1 for this ℓ and $t = 1$. Observe $\hat{\ell} = 2$ works for the recoded map $\hat{\varphi}$.

The same result follows for left-resolving and left-closing factor maps by concentrating on blocks $u = [u_1, \ldots, u_\ell], v = [v_1, \ldots, v_\ell] \in \mathcal{W}(A, \ell)$ with $u_\ell = v_\ell$.   □

Next we see the closing property is important for another special class of maps. Let $\varphi : \Sigma_A \to \Sigma_B$ be a finite-to-one factor map between irreducible subshifts of finite type. We say φ is *constant-to-one* if each point in Σ_B has the same number of preimages under φ. If the degree of φ is d we will also say φ is *exactly or uniformly d-to-one*. Let $z \in \Sigma_B$. Two points $x, y \in \varphi^{-1}(z)$ are *totally or mutually separated* if there is a distance $\delta > 0$ and $d(\sigma^n(x), \sigma^n(y)) \geq \delta$ for all $n \in \mathbb{Z}$. Observe that all the points in every $\varphi^{-1}(z)$ are mutually separated if and only if the factor map is both left and right-closing. The third application of the recoding construction demonstrates the connection between these ideas.

Proposition 4.3.4 *Let $\varphi : \Sigma_A \to \Sigma_B$ be a finite-to-one factor map between irreducible subshifts of finite type. Then φ is constant-to-one if and only it is left and right-closing.*

Proof. Suppose φ is a one-block, left and right-closing map. We can use the recoding construction to produce a map that is simultaneously left and right-resolving. There is an ℓ_m and an ℓ_a in $\mathbb{N}$ so that if $u = [u_{-\ell_m}, \ldots, u_{\ell_a}]$, $[v_{-\ell_m}, \ldots, v_{\ell_a}] \in \mathcal{W}(A, \ell_m + \ell_a + 1)$ with $\varphi(u) = \varphi(v)$ and $u_0 = v_0$ then $u_1 = v_1$ and $u_{-1} = v_{-1}$. Apply Construction 4.3.1 with $\ell = \ell_m + \ell_a + 1$ and $t = \ell_m + 1$. The recoded factor map $\hat{\varphi}$ is left and right-resolving. Every symbol in $\Sigma_{\hat{B}}$ has the same number of preimages so every symbol is magic and $\hat{\varphi}$ is constant-to-one.

Next suppose φ is a one-block, exactly d-to-one map which is not right-closing. Then there is a $y \in \Sigma_B$ with $\varphi^{-1}(y) = \{x^1, \ldots, x^d\}$ and $x_i^1 = x_i^2$ for all $i \leq 0$. This means there is an $N \geq 0$ so that

$$|\{j \in L_A : j = x_0 \text{ for some } [x_{-N}, \ldots, x_N] \in \varphi^{-1}([y_{-N}, \ldots, y_N])\}| \leq d - 1.$$

This contradicts φ being d-to-one everywhere. So φ is left and right-closing.   □

Next we prove a simple property of finite-to-one factor maps.

Proposition 4.3.5 *Let* $\varphi : \Sigma_A \to \Sigma_B$ *be a finite-to-one factor map between irreducible subshifts of finite type. If φ is not a conjugacy, then there is a periodic point in Σ_B with more than one preimage under φ.*

Proof. Let $\varphi : \Sigma_A \to \Sigma_B$ be a finite-to-one, one-block factor map between irreducible subshifts of finite type. If the degree of φ is greater than one then every point has more than one preimage. Assume the degree is one. If the map is not a conjugacy there is a point $z \in \Sigma_B$ with more than one preimage. If $x \neq y \in \varphi^{-1}(z)$, then there is an $N \in \mathbb{Z}$ and $x_i \neq y_i$ for either $i \geq N$ or $i \leq N$ (or both). If this were not true then either φ would have a diamond or x and y would be the same point. Assume $x_i \neq y_i$ for $i \geq N$ and consider the sequence of pairs $\{(x_i, y_i)\}$ for $i \geq N$. There must be at least two pairs, (x_j, y_j) and (x_k, y_k), for $N \leq j < k$ that agree. Then the two blocks $[x_j, \ldots, x_k]$ and $[y_j, \ldots, y_k]$, each repeated infinitely often, define distinct periodic points in Σ_A which map to the same point. $\qquad\square$

Next we use resolving maps for an interesting and useful construction. The construction illustrates another reason resolving maps are so useful.

Construction 4.3.6 Let Σ_B be an irreducible subshift of finite type with positive entropy and z a point of least period p in Σ_B. Then there is an irreducible subshift of finite type Σ_A and a one-to-one almost everywhere, right-resolving factor map $\varphi : \Sigma_A \to \Sigma_B$ so that every periodic point in Σ_B, except those in the orbit of z, has a single preimage in Σ_A and each point in the orbit of z has exactly two preimages, each of period $2p$, in Σ_A.

Proof. By going to a higher block presentation for Σ_B we may assume the periodic point $z \in \Sigma_A$ is $(\ldots, z_0, z_1, \ldots, z_{p-1}, z_0, \ldots)$ with all the symbols $z_0, \ldots, z_{p-1}$ distinct. This is equivalent to saying z is described by a simple cycle in G_B. A cycle is simple when no vertex is used more than once. By going to a still higher block presentation we can also assume that the only edges in G_B connecting the z_k are the edges from z_k to $z_{k+1} \mod p$. We define a new graph G_A. The new graph has vertices

$$V_A = (V_B - \{z_0, z_1, \ldots, z_{p-1}\}) \cup \{z_0^1, \ldots, z_{p-1}^1, z_0^2, \ldots, z_{p-1}^2\}.$$

We replace each symbol, z_k, in the cycle describing z by two new symbols, z_k^1 and z_k^2. The edges in the new graph are defined by the following rules:

(i)   if $i, j \in V_A$ and $i, j \notin \{z_0, \ldots, z_{p-1}\}$ then there is an edge from i to j in G_A if and only if there is an edge in G_B;

(ii)   if $i \in V_A$ and $i \notin \{z_0, \ldots, z_{p-1}\}$ then there is an edge from i to z_k^1 in G_A
       if and only if there is an edge from i to z_k in G_B;

(iii) if $j \in V_A$ and $j \notin \{z_0, \ldots, z_{p-1}\}$ then there is an edge from z_k^1 to j and
       from z_k^2 to j in G_A if and only if there is an edge from z_k to j in G_B;

(iv)   there is an edge from z_k^1 to z_{k+1}^1 for $k = 0, \ldots, p - 2$, there is an edge
       from z_{p-1}^1 to z_0^2, there is an edge from z_k^2 to z_{k+1}^2 for $k = 0, \ldots, p-2$ and
       there is an edge from z_{p-1}^2 to z_0^1 in G_B.

The cycle of length p made up of the z_k's in G_B is blown up to a cycle of
length $2p$ in G_A. The edges in G_B not involving the z_k's are left alone and
connections between these two parts of the graph in G_A are made so that
the resulting map will have the desired properties. The graph G_A is strongly
connected so Σ_A will be irreducible.

The map from G_A to G_B is defined by the way it maps vertices. It is the
identity except on the z_k^i. It maps z_k^i to z_k. It is easily checked that this defines
a right-resolving factor map from Σ_A to Σ_B. The map is one-to-one almost
everywhere since all symbols except the z_k are magic symbols. There are other
symbols since Σ_B was assumed to have positive entropy. Every cycle except
the cycle of all z_k's in G_B has a single cycle of the same length as its preimage
since every such cycle contains a vertex that is not a z_k.   □

This construction is illustrated in Example 4.3.7. It highlights the role periodic points play in finite-to-one factor maps as described in Proposition 4.3.5
We could have made the map in the construction left-resolving by focusing on
predecessors instead of successors.

Example 4.3.7 Let Σ_B be described by the graph on the left in Figure 4.3.1.
The point $z = (\ldots z_1.z_0 z_1 z_0 \ldots)$ has period two . The graph on the right is the
graph of Σ_A produced by Construction 4.3.6.

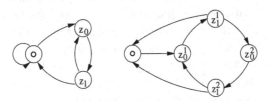

Figure 4.3.1

The next example is of a one-to-one almost everywhere factor map that is
neither left nor right-resolving. We will see it is indecomposable.

Example 4.3.8 Let Σ_A be the subshift of finite type described by the graph on the left in Figure 4.3.2.

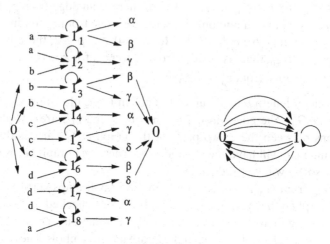

Figure 4.3.2

It has nine vertices, $\{0, 1_1, 1_2, \ldots, 1_8\}$. There is one edge from 0 to 1_k for k odd and two edges from 0 to 1_k for k even. These are the edges going into the 1_k from the left. There is a self-loop at each of the 1_k. There is one edge from 1_k to zero for k even and two edges from 1_k to 0 for k odd. These are the edges going out of the 1_k to the right. The symbols for Σ_A are the edge labels of the graph. Define Σ_B by the graph on the right. It has two vertices 0 and 1. There are four edges from 0 to 1 labelled a, b, c, d, four edges labelled $\alpha, \beta, \gamma, \delta$ from 1 to 0 and a self-loop labelled m at 1. Here also, the symbols for Σ_B are the edge labels. The factor map $\varphi : \Sigma_A \to \Sigma_B$ is defined by a graph homomorphism of G_A to G_B. It sends the vertex 0 to 0 and the vertices $1_1, \ldots 1_8$ to 1. It is defined on the edges by dropping the subscripts on the symbols. The fixed point of all m's in Σ_B has eight preimages. Every other periodic point has a single preimage. To see this, observe that a block of the form $[i, m, \ldots, m, j]$ for $i \in \{a, b, c, d\}$ and $j \in \{\alpha, \beta, \gamma, \delta\}$ has exactly one preimage. Any point in Σ_B that neither begins nor ends with all m's has exactly one preimage. Suppose there is a subshift of finite type Σ_C and two factor maps $\varphi_1 : \Sigma_A \to \Sigma_C$ and $\varphi_2 : \Sigma_C \to \Sigma_B$ with $\varphi = \varphi_2 \circ \varphi_1$. We will prove that either φ_1 or φ_2 is a conjugacy. When a factor map φ has this property we say it is *indecomposable*.

Suppose we have Σ_C, φ_1 and φ_2 as stated. Proposition 4.3.5 shows that if φ_1 is not a conjugacy it must identify some of the fixed points in Σ_A. Assume it identifies the fixed points $x = (m_1)^\infty$ and $y = (m_2)^\infty$. Let

$$x^n = \leftarrow b_2(m_1)^n.(m_1)^\infty$$
$$y^n = (m_2)^\infty.(m_2)^n \beta_1 \rightarrow$$

where the left arrow means any admissible sequence not beginning with a string of all m_k's and the right arrow means any admissible sequence not ending with a string of all m_k's. Since $\{\varphi_1(x^n)\}$ and $\{\varphi_1(y^n)\}$ both converge to $\varphi_1(x) = \varphi_1(y)$ as $n \to \infty$ we know that for all sufficiently large n, the points $[\varphi_1(x^n), \varphi_1(y^n)]$ exist in Σ_C. Recall the bracket of two points in a subshift of finite type is the point

$$[w, v]_i = \begin{cases} w_i & \text{for} \quad i \leq 0 \\ v_i & \text{for} \quad i \geq 0. \end{cases}$$

Let $z^n \in \Sigma_A$ be a point with $\varphi_1(z^n) = [\varphi_1(x^n), \varphi_1(y^n)]$. Then

$$\begin{aligned} \varphi(z^n) &= \varphi_2([\varphi_1(x^n), \varphi_1(y^n)]) \\ &= [\varphi_2 \circ \varphi_1(x^n), \varphi_2 \circ \varphi_1(y^n)] \\ &= [\varphi(x^n), \varphi(y^n)]. \end{aligned}$$

The point

$$[\varphi(x^n), \varphi(y^n)] = \quad \leftarrow bm^n.m^n\beta \rightarrow$$

has one preimage under φ since it neither begins or ends with all m's. Consequently, z^n must be of the form

$$z^n = \quad \leftarrow b_3(m_3)^n.(m_3)^n\beta_3 \rightarrow .$$

Since the points $\varphi_1(z^n)$ converge to $\varphi_1(x) = \varphi_1(y)$ and the points z^n converge to the point $(m_3)^\infty$ we see that φ_1 identifies the point $(m_3)^\infty$ with the points $(m_1)^\infty$ and $(m_2)^\infty$. Continuing in the same manner shows that all the fixed points in Σ_A are identified by φ_1. Then Proposition 4.3.5 shows that the map φ_2 is a conjugacy. The same reasoning applies if we assume any other pair of fixed points are identified by φ_1. □

Exercises

For the following exercises let Σ_A and Σ_B be irreducible subshifts of finite type and $\varphi : \Sigma_A \to \Sigma_B$ a factor map.

1. Prove that φ is infinite-to-one if and only if there exists a $z \in \Sigma_B$ and $x, y \in \varphi^{-1}(z)$ with x and y forwardly and backwardly asymptotic.

2. Show that if φ is finite-to-one, $z \in \Sigma_B$ and $x, y \in \varphi^{-1}(z)$ then x and y are either mutually separated, forwardly asymptotic or backwardly asymptotic.

3. Prove that if φ is d-to-one a.e. and $z \in \Sigma_B$ then $\varphi^{-1}(z)$ contains d mutually separated points.

4. Suppose $\varphi : \Sigma_A \to \Sigma_B$ is a d-to-one a.e. factor map between irreducible subshifts of finite type and there is a $z \in \Sigma_B$ with more than d preimages. Prove that the set of points with more than d preimages is dense in Σ_B.

5. Show that if $\varphi : \Sigma_A \to \Sigma_B$ is a finite-to-one factor map, then it is an open mapping if and only if it is left and right-closing. Recall that a mapping is open if the image of every open set is open. Conclude that it is an open mapping if and only if it is constant-to-one.

6. Show that if $\varphi : X_A \to X_B$ is a finite-to-one factor map, then it is an open mapping if and only if it is right-closing.

7. Prove the map described in Example 4.1.3 is indecomposable.

§ 4.4 Subsystems and Infinite-to-One Factor Maps

In this section we will examine two seemingly unrelated problems. The first problem is to determine if one subshift of finite type can be embedded in another. The second problem is to determine when an infinite-to-one factor map exists between two subshifts of finite type. Surprisingly, the same techniques provide complete answers to both questions.

A subshift of finite type Σ_B can be embedded in a subshift of finite type Σ_A if Σ_A contains a subshift of finite type conjugate to Σ_B. Our first problem is to find necessary and sufficient conditions for embedding one subshift of finite type into another. One case is when Σ_B is conjugate to Σ_A. The other case is when Σ_B is conjugate to $\Sigma \subsetneq \Sigma_A$. This is called a *proper embedding*. In this section we are interested in proper embeddings. Let Σ_A be an irreducible subshift of finite type and suppose Σ is a subshift of finite type properly contained in Σ_A. By going to a higher block presentation of Σ_A we may assume Σ is obtained from Σ_A by deleting some of the symbols from the alphabet of Σ_A. The Perron-Frobenius Theorem tells us that the entropy of Σ is strictly less than the entropy of Σ_A. Let $Per(\Sigma_A, n)$ denote the set of periodic points with least period n in Σ_A. When $\Sigma \subseteq \Sigma_A$, $Per(\Sigma, n) \subseteq Per(\Sigma_A, n)$ and so $|Per(\Sigma, n)| \leq |Per(\Sigma_A, n)|$. In order to properly embed one subshift of finite type into another the condition on entropy must be met and the condition on periodic points must be met. These two conditions are not only necessary but sufficient.

Theorem 4.4.1 *Let Σ_A be an irreducible and aperiodic subshift of finite type. An irreducible subshift of finite type, Σ_B, can be properly embedded in Σ_A if and only if:*

(i) $h(\Sigma_B, \sigma) < h(\Sigma_A, \sigma)$;

(ii) $|Per(\Sigma_B, n)| \leq |Per(\Sigma_A, n)|$, *for every n.*

Proof. We have already seen that the two conditions are necessary for a proper embedding.

Now we outline the proof of the converse There are five steps. First we choose markers in Σ_A. This is easy. We have more freedom when making constructions in Σ_A than we do in Σ_B because the larger entropy allows more flexibility. The second step is to choose markers in Σ_B. This is more delicate. In order to insure that the final map is continuous we need to choose markers which will occur with bounded gaps between them. In order to take advantage of the extra entropy of Σ_A when constructing the map we must keep the markers from occurring too close together. In the third step we describe the way markers are used in Σ_B to mark a point. In the fourth step we make assignments between points of low period in Σ_B and Σ_A. Then we make assignments between blocks that occur between markers in Σ_B and blocks that occur between markers in Σ_A. The fifth and final step is to put the markers and block assignments together to construct the embedding.

Step 1. Constructing markers in Σ_A. First generalize Lemma 3.3.17 to find a higher block presentation of Σ_A with a distinguished symbol M so that the subshift of finite type obtained be deleting M is irreducible, aperiodic and still still has entropy strictly larger than the entropy of Σ_B. Then choose six distinct blocks $\bar{u}^1, \bar{u}^2, \bar{v}^1, \bar{v}^2, \bar{w}^1$ and $\bar{w}^2$ in Σ_A so that:

(i)   all six blocks have the same length;
(ii) $\bar{u}^1, \bar{v}^1$ and $\bar{w}^1$ begin with M and contain M nowhere else;
(iii) $\bar{u}^2, \bar{v}^2$ and $\bar{w}^2$ end with M and contain M nowhere else.

Step 2. Constructing markers in Σ_B. Fix two integers $0 < q < r$. The values for q and r will be chosen later. There will be two kinds of markers in Σ_B. The first type is made up of long blocks coming from points of small period. Let

$$P(q, k) = \{[x_0, \dots, x_{k-1}] : \sigma^j(x) = x \quad \text{for some} \quad j < q\}.$$

Then let

$$P = \bigcup_{k \geq r} P(q, k).$$

The blocks in P are the first kind of marker. The blocks chosen to be the second type of marker are chosen to avoid the first kind of periodic marker and each other but insure markers occur with bounded gaps between them. Let

$$\mathcal{N}(q, r) = \{u^1, \dots u^\alpha\} = \{u \in \mathcal{W}(B, 2r - 1) : \text{no subblock of } u \text{ is in} \quad P(q, r)\}.$$

Blocks in $\mathcal{N}$ avoid blocks in P to a sufficient degree and if taken to be markers insure that markers occur with bounded gaps between them. But they do not avoid each other. We must build this in. For each i,

$$[u^i]_{-r+1} = \{x : [x_{-r+1}, \dots, x_0, \dots, x_{r-1}] = u^i\}.$$

Let

$$E_1 = [u^1]_{-r+1}.$$

Note $E_1 \cap \sigma^j(E_1) = \phi$ for $0 < j < q$. For $1 \leq i < \alpha$, let

$$E_{i+1} = E_i \bigcup \{[u^{i+1}]_{-r+1} - \bigcup_{-q<j<q} \sigma^j(E_i)\}.$$

This adds to E_i the part of $[u^{i+1}]_{-r+1}$ which will preserve the condition $E_i \cap \sigma^j(E_i) = \phi$ for $0 < j < q$. Let $E = E_\alpha$. Note

(i) $$E \subseteq \bigcup_i [u^i]_{-r+1},$$

(ii) $$\bigcup_i [u^i]_{-r+1} \subseteq \bigcup_{-q<j<q} \sigma^j(E),$$

(iii) $$E \cap \sigma^j(E) = \phi \text{ for } 0 < j < q.$$

Let $\mathcal{N}$ be the finite set of disjoint cylinders which make up E. These are the other markers. We have two types of markers in Σ_B, the long blocks of small period in $\mathcal{P}$ and the blocks which insure the markers occur with bounded gaps in $\mathcal{N}$.

Step 3. Marking a point in Σ_B. Let $x \in \Sigma_B$. We scan the point x and mark two things:

(i)   the entire occurrence of a marker of type $\mathcal{P}$ (making the length maximal);
(ii)   the center coordinate of a marker of type $\mathcal{N}$ (the center of the u^i).

This is illustrated in Figure 4.4.1.

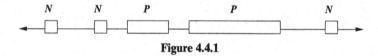

Figure 4.4.1

Note the following.

(i)   A coordinate marked $\mathcal{N}$ cannot occur inside a block marked $\mathcal{P}$ because it's the center of a u^i block.
(ii)   Two coordinates marked $\mathcal{N}$ can occur no closer than q by the construction of E.
(iii) Two blocks marked $\mathcal{P}$ may overlap but the overlap must be of length less than q because of the periodicity of the blocks.

Next is an important observation. The gap between any two marked coordinates is bounded. To see this, first note that if a coordinate is in a $\mathcal{P}(q, r)$ block then it is in a block marked $\mathcal{P}$. Then note that if a coordinate in a point is the center coordinate of a u^i block there will be a marked coordinate within q coordinates. This follows from the choice of the u^i and condition (ii) on E. Fix a marked point $x \in \Sigma_B$ and focus on x_0. Suppose none of the coordinates $x_0, x_1, \ldots, x_{r-1}$ are in a $\mathcal{P}(q, r)$ block. Then x_{r-1} is the center coordinate of a u^i

block. This means at least one of the coordinates $x_0, x_1, \ldots, x_{q+r-1}$ is marked. This is the property that will make the embedding continuous.

Step 4. Assignments using markers Let $\bar{A}$ be the transition matrix for the subshift of finite type obtained from Σ_A by deleting the symbol M. Let $a(n) = min_{i,j}\{(\bar{A}^{n-2m})_{ij}\}$ where m is the length of the markers in Σ_A. Choose q so that $a(n) > \Sigma_{i,j}(B^n)_{ij}$ for all $n \geq q$. We will assign blocks that occur between markers in Σ_B to blocks that occur between markers in Σ_A. There are five kinds of assignments necessary. They are

(i)   between coordinates marked $\mathcal{N}$,
(ii)   inside a $\mathcal{P}$ block,
(iii) between $\mathcal{P}$ blocks,
(iv)   from a coordinate marked $\mathcal{N}$ to a $\mathcal{P}$ block and
(v)   from a $\mathcal{P}$ block to a coordinate marked $\mathcal{N}$.

Assignments from a coordinate marked $\mathcal{N}$ to coordinate marked $\mathcal{N}$. This is the simplest case. For each ℓ consider the blocks of that length in Σ_B which can occur between two markers of type $\mathcal{N}$. Such a block occurs with its first and last symbols marked $\mathcal{N}$ and no other symbol marked. Make a one-to-one correspondence between these blocks of length ℓ in Σ_B and some subset of the blocks of length ℓ in Σ_A that begin with $\bar{u}^1$, end with $\bar{u}^2$ and do not contain M except at the beginning and end. The choice of q guarantees there are a sufficient number of such blocks in Σ_A.

Assignments inside a $\mathcal{P}$ block. This is the most complicated case. First match each periodic point in Σ_B of period less than q to a point of the same period in Σ_A. Do this so that no two points in Σ_B are matched to the same point in Σ_A and so that the matching commutes with the shifts. For each symbol $i \in L_{\bar{A}}$ choose a block, $\bar{v}^{(1,i)}$, of length $2q$ in Σ_A that begins with $\bar{v}^1$, ends in i and doesn't contain the symbol M except as the first symbol. This guarantees that $\bar{v}^{(1,i)}$ contains no marker other than $\bar{v}^1$ and $\bar{v}^i$ occurs only at the beginning. It also guarantees that $\bar{v}^{(1,i)}$ cannot be contained in any periodic point of period less than q. Similarly, choose for each $j \in L_{\bar{A}}$ a block, $\bar{v}^{(2,j)}$, of length $2q$ that begins with j ends with $\bar{v}^2$ and contains M only as the last symbol. A $\mathcal{P}$ block in Σ_B gets assigned to a block in Σ_A as indicated in Figure 4.4.2.

A $\mathcal{P}$ block of length ℓ gets assigned to a block of length $\ell - 2q$. The $\mathcal{P}$ block comes from a point of period less than q which was assigned to a point of the same period in Σ_A. The $\mathcal{P}$ block is assigned to a block that preserves this assignment on the central $\ell - 6q$ block, begins with the appropriate $\bar{v}^{(1,i)}$ and ends with the appropriate $\bar{v}^{(2,j)}$. This is done to insure we can recognize in the image of Σ_B an intended occurrence of the markers $\bar{v}^1$ and $\bar{v}^2$. Some of the points of period less than q in Σ_A may contain the markers. That is why we need the blocks $\bar{v}^{(1,i)}$ and $\bar{v}^{(2,j)}$. They are chosen long enough to insure we will be able to identify a true occurrence of a marker in the image of Σ_B. Then

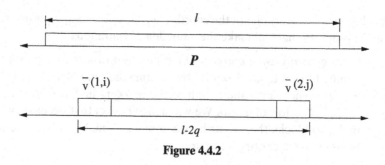

Figure 4.4.2

if ℓ is bigger than $7q$ we will be able to recover the $\mathcal{P}$ block from the assigned block in Σ_A.

Assignments between $\mathcal{P}$ blocks. Two blocks marked $\mathcal{P}$ may overlap, but the overlap must be by less than q entries. A $\mathcal{P}$ block gets assigned to a block in Σ_A that is q entries shorter at each end. This means two coordinates assigned from different $\mathcal{P}$ blocks will be separated by at least q unassigned entries. This gives the required freedom. Match the blocks that can occur in Σ_B between $\mathcal{P}$ blocks to blocks of the appropriate length in Σ_A that begin with $\bar{w}^1$, end with $\bar{w}^2$ and contain M nowhere else. Make sure no two blocks from Σ_B are matched to the same block in Σ_A

Assignments from a coordinate marked $\mathcal{N}$ to a $\mathcal{P}$ block. This is just like (iii). Consider the blocks in Σ_B that can occur with the first symbol marked $\mathcal{N}$, the last symbol marked $\mathcal{P}$ and no other symbol marked. A coordinate marked $\mathcal{N}$ can occur next to a block marked $\mathcal{P}$ but we still have the required freedom since a $\mathcal{P}$ block is matched to a block in Σ_A that is q entries shorter at each end. Match the blocks in Σ_B that can occur to blocks of the appropriate length in Σ_A that begin with $\bar{u}^1$, end with $\bar{w}^2$ and do not contain M except at the beginning and end. Make sure no two blocks from Σ_B are matched to the same block in Σ_A.

The last assignment is for blocks that begin with a coordinate marked $\mathcal{P}$ and end with a coordinate marked $\mathcal{N}$. Make these assignments just as is in the last case but choose blocks in Σ_A that begin with $\bar{w}^1$ and end with $\bar{u}^2$.

Step 5. The map. Set $r = 8q$. Consider the marker structure for an $x \in \Sigma_B$. It will map to a point $\varphi(x) \in \Sigma_A$ with a marker structure as illustrated in Figure 4.4.3. The marker structure of $\varphi(x)$ allows you to reconstruct where markers occurred in x and whether they were type $\mathcal{P}$ or $\mathcal{N}$.

There are bounds on the distances

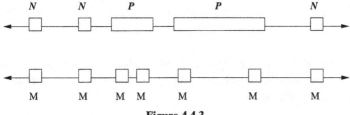

Figure 4.4.3

$$q \leq d_1 \leq 2r + 2q = 18q$$
$$q \leq d_2 \leq 2r + 3p = 19q$$
$$2q \leq d_3$$
$$q \leq d_4 \leq 2r + 4q = 20q$$
$$2q \leq d_5$$
$$q \leq d_6 \leq 2r + 3q = 19q.$$

The blocks between the markers can be filled using the assignments in Step 4. In Figure 4.4.3, moving from left to right use the assignment between coordinates marked $\mathcal{N}$, next use the assignment from a coordinate marked $\mathcal{N}$ to a $\mathcal{P}$ block, then the assignment inside a $\mathcal{P}$ block, then the assignment between $\mathcal{P}$ blocks, then inside a $\mathcal{P}$ block and finally the assignment from a $\mathcal{P}$ block to a coordinate marked $\mathcal{N}$.

This method, first putting down the marker structure of the image point and then the fillers, produces a continuous map since all choices depend on a bounded number of coordinates. The map, φ, commutes with the shift because the construction does not take into account the position of the block. The map is invertible because of the choice of markers and fillers in Σ_A. □

Corollary 4.4.2 *Let Σ_A be an irreducible and aperiodic subshift of finite type. If Σ_B is another irreducible subshift of finite type with $h(\Sigma_B, \sigma) < h(\Sigma_A, \sigma)$ then there is a subshift of finite type $\Sigma_{\bar{B}} \subseteq \Sigma_A$ and a right-closing factor map $\varphi : \Sigma_{\bar{B}} \to \Sigma_B$*

Proof. The only obstruction to embedding Σ_B in Σ_A might be the number of periodic points. The Perron-Frobenius Theorem says that asymptotically $|Per(A, n)|$ grows exponentially faster than $|Per(B, n)|$. Using Construction 4.3.6 we can construct a subshift of finite type $\Sigma_{\bar{B}}$ with a right-resolving factor map $\varphi : \Sigma_{\bar{B}} \to \Sigma_B$ such that $\Sigma_{\bar{B}}$ meets the condition on periodic points for embedding into Σ_A. Then apply Theorem 4.4.1 to $\Sigma_{\bar{B}}$. □

Corollary 4.4.3 *Let Σ_A be an irreducible and aperiodic subshift of finite type. If Σ_B is another irreducible subshift of finite type with $h(\Sigma_B, \sigma) < h(\Sigma_A, \sigma)$ then* *there is a subshift of finite type $\Sigma_{\bar{B}}$ contained in Σ_A with the same entropy as Σ_B.*

The problem of embedding one-sided subshifts of finite type is very different from the two-sided problem. At present there are no known necessary and sufficient conditions.

Problem 4.4.4 Find necessary and sufficient conditions for properly embedding one one-sided subshift of finite type in another (See Problem 4 in this section).

Next we turn to the problem of determining when there is an infinite-to-one factor map between two subshifts of finite type. We saw in Section 4.2 that there are no necessary and sufficient conditions known for the existence of a finite-to-one factor map between two subshifts of finite type. The infinite-to-one case is different. Suppose Σ_A and Σ_B are irreducible subshifts of finite type and there is an infinite-to-one factor map from Σ_A to Σ_B. Then $h(\Sigma_A, \sigma) > h(\Sigma_B, \sigma)$ by Corollary 4.1.8. and each periodic point in Σ_A must be mapped to a periodic point in Σ_B where the period of the image divides the period of the original point. These two conditions are necessary and sufficient for the existence of an infinite-to-one factor map.

Theorem 4.4.5 *Let Σ_A and Σ_B be irreducible and aperiodic subshifts of finite type. There is an infinite-to-one factor map from Σ_A to Σ_B if and only if*

(i) $h(\Sigma_A, \sigma) > h(\Sigma_B, \sigma)$ *and*
(ii) *for every periodic point in Σ_A there is a periodic point in Σ_B whose period divides the period of the point in Σ_A.*

Proof. We have already seen that the two conditions are necessary for the existence of an infinite to one factor map.

The proof of the converse uses Corollary 4.4.2 and then follows almost exactly the proof of Theorem 4.4.1. There are five steps in each proof. The only real difference is in the first step. Steps two through five are very similar. For the first step in this proof we use Corollary 4.4.2 to find a special subshift of finite type contained in Σ_A and a factor map from this shift onto Σ_B. In steps two through four we use markers and block assignments to extend the map to a factor map from all of Σ_A to Σ_B.

Step 1. Apply Corollary 4.4.2 to construct $\Sigma \subseteq \Sigma_A$ with a factor map $\psi : \Sigma \to \Sigma_B$. Since Σ_B is irreducible and aperiodic there is an N_B such that Σ_B contains a point of period p for every $p \geq N_B$. For any fixed $q \geq N_B$ let Σ' be the union of Σ and all points in Σ_A with period less than q. Observe that Σ' is a subshift of finite type which may be reducible and by the condition on periodic points (ii) of Σ_A and Σ_B we can extend ψ to a new factor map $\psi' : \Sigma \to \Sigma_B$. By going to a higher block presentation of Σ_A and Σ' we can assume Σ' is defined by a transition matrix $\bar{B}$ obtained by setting some of $A's$ entries equal to zero and ψ' is a one-block map. We denote this situation by $\bar{\varphi} : \Sigma_{\bar{B}} \to \Sigma_B$. In the next four steps we extend $\bar{\varphi}$ to a factor map $\varphi : \Sigma_A \to \Sigma_B$.

Step 2. Constructing markers in Σ_A. In this step we construct markers in Σ_A almost exactly as we constructed markers for Σ_B in step 2 of the proof of Theorem 4.4.1. The only difference is that we use long blocks from $\Sigma_{\bar{B}}$ instead of long blocks from points with low period. Fix two integers $0 < q < r$. The values for q and r will be chosen later. There will be two kinds of markers in Σ_B. The first type is made up of long blocks coming from $\Sigma_{\bar{B}}$ Then let

$$M = \bigcup_{k \geq r} W(\bar{B}, k).$$

The blocks in M are the first kind of marker. The blocks chosen to be the second type of marker are chosen to avoid the first kind of Σ_B marker and each other, but insure that markers occur with bounded gaps between them. Let

$$\begin{aligned} N(q,r) &= \{u^1, \ldots u^\alpha\} \\ &= \{u \in W(A, 2r-1) : \text{no subblock of } u \text{ is in } \quad W(\bar{B}, r)\}. \end{aligned}$$

Blocks in N avoid blocks in M to a sufficient degree and if taken to be markers insure that markers occur with bounded gaps between them. But they do not avoid each other. We must build this in. For each i, let

$$[u^i]_{-r+1} = \{x : [x_{-r+1}, \ldots, x_0, \ldots, x_{r-1}] = u^i\}.$$

Let

$$E_1 = [u^1]_{-r+1}.$$

Note $E_1 \cap \sigma^j(E_1) = \phi$ for $0 < j < q$ since $\Sigma_{\bar{B}}$ contains all points in Σ_A with period less than q. For $1 \leq i < \alpha$, let

$$E_{i+1} = E_i \bigcup \{[u^{i+1}]_{-r+1} - \bigcup_{-q<j<q} \sigma^j(E_i)\}.$$

This adds to E_i the part of $[u^{i+1}]_{-r+1}$ which will preserve the condition $E_i \cap \sigma^j(E_i) = \phi$ for $0 < j < q$. Let $E = E_\alpha$.

Note

(i) $$E \subseteq \bigcup_i [u^i]_{-r+1},$$

(ii) $$\bigcup_i [u^i]_{-r+1} \subseteq \bigcup_{-q<j<q} \sigma^j(E),$$

(iii) $$E \cap \sigma^j(E) = \phi \text{ for } 0 < j < q.$$

Let $\mathcal{N}$ be the finite set of disjoint cylinders that make up E. These are the other markers. We have two types of markers in Σ_B, the long blocks of small period in $\mathcal{M}$ and the blocks which insure the markers occur with bounded gaps in $\mathcal{N}$.

Step 3. Marking a point in Σ_A. In this step we mark a point from Σ_A as we marked a point from Σ_B in step 3 of the proof of Theorem 4.4.1. Let $x \in \Sigma_A$. We scan the point x and mark two things:

(i)   the entire occurrence of a marker of type $\mathcal{M}$ (making the length maximal);
(ii) the center coordinate of a marker of type $\mathcal{N}$ (the center of the u^i).

This is illustrated in Figure 4.4.4.

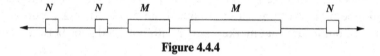

Figure 4.4.4

Note the following.

(i)   A coordinate marked $\mathcal{N}$ cannot occur inside a block marked $\mathcal{M}$ because it's the center of a u^i block.
(ii) Two coordinates marked $\mathcal{N}$ can occur no closer than q by the construction of E.
(iii) Two blocks marked $\mathcal{M}$ cannot overlap.

Next is an important observation. The gap between any two marked coordinates is bounded. To see this, first note that if a coordinate is in a $\mathcal{W}(\bar{B}, r)$ block then it is in a block marked $\mathcal{M}$. Then note that if a coordinate in a point is the center coordinate of a u^i block there will be a marked coordinate within q coordinates. This follows from the choice of the u^i and condition (ii) on E. Fix a marked point $x \in \Sigma_A$ and focus on x_0. Suppose none of the coordinates $x_0, x_1, \ldots, x_{r-1}$ are in a $\mathcal{W}(\bar{B}, r)$ block. Then x_{r-1} is the center coordinate of a u^i block. This means at least one of the coordinates $x_0, x_1, \ldots, x_{q+r-1}$ is marked. This is the property that will make the embedding continuous.

Step 4. Assignments using markers. Once again we mimic as closely as possible the proof of Theorem 4.4.1. In this step we make assignments between blocks in Σ_A and blocks in Σ_B. Choose q larger than the value of N_B in Step 1 and so that B^q is a strictly positive matrix. The necessary assignments are

(i)   between coordinates marked $\mathcal{N}$,
(ii) inside an $\mathcal{M}$ block
(iii) between $\mathcal{M}$ blocks,
(iv) from a coordinate marked $\mathcal{N}$ to a $\mathcal{M}$ block and
(v)   from a $\mathcal{M}$ block to a coordinate marked $\mathcal{N}$.

Assignments from a coordinate marked $\mathcal{N}$ to coordinate marked $\mathcal{N}$. This is the simplest case. Fix a symbol M in Σ_B. For each $\ell \geq q$ consider the blocks of that length in Σ_A which can occur between two markers of type $\mathcal{N}$. This means the block occurs with its first and last symbols marked $\mathcal{N}$ and no other symbol marked. Assign each of these blocks to a block of length ℓ in Σ_B that begins and ends with M.

Assignments inside an $\mathcal{M}$ block. Fix $r = 3q$. For $\ell \geq r$, match each block of length ℓ in $\mathcal{M}$ to the $\bar{\varphi}$ image of its middle $\ell - 2q$ block in Σ_B.

Assignments between $\mathcal{M}$ blocks. Two blocks marked $\mathcal{M}$ may not overlap. An $\mathcal{M}$ block gets assigned to a block in Σ_B that is q entries shorter at each end. Match each block of length ℓ in Σ_A that can occur between two $\mathcal{M}$ blocks with no other marked coordinates to a block of length $\ell + 2q$ in Σ_B that has the appropriate symbol at each end. The matching must take into account the $\bar{\varphi}$ image of the $\mathcal{M}$ blocks that are connected.

Assignments from a coordinate marked $\mathcal{N}$ to a $\mathcal{M}$ block. Consider the blocks in Σ_A that can occur between a coordinate marked $\mathcal{N}$ and an $\mathcal{M}$ block with no other marked coordinates. A coordinate marked $\mathcal{N}$ can occur next to a block marked $\mathcal{M}$ but we still have the required freedom since an $\mathcal{M}$ block is matched to a block in Σ_B that is q entries shorter at each end. Match the blocks in Σ_A that can occur to blocks of the appropriate length in Σ_B that have an M at the beginning, are of the correct length and connect to the $\bar{\varphi}$ image of the $\mathcal{M}$ block.

The fifth type of assignment is the same as the forth with the ends reversed.

Step 5. The map. Consider the marker structure for an $x \in \Sigma_A$. It will map to a point $\varphi(x) \in \Sigma_B$. The symbols in x marked $\mathcal{N}$ are sent to the symbol M in Σ_B and the middles of the blocks marked $\mathcal{M}$ are sent to their $\bar{\varphi}$ images. This is illustrated in Figure 4.4.5.

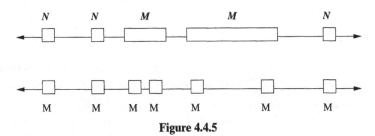

Figure 4.4.5

Then use the assignments from Step 4 to fill in the images of the blocks connecting the marked coordinates.

This method, first putting down the marker structure of a point, mapping the markers and then filling in the images of the connecting blocks results in a continuous map of Σ_A into Σ_B since the marked coordinates occur with bounded gaps between them. The map φ commutes with the shift because the construction does not take into account the position of the block. The map agrees with $\bar{\varphi}$ on $\Sigma_{\bar{B}}$ and so maps Σ_A onto Σ_B.   □

Exercises

For exercises 1, 2 and 3 let A and B be irreducible transition matrices. Let X_A and X_B be the one-sided subshifts of finite type they define and Σ_A and Σ_B the two-sided subshifts of finite type they define.

1. Show that there exists a factor map from X_A to X_B if and only if there is a factor map from Σ_A to Σ_B.
2. Let $\bar{\varphi} : \Sigma_A \to \Sigma_B$ be a conjugacy. Show that $\bar{\varphi}$ can be used to define a finite-to-one factor map $\varphi : X_A \to X_B$ but X_A and X_B need not be conjugate.
3. Let $\varphi : X_A \to X_B$ be a factor map and $\bar{\varphi} : \Sigma_A \to \Sigma_B$ the factor map induced by φ on the two-sided shifts. Show that φ is finite-to-one if and only if $\bar{\varphi}$ is finite-to-one and that φ is infinite-to-one if and only $\bar{\varphi}$ is inifinite-to-one.
4. Give a counterexample to the embedding theorem, Theorem 4.4.1, for one-sided subshifts of finite type. Hint: use the total column amalgamation discussed in Section 2.
5. Show that Corollaries 4.4.2 and 4.4.3 hold for one-sided subshifts of finite type.
6. Show that the unequal entropy factor theorem, Theorem 4.4.5, holds for one-sided subshifts of finite type.

Notes

Section 4.1

Continuous shift-commuting maps from the two-sided two-shift into itself were first explicitly studied by G.A. Hedlund and his coworkers at the Institute for Defense Analysis in the early 1960's [H1], [H2]. They discovered many of the basic properties of factor maps in this setting. Many of these results were extended to subshifts of finite type in the early 1970's by E. Coven and M.Paul

[CP1]. Resolving maps were used by B. McMillan [Mc] in 1953 where he termed such maps *unifiler*. The properties of these maps were used implicitly by R. Adler and B. Weiss in their 1967 paper [AW1] and were singled out and called resolving by R. Adler and B. Marcus in their 1979 paper [AM]. Proposition 4.1.6 which shows the entropy of a factor is less than or equal to the entropy of the domain subshift of finite type is a special case of a general theorem about compact dynamical systems. The idea of a map having a diamond and the contents of Theorem 4.1.7 go back to Hedlund's papers and were also used and expanded by Coven and Paul.

Section 4.2

As noted before, finite-to-one maps from the two-shift to the two-shift were studied by Hedlund in [H1], [H2] and finite-to-one maps from an irreducible subshift of finite type to itself were studied by Coven and Paul [CP1]. They proved that every finite-to-one map has a degree and proved Theorem 4.2.1 in this setting. There are restrictions on the possible degrees of factor maps between two subshifts of finite type. This was first observed by Hedlund in [H2] when the domain and image are both the full two-shift. This was extended to irreducible subshifts of finite type by M. Boyle [By2] and by P. Trow [Tw2]. Theorem 4.2.4 (and Exercise 3) which says a finite-to-one factor map between irreducible and aperiodic subshifts of finite type can be replaced by a one-to-one almost everywhere factor map is due to J. Ashley [Ay2] in 1990. Necessary and sufficient conditions for the existence of a finite-to-one factor map from a subshift of finite type onto a full n-shift are given in Theorem 4.2.5 which was proved by B. Marcus in 1978 [Ma1]. The Road Coloring Problem 4.2.6 originated in [AW1] and was stated explicitly by R. Adler, L.W. Goodwyn and B. Weiss in their 1977 paper [AGW]. There are a number of partial results such as the one due to G.L. O'Brien in [O'B] and the one due to J. Friedman in [Fn]. The lattice in 4.2.7 was constructed by B. Kitchens in 1980 and a probabilistic version can be found in [K1]. Theorem 4.2.8 concerning the relationship between the Jordan forms away from zero was proved by B. Kitchens in 1981 [K1]. Corollaries 4.2.9 and 4.2.11 were proved prior to Theorem 4.2.8 using very different methods. Corollary 4.2.9 is due to M. Nasu in [Na1] and Corollary 4.2.11 is due to B. Marcus in [Ma1]. Theorem 4.2.16 and Proposition 4.2.19 were proved by B. Kitchens, B. Marcus and P. Trow in 1989 [KMT]. Corollary 4.2.17 to Theorem 4.2.16 is the 'only if' part of Theorem 4.2.18. Theorem 4.2.18 states that two aperiodic subshifts of finite type are factors of each other if and only if they are shift equivalent. The 'if' part of the theorem has a longer history. P. Trow tried to extend B. Marcus' Theorem 4.2.5 for full shifts to a larger class of subshifts of finite type. In his 1986 paper [Tw1] he proved it could be done in some cases if the map between the subshifts of finite type commuted with a power of the shifts rather than the shifts themselves. In 1987 M. Boyle, B. Marcus and P. Trow [BMT] generalized Trow's theorem and proved the 'if' part of Theorem 4.2.18 except the maps constructed commuted with a power of the shift and not the shift. Finally, J. Ashley in 1989 [Ay3] proved the maps could be constructed to be true factor maps.

Section 4.3

Construction 4.3.1 was made by B. Kitchens in 1980 to prove Proposition 4.3.3. The topological property of closing was singled out to be the analogue of the combinatorial property of resolving. In 1982 M. Nasu proved Proposition 4.3.4 [Nu2]. Proposition 4.3.5 goes back to the work of Hedlund and W. Krieger made Construction 4.3.6 to investigate the possible zeta functions for subshifts of finite type with a specified entropy in his 1979 paper [Kg4]. B. Kitchens discovered indecomposable factor maps in the early 1980's and recently M. Boyle [B] proved that any finite-to-one factor map can be decomposed into a composition of indecomposable factor maps in only a finite number of ways.

Section 4.4

In the early 1980's W. Krieger developed the marker-filler method of constructing continuous maps and used it to prove Theorem 4.4.1 [Kg5]. Soon afterwards M. Boyle used these methods to prove Theorem 4.4.4 [By1]. The simplicity of the sufficient conditions in both these theorems came as a surprise. Before then the conjugacy and the finite-to-one factor problems had recieved much more attention and in those problems sufficient conditions seemed very difficult to formulate.

References

[AGW] R. Adler L.W. Goodwyn and B. Weiss, *Equivalence of Topological Markov Shifts*, Israel Journal of Mathematics no. 27 (1977), 49–63.

[AM] R. Adler and B. Marcus, *Topological Entropy and Equivalence of Dynamical Systems*, Memoirs of the American Mathematical Society no. 219 (1979).

[AW1] R. Adler and B. Weiss, *Entropy, a Complete Metric Invariant for Automorphisms of the Torus*, Proceedings of the National Academy of Sciences, USA no. 57 (1967), 1573–1576.

[Ay2] J. Ashley, *Bounded-to-1 Factors of an Aperiodic Shift of Finite Type Are 1-to-1 Almost Everywhere Factors Also*, Ergodic Theory and Dynamical Systems **10** (1990) 615–625.

[Ay3] J. Ashley, *Resolving Factor Maps for Shifts of Finite Type with Equal Entropy*, Ergodic Theory and Dynamical Systems **11** (1991), 219–240.

[By1]   M. Boyle, *Lower Entropy Factors of Sofic Systems* Ergodic Theory and Dynamical Systems, **4** (1984), 541–557.

[By2]   M. Boyle, *Constraints on the Degree of Sofic Homomorphisms and the Induced Multiplication of Measures on Unstable Sets,* Israel Journal of Mathematics **53** (1986), 52–68.

[By3]   M. Boyle, *Factoring Factor Maps,* preprint.

[BMT]   M. Boyle, B. Marcus and P. Trow, *Resolving Maps and the Dimension Group for Shifts of Finite Type,* Memoirs of the American Mathematical Society no. 377 (1987).

[CP1]   E. Coven and M. Paul, *Endomorphisms of Irreducible Subshifts of Finite Type,* Mathematical Systems Theory **8** (1974), 167–175.

[Fn]   J. Friedman, *On the Road Coloring Problem,* Proceedings of the American Mathematical Society **110** (1990), 1133–1135.

[H1]   G.A. Hedlund, *Transformations Commuting with the Shift,* Topological Dynamics (J. Auslander and W. Gottschalk, eds.), W.A. Benjamin, 1968.

[H2]   G.A. Hedlund, *Endomorphisms and Automorphisms of the Shift Dynamical System,* Mathematical Systems Theory **3** no. 4 (1969), 320–375.

[K1]   B. Kitchens, *An Invariant for Continuous Factors of Markov Shifts,* Proceedings of the American Mathematical Society **83** (1981), 825–828.

[KMT]   B. Kitchens, P. Trow and B. Marcus, *Eventual Factor Maps and Compositions of Closing Maps,* Ergodic Theory and Dynamical Systems **11** (1991), 857–113.

[Kg4]   W. Krieger, *On the Periodic Points of Topological Markov Chains,* Mathematische Zeitschrift **169** (1979), 99–104.

[Kg5]   W. Krieger, *On Subsystems of Topological Markov Chains,* Ergodic Theory and Dynamical Systems **2** (1982), 195–202.

[Mc]   B. McMillan, *The Basic Theorems of Information Theory,* Annals of Mathematical Statistics **24** (1953), 196–219.

[Mr1]   B. Marcus, *Factors and Extensions of Full Shifts,* Monatshefte fü Mathematik **88** (1979), 239–247.

[Nu1]   M. Nasu, *Uniformly Finite-to-one and Onto Extensions of Homomorphisms Between Strongly Connected Graphs,* Discrete Mathematics no. 39 1982, 171–197.

[O'B]   G.L. O'Brien, *The Road Coloring Problem,* Israel Journal of Mathematics 39 (1981), 145–154.

[Tw1]   P. Trow, *Resolving Maps which Commute with a Power of the Shift,* Ergodic Theory and Dynamical Systems **6** (1986), 281–293.

[Tw2]   P. Trow, *Degrees of Finite-to-one Factor Maps,* Israel Journal of Mathematics **71** (1990), 229–238.

Chapter 5. Almost-Topological Conjugacy

In this chapter we investigate two equivalence relations between subshifts of finite type. The first is finite equivalence and the second is almost-topological conjugacy. These equivalence relations are weaker than topological conjugacy which we discussed in Chapter 2. They are meant to capture the "typical" behavior of points in a subshift of finite type.

In the first section the structure of reducible subshifts of finite type is briefly discussed. In the second section we define the fibered product of two factor maps. Next we define finite equivalence and almost-topological conjugacy. Then we conclude with necessary and sufficient conditions for finite equivalence and almost-topological conjugacy.

§5.1 Reducible Subshifts of Finite Type

To this point we have dealt almost exclusively with irreducible subshifts of finite type. Here we briefly discuss the structure of arbitrary subshifts of finite type.

In Section 1.4 a point $x \in X_A$ or Σ_A is defined to be *wandering* if there exists an open set $U \ni x$ and $\sigma^i(U) \cap U = \phi$ for every $i \neq 0$ in $\mathbb{N}$ or $\mathbb{Z}$. A point is *nonwandering* if for every open set U containing it there is an $i \neq 0$ with $\sigma^i(U) \cap U \neq \phi$. Denote the set of nonwandering points in a subshift of finite type X_A or Σ_A by Ω_A. A subshift of finite type is nonwandering if there are no wandering points.

In Section 1.1 a point is defined to be *transitive* in either X_A or Σ_A if it has a dense orbit. A subshift of finite type is *transitive* if it contains a transitive point.

In Section 1.4 we showed that a one-sided subshift of finite type is transitive if and only if it has an irreducible transition matrix. We also showed that a two-sided subshift of finite type is transitive and nonwandering if and only if it has an irreducible transition matrix.

Let A be an arbitrary transition matrix and x a point in X_A or Σ_A. A point y is in the *positive limit set of* x if there is a sequence $\{n_i\}$ going to positive infinity and the sequence $\{\sigma^{n_i}(x)\}$ converges to y. If $x \in \Sigma_A$ a point y is in the *negative limit set of* x if there is a sequence $\{m_i\}$ going to negative infinity and the sequence $\{\sigma^{m_i}(x)\}$ converges to y.

We examine the structure of a subshift of finite type by applying Theorem 1.3.10 to its transition matrix. Theorem 1.3.10 shows that a state for a transition matrix is either transient or belongs to an irreducible component of A. An irreducible component of A is a *maximal component* if its Perron value is equal to the spectral radius of A. Corollary 1.3.11 asserts that a transition matrix must have at least one maximal component. For a transition matrix A with irreducible components $A_1, \ldots, A_k$ we say $X_{A_1}, \ldots, X_{A_k} \subseteq X_A$ and $\Sigma_{A_1}, \ldots, \Sigma_{A_k} \subseteq \Sigma_A$ are the *irreducible components of* X_A and Σ_A. If A_j is a maximal component of A then $X_{A_j} \subseteq X_A$ and $\Sigma_{A_j} \subseteq \Sigma_A$ are *maximal components of* X_A and Σ_A. For one-sided subshifts of finite type we can describe the nonwandering structure.

Observation 5.1.1 *Let A be an arbitrary transition matrix and X_A the one-sided subshift of finite type it defines. For $x \in X_A$*

(i)   *the positive limit set of x is contained in one of the irreducible components of X_A,*

(ii) *x is nonwandering if and only if every x_i is a state in the same irreducible component of A and*

(iii) *x is nonwandering if and only if it belongs to an irreducible component of X_A.*

Proof. Statement (i) follows immediately from Theorem 1.3.10. For statement (ii) suppose every x_i is contained in the same irreducible component of A. The periodic points are dense in each irreducible component so x is nonwandering. For the converse suppose x_n is a state for an irreducible component for A and x_{n-1} is not a state for the same component. Then $[x_{n-1}, x_n] \neq [x_{i-1}, x_i]$ for every $i \neq n$. The cylinder set $[x_0, \ldots, x_n]_0$ is open, contains x and $\sigma^i([x_0, \ldots, x_n]_0) \cap [x_0, \ldots, x_n]_0 = \phi$ for every $i \neq 0$. Consequently, x is wandering. The statements (ii) and (iii) are clearly equivalent.   □

The analog for two-sided subshifts of finite type follows.

Observation 5.1.2 *Let A be any arbitrary transition matrix and Σ_A the two-sided subshift of finite type it defines. For $x \in \Sigma_A$*

(i)   *the positive and negative limit sets of x are each contained in an irreducible component of Σ_A,*

(ii) *x is nonwandering if and only if every x_i is a state in the same irreducible component of A,*

(iii) x is nonwandering if and only if it belongs to an irreducible component of Σ_A and

(iv) x is nonwandering if and only if its positive and negative limit sets are subsets of the same irreducible component of Σ_A.

Proof. Statement (i) follows immediately from Theorem 1.3.10. The proof of (ii) is almost identical to the proof of Observation 5.1.1 (ii). If every x_i is a state for the same irreducible component of A then x is nonwandering. Suppose x_n is a state for an irreducible component of A and x_{n-1} is not a state for the same irreducible component. Then let m be the larger of $|n|$ and $|n - 1|$. The cylinder set $[x_{-m}, \ldots, x_m]_{-m}$ is open, contains x and $\sigma^i([x_{-m}, \ldots, x_m]_{-m}) \cap [x_{-m}, \ldots, x_m]_{-m} = \phi$ for all $i \in \mathbb{Z}$. This shows x is wandering. Statements (ii), (iii) and (iv) are clearly equivalent.   □

A corollary to these observations characterizes nonwandering subshifts of finite type.

Corollary 5.1.3 *Let A be a transition matrix. The following are equivalent:*

(i) X_A *is nonwandering;*
(ii) Σ_A *is nonwandering;*
(iii) *every state of A belongs to an irreducible component of A and there are no transitions between the components.*

We end this section with a lemma and corollary which will be used later.

Lemma 5.1.4 *Suppose A is a $\{0, 1\}$ transition matrix, B is an irreducible $\{0, 1\}$ transition matrix and there is a right-resolving factor map $\varphi : \Sigma_A \to \Sigma_B$. Then there are no transitions out of a maximal component of A.*

Proof. Let A' be a maximal component of A. The restriction of φ to $\Sigma_{A'}$ defines a one-block factor map $\varphi' : \Sigma_{A'} \to \Sigma_B$. The spectral radius of A is equal to the Perron values of A' and B. Suppose i is a state for A' and there is a transition in A from i to a state j not in A'. Since φ on Σ_A is right-resolving there is no state $j' \neq j$ with $j' \in f(i)$ and $\varphi(j') = \varphi(j)$. Since φ maps $\Sigma_{A'}$ onto Σ_B there is some state j' for A' with $\varphi'(j') = \varphi(j)$. Add the transition from i to j' into A'. This raises the Perron value for A' but does not cause a diamond in the map $\varphi' : \Sigma_{A'} \to \Sigma_B$. That contradicts Theorem 4.1.7   □

Corollary 5.1.5 *Suppose A is a $\{0, 1\}$ transition matrix, B and C are irreducible $\{0, 1\}$ transition matrices, there is a right-resolving factor map $\varphi : \Sigma_A \to \Sigma_B$ and there is a left-resolving factor map $\psi : \Sigma_A \to \Sigma_C$. Then Σ_A is nonwandering and every component is maximal.*

Proof. Since φ is right-resolving, there are no transitions out of any maximal component of A. Since ψ is left-resolving there can similarly be no transitions into a maximal component of A. If i is any state for A and φ is right-resolving there must be a path of transitions from i into a maximal component. This means i is in the maximal component to start. Every irreducible component of A is maximal and there are no transitions between them. Corollary 5.1.3 shows Σ_A is nonwandering. $\square$

§ 5.2 Almost-topological Conjugacy

In Chapter 2 we examined the question of conjugacy between subshifts of finite type and produced many algebraic invariants for conjugacy but found no satisfactory sufficient condition. In this section we will define another type of equivalence between subshifts of finite type. This equivalence is meant to preserve the typical dynamical behavior of points while ignoring the nontypical behavior. For this type of equivalence simple necessary and sufficient conditions are known. We restrict our attention to two-sided subshifts of finite type and define two equivalence relations. First we need a construction.

Fibered Product Construction 5.2.1 Suppose $\varphi : \Sigma_A \to \Sigma_C$ and $\psi : \Sigma_B \to \Sigma_C$ are finite-to-one factor maps between irreducible subshifts of finite type. We construct an irreducible subshift of finite type $\Sigma_{\bar{C}}$ with finite-to-one factor maps $\bar{\psi} : \Sigma_{\bar{C}} \to \Sigma_A$ and $\bar{\varphi} : \Sigma_{\bar{C}} \to \Sigma_B$. Moreover, $\bar{\varphi}$ mirrors the properties of φ and $\bar{\psi}$ mirrors the properties of ψ.

Suppose φ and ψ are one-block maps. Define a subshift of finite type $\Sigma_{C'}$ as follows. Let

$$L_{C'} = \{(i,j) : i \in L_A, j \in L_B \text{ and } \varphi(i) = \psi(j)\}.$$

Define transitions by $(i,j) \to (i',j')$ if $i \to i'$ in Σ_A and $j \to j'$ in Σ_B. Note that $\Sigma_{C'}$ contains all points in $(x,y) \in \Sigma_A \times \Sigma_B$ with $\varphi(x) = \psi(y)$. Next define one-block factor maps $\psi' : \Sigma_{C'} \to \Sigma_A$ and $\varphi' : \Sigma_{C'} \to \Sigma_B$ by $\psi'((i,j)) = i$ and $\varphi'((i,j)) = j$. Both maps are easily seen to be onto. Since φ and ψ are finite-to-one they have no diamonds by Theorem 4.1.7. Consequently, ψ' and φ' have no diamonds and are finite-to-one. Let $\Sigma_{\bar{C}}$ be a maximal component of $\Sigma_{C'}$ with $\bar{\psi}$ and $\bar{\varphi}$ the maps ψ' and φ' restricted to it. Since $\Sigma_{\bar{C}}$ is a maximal component and $\bar{\psi}$ and $\bar{\varphi}$ are finite-to-one Theorem 4.1.7 shows they are onto. We say $\Sigma_{\bar{C}}$ together with the factor maps $\bar{\psi}$ and $\bar{\varphi}$ is the *fibered product* of the factor maps $\varphi : \Sigma_A \to \Sigma_C$ and $\psi : \Sigma_B \to \Sigma_C$.

Observation 5.2.2 *Let* $\varphi : \Sigma_A \to \Sigma_C$ *and* $\psi : \Sigma_B \to \Sigma_C$ *be finite-to-one factor maps between irreducible subshifts of finite type and* $\Sigma_{\bar{C}}$ *with* $\bar{\psi}$ *and* $\bar{\varphi}$ *their fibered product. Then*

(i)   *if φ is d-to-one a.e. then $\bar{\varphi}$ is d-to-one a.e.,*
(ii) *if φ is left-resolving then $\bar{\varphi}$ is left-resolving and*
(iii) *if φ is right-resolving then $\bar{\varphi}$ is right-resolving.*

The same is true for ψ and $\bar{\psi}$.

Proof. For (*i*) suppose φ is d-to-one a.e. and x is a doubly transitive point in Σ_C. Then $\varphi^{-1}(x) = \{x^1, \dots, x^d\}$ with each x^i a doubly transitive point in Σ_A. Choose a point $y \in \psi^{-1}(x)$. It is doubly transitive in Σ_B. Obesrve $\bar{\varphi}^{-1}(y) = \{(x^1, y), \dots, (x^d, y)\} \subseteq \Sigma_{\bar{C}}$. So $\bar{\varphi}$ is d-to-one on the doubly transitive points.

For (*ii*) let $(i, j) \in L_{\bar{C}}$. For each $j' \in p_B(j)$ there is a unique $i' \in p_A(i)$ with $\varphi(i') = \psi(j')$. Then (i', j') is the unique element of $p_{\bar{C}}((i, j))$ whose $\bar{\varphi}$ image is j'. This means $\bar{\varphi}$ is left-resolving.

The proof of (*iii*) is the same as the proof of (*ii*) and the arguments are symmetric between φ and ψ. □

Definition 5.2.3 Two irreducible subshifts of finite type, Σ_A and Σ_B, are said to be *finitely equivalent* if there is a third irreducible subshift of finite type Σ_C and both Σ_A and Σ_B are finite-to-one factors of Σ_C.

Definition 5.2.4 Two irreducible subshifts of finite type, Σ_A and Σ_B, are said to be almost-topologically conjugate if there is a third irreducible subshift of finite type Σ_C and both Σ_A and Σ_B are one-to-one a.e. factors of Σ_C.

First we need to see that these are equivalence relations. Only transitivity for the relation needs to be checked. Suppose Σ_{A_1} is related to Σ_{A_2} and Σ_{A_2} is related to Σ_{A_3}. This means we have the diagram in Figure 5.2.1.

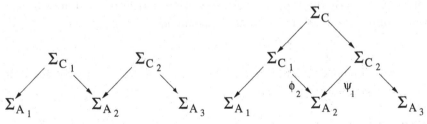

Figure 5.2.1   **Figure 5.2.2**

Using Construction 5.2.1 we build the fibered product over $\varphi_2 : \Sigma_{C_1} \to \Sigma_{A_2}$ and $\psi_2 : \Sigma_{C_2} \to \Sigma_{A_2}$ to get the diagram in Figure 5.2.2.

Then apply Observation 5.2.2 to see that both finite equivalence and almost-topological conjugacy describe equivalence relations between subshifts of finite type.

Next we find necessary conditions for the equivalence relations to hold between subshifts of finite type. Theorem 4.1.7 shows that two finitely equivalent irreducible subshifts of finite type have the same topological entropy. Then Observation 4.2.3 shows two almost-topologically conjugate irreducible subshifts of finite type must have the same period as well as the same entropy. We will show these conditions are sufficient.

Lemma 5.2.5 *Suppose A and B are square, nonnegative integer matrices. Further suppose they are irreducible. There exists a positive integer matrix F with $AF = FB$ if and only if A and B have the same Perron eigenvalue.*

Proof. The condition that A and B have the same Perron eigenvalue is necessary since $AF = FB$ implies $A^k F = FB^k$ for each k. The Perron-Frobenius Theorem says the nonzero entries of A^k and B^k grow exponentially with their Perron values as base.

To see that the condition is sufficient for the existence of F suppose $\lambda_A = \lambda_B = \lambda$. Let r_A be a right Perron eigenvector for A and ℓ_B a left Perron eigenvector for B. Both are strictly positive. Consider the rectangular matrix $[r_A \ell_B]$ and note $A[r_A \ell_B] = \lambda[r_A \ell_B] = [r_A \ell_B]B$.

If $[r_A \ell_B]$ is $m \times n$ then we can consider $[r_A \ell_B]$ as a vector in $\mathbb{R}^{m \times n}$. For any $\varepsilon > 0$ there will be a positive real number t so that $t[r_A \ell_B]$ has distance less than ε from a $\mathbb{Z}^{m \times n}$ lattice point in $\mathbb{R}^{m \times n}$. For ε and t write $t[r_A \ell_B] + E = F$ where each entry of E is less than ε in absolute value and F is a positive integer matrix. Then

$$AF - FB = A(t[r_A \ell_B] + E) - (t[r_A \ell_B] + E)B = AE - EB.$$

The equation $AF - FB$ involves only integer matrices so the resulting matrix has integer entries. By choosing ε sufficiently small we can insure that the absolute values of the entries of $AE - EB$ are less than one. This means $AF - FB = 0$. □

We use Lemma 5.2.5 to prove the sufficiency of the entropy condition in the next theorem.

Theorem 5.2.6 *Let Σ_A and Σ_B be irreducible subshifts of finite type. Then Σ_A and Σ_B are finitely equivalent if and only if they have the same topological entropy.*

Proof. Corollary 4.1.8 shows equal entropy to be necessary for finite equivalence.

Suppose Σ_A and Σ_B have the same topological entropy. Then the transition matrices A and B have the same Perron value. Suppose A is an $m \times m$, $\{0, 1\}$

transition matrix and B is an $n \times n$, $\{0,1\}$ transition matrix. Use Lemma 5.2.5 to choose a positive integer matrix F satisfying $AF = FB$.

We use the matrix F to define a new $\{0,1\}$ transition matrix C (not necessarily irreducible), a left-resolving factor map $\varphi_A : \Sigma_C \to \Sigma_A$ and a right-resolving factor map $\varphi_B : \Sigma_C \to \Sigma_B$.

Define the symbol set for Σ_C to be

$$L_C = \{(i,j,k) : i \in L_A, j \in L_B, 1 \leq k \leq F_{ij}\}.$$

To define the transitions among the states in L_C we construct a *tableau associated with A, B and F*. For each $i \in L_A$ there is a page in the tableau and for each $j \in L_B$ there is a paragraph on each page. Fix an $i \in L_A$. Let the follower set of j be $f(j) = \{j_1, j_2, \ldots, j_{|f(j)|}\}$. To construct the page for i we lay out a list of partially filled assignments as follows.

$$
\begin{array}{ccccc}
(i,1,1) & \to & (\cdot, 1_1, \cdot) & \cdots & (\cdot, 1_{|f(1)|}, \cdot) \\
\vdots & \to & \vdots & \cdots & \vdots \\
(i,1,F_{i1}) & \to & (\cdot, 1_1, \cdot) & \cdots & (\cdot, 1_{|f(1)|}, \cdot) \\
\vdots & \to & \vdots & \cdots & \vdots \\
(i,j,1) & \to & (\cdot, j_1, \cdot) & \cdots & (\cdot, j_{|f(j)|}, \cdot) \\
\vdots & \to & \vdots & \cdots & \vdots \\
(i,j,F_{ij}) & \to & (\cdot, j_1, \cdot) & \cdots & (\cdot, j_{|f(j)|}, \cdot) \\
\vdots & \to & \vdots & \cdots & \vdots \\
(i,n,1) & \to & (\cdot, n_1, \cdot) & \cdots & (\cdot, n_{|f(n)|}, \cdot) \\
\vdots & \to & \vdots & \cdots & \vdots \\
(i,n,F_{in}) & \to & (\cdot, n_1, \cdot) & \cdots & (\cdot, n_{|f(n)|}, \cdot)
\end{array}
$$

The paragraph corresponding to $j \in L_B$ is composed of the rows where j appears as the middle symbol of the left most entry.

If we see in the tableau $(i,j,k) \to (\cdot, j', \cdot)$ we have by definition $j' \in f(j)$. We want to fill in the left blank with an $i' \in f(i)$ and the right blank with a k', $1 \leq k' \leq F_{i'j'}$ so that the resulting (i',j',k') appears only once on the page for i. We also require that each (i',j',k') that can appear on the page does appear.

The equation $AF = FB$ shows that this can be done. To see this compute

$$(AF)_{ij'} = \sum_{i'} A_{ii'} F_{i'j'} = \text{number of fillers with } j' \text{ in the center}$$

and

$$(FB)_{ij'} = \sum_{j} F_{ij} B_{jj'} = \text{number of spaces with } j' \text{ in the center.}$$

The number of fillers and the number of spaces are equal. Fill the tableau in a way which obeys the two rules. The resulting filled tableau defines the transition matrix C.

The map $\varphi_A : \Sigma_C \to \Sigma_A$ is defined by sending the symbol $(i,j,k) \in L_C$ to $i \in L_A$. To see it is left-resolving fix a pair $[i,i'] \in \mathcal{W}(A,2)$. Observe that by construction any (i',j',k') occurs exactly once on the page for i. So there is exactly one (i,j,k) which can precede (i',j',k'). The map φ_A is left-resolving and hence finite-to-one. The map $\varphi_B : \Sigma_C \to \Sigma_B$ is defined by sending the symbol $(i,j,k) \in L_C$ to $j \in L_B$. It is right-resolving. To see this fix a pair $[j,j'] \in \mathcal{W}(B,2)$ and observe that by construction any row on a page with (i,j,k) as the left most entry has exactly one (i',j',k') in its row.

The resulting matrix C may be reducible but by Corollary 1.3.11 C must have a maximal component C'. Then φ_A and φ_B restricted to $\Sigma_{C'}$ map it into Σ_A and Σ_B. Theorem 4.1.7 shows φ_A maps $\Sigma_{C'}$ onto Σ_A and φ_B maps $\Sigma_{C'}$ onto Σ_B.   □

There are several things to notice about the construction. First there are many choices for F. Then once F is fixed there may be many ways to fill the tableau. The filled tableau defines a transition matrix C which need not be irreducible.

Our final theorem in this section will be the next one. Its proof will require several lemmas.

Theorem 5.2.7 *Let Σ_A and Σ_B be irreducible subshifts of finite type. Then Σ_A and Σ_B are almost-topologically conjugate if and only if they have the same topological entropy and the same period.*

The strategy for the proof of Theorem 5.2.7 is to first take A and B to be aperiodic. Then carefully fill the tableau to produce an irreducible and aperiodic component in C. Corollary 5.1.5 guarantees that this component will be maximal. Then we can take Σ_C irreducible and aperiodic and the maps φ_A and φ_B finite-to-one. This allows us to apply Theorem 4.2.4 to replace φ_A and φ_B with one-to-one a.e. factor maps. Finally we reduce the periodic case to the aperiodic case which will complete the proof.

Lemma 5.2.8 *Let Σ_A and Σ_B be irreducible and aperiodic subshifts of finite type with the same entropy. Then Σ_A and Σ_B are almost-topologically conjugate.*

Proof. We know equal entropy is necessary. Suppose Σ_A and Σ_B are as stated and have equal entropy. If the entropy is zero both Σ_A and Σ_B consist of a single fixed point so we may assume the entropy is positive.

Fix special symbols $i_0 \in L_A$ with at least two predecessors and $j_0 \in L_B$ with at least two followers. Since A and B are aperiodic we can find a large

p so that i_0 and j_0 both occur in periodic points with least period p and in periodic points with least period $p + 1$. For p and $p + 1$ fix periodic blocks $[i_0, i_1, \dots, i_{p-1}, i_0]$ and $[i_0, \bar{i}_1, \dots, \bar{i}_p, i_0]$ in $\mathcal{W}(A)$ with $i_{p-1} \neq \bar{i}_p$. Similarly, fix periodic blocks $[j_0, j_1, \dots, j_{p-1}, j_0]$ and $[j_0, \bar{j}_1, \dots, \bar{j}_p, j_0]$ in $\mathcal{W}(B)$ with $j_1 \neq \bar{j}_1$.

Apply Lemma 5.2.5 to produce a positive, integer matrix F with $AF = FB$. If necessary multiply F by a constant to make its smallest entry greater than $2p$.

Next write down the partial assignment for the tableau associated to A, B, and F. We will fill it so there is an irreducible and aperiodic component in the resulting transition matrix C.

We do this using the periodic blocks we fixed containing i_0 and j_0. Start filling the tableau by making the transitions

$$
\begin{aligned}
(i_r, j_r, r+1) &\to (i_{r+1}, j_{r+1}, r+2) && \text{for } 0 \le r < p-1 \\
(i_{p-1}, j_{p-1}, p) &\to (i_0, j_0, 1) \\
(i_0, j_0, 1) &\to (\bar{i}_1, \bar{j}_1, p+1) \\
(\bar{i}_r, \bar{j}_r, p+r) &\to (\bar{i}_{r+1}, \bar{j}_{r+1}, p+r+1) && \text{for } 0 < r < p \\
(\bar{i}_p, \bar{j}_p, p) &\to (i_0, j_0, 1).
\end{aligned}
$$

It is possible to make these assignment since $i_{p-1} \neq \bar{i}_p$, $j_1 \neq \bar{j}_1$ and the progressively higher k's occurring in the third entry of each symbol insure there are no other conflicts. Fill the rest of the tableau.

The point is that the symbol $(i_0, j_0, 1)$ occurs in an irreducible component of C containing cycles of length p and $p + 1$. This means the component is aperiodic. Let this component be C'.

By Corollary 5.1.5 this is a maximal component of C. The maps φ_A and φ_B restricted to $\Sigma_{C'}$ define finite-to-one factor maps onto Σ_A and Σ_B.

Replace φ_A and φ_B with one-to-one a.e. factor maps by applying Theorem 4.2.4. $\qquad\square$

The final step in the proof of Theorem 5.2.7 is to reduce the periodic case to the aperiodic case. In order to do this we need to examine the structure of irreducible periodic subshifts of finite type. Let Σ_A be an irreducible subshift of finite type with period p. Let $C_0, \dots, C_{p-1}$ be the cyclic subsets of the states as defined in Section 1.3. Suppose $C_0 = \{1, \dots, m\}$. Define A_0 to be the submatrix of A^p with rows and columns indexed by the states in C_0. The matrix A_0 is irreducible and aperiodic. Define a new transition matrix with states $L_{\bar{A}} = \{1_r, \dots, m_r : 0 \le r \le p-1\}$. Define transitions $i_r \to i_{r+1}$ for $0 \le r < p-1$

and put $(A^p)_{ij}$ transitions from i_{p-1} to j_0. The transition matrix will be

(†)
$$\bar{A} = \begin{bmatrix} 0 & I & 0 & \cdots & 0 \\ 0 & 0 & I & \cdots & 0 \\ & & \ddots & \ddots & \\ 0 & 0 & 0 & \cdots & I \\ A_0 & 0 & \cdots & & 0 \end{bmatrix}$$

where I is the $m \times m$ identity matrix.

Define a conjugacy $\pi : \Sigma_A \to \Sigma_{\bar{A}}$ as follows. For each pair $i, j \in C_0$, $(A^p)_{ij} = (\bar{A}^p)_{i_0 j_0}$. For each pair $i, j \in C_0$, match the blocks of length p from i to j in $\mathcal{W}(A)$ to the blocks of length p from i_0 to j_0 in $\mathcal{W}(\bar{A})$ in a one-to-one way. Define π by saying it scans a point in Σ_A and sends each block connecting two consecutive occurrences of states in C_0 to the block it is matched to in $\mathcal{W}(\bar{A})$. This defines the conjugacy.

Proof of Theorem 5.2.7 Suppose Σ_A and Σ_B are irreducible subshifts of finite type with the same topological entropy and period. Let the period be p. By the previous discussion about periodic matrices we can assume both A and B have the form shown in (†) with A_0 and B_0 in the lower left corners. The matrices A_0 and B_0 are aperiodic so we can apply Lemma 5.2.8 to construct an almost-topological conjugacy with $\varphi_{A_0} : \Sigma_{C_0} \to \Sigma_{A_0}$ and $\varphi_{B_0} : \Sigma_{C_0} \to \Sigma_{B_0}$. The matrix C_0 is aperiodic and the maps φ_{A_0} and φ_{B_0} are one-to-one a.e. one-block maps. Using C_0 construct an irreducible matrix C with period p as in (†). Using the notation for states as in the previous discussion we consider Σ_{A_0} to have states $\{1, \ldots, m\}$, Σ_A to have states $\{1_r, \ldots, m_r : 0 \le r < p\}$, Σ_{C_0} to have states $\{1, \ldots, \ell\}$ and Σ_C to have states $\{1_r, \ldots, \ell_r : 0 \le r < p\}$. Define a one-block factor map $\varphi_A : \Sigma_C \to \Sigma_A$ by $\varphi_A(i_r) = (\varphi_{A_0}(i))_r$. The map φ_{A_0} is one-to-one a.e. so the new map φ_A will be one-to-one a.e. also. Define the map $\varphi_B : \Sigma_C \to \Sigma_B$ in the same way.   □

Exercises

1.   Suppose the irreducible subshifts of finite type Σ_A and Σ_B have entropy $\log n$ for an integer n. Use Theorem 4.2.4 and the fibered product of Construction 5.2.1 to show Σ_A and Σ_B are finitely equivalent.

2.   Let $A = \begin{bmatrix} 1 & 1 \\ 1 & 0 \end{bmatrix}$ and $B = \begin{bmatrix} 0 & 1 & 1 \\ 1 & 0 & 1 \\ 0 & 1 & 0 \end{bmatrix}$.

   a.   Compute a positive integral matrix F so $AF = FB$.
   b.   Construct the tableau associated with A, B and F.
   c.   Fill in the tableau to produce an irreducible and aperiodic transition matrix C and one-to-one a.e. factor maps $\varphi_A : \Sigma_C \to \Sigma_A$ and $\varphi_B : \Sigma_C \to \Sigma_B$.

Notes

Section 5.2

One of the most important theorems of ergodic theory is the Isomorphism Theorem for Bernoulli shifts. It states that two Bernoulli shifts are measurably isomorphic if and only if they have the same measure-theoretic entropy. Equal measure-theoretic entropy was shown to be necessary for isomorphism by A. Kolmogorov and Y. Sinai in 1959 [Ko], [Si2]. Equal measure-theoretic entropy was shown to be sufficient by D. Ornstein in 1970 [Or]. See [Ru] for an introduction to this theory. Theorem 5.2.7 can be viewed as an analogous theorem in the category of subshifts of finite type.

The idea of almost-topological conjugacy was first used to investigate the dynamics of automorphisms of the two-dimensional torus. This was done by R. Adler and B. Weiss in 1967 [AW1], [AW2]. Two automorphisms of the torus are topologically conjugate if and only if they are algebraically conjugate. Algebraic conjugacy is a very strong condition and is necessary because of the topology of the torus. Adler and Weiss investigated the relationship between automorphisms with the same topological entropy. K. Berg in 1967 [Be] and Adler and Weiss constructed Markov partitions for hyperbolic automorphisms of the two-dimensional torus. This allowed them to model the dynamics of the automorphisms by subshifts of finite type. Example 1.2.7 illustrates this construction. Adler and Weiss then showed that when two automorphisms have the same topological entropy the associated subshifts of finite type are almost-topologically conjugate. This means the "typical" dynamics of the automorphisms are the same.

The idea of almost-topological conjugacy between dynamical systems was formalized as an equivalence relation by R. Adler, L.W. Goodwyn and B. Weiss in 1977 [AGW]. In their paper they proved that two irreducible and aperiodic subshifts of finite type with topological entropy $\log n$, for n an integer, are almost-topologically conjugate. This paper is also where the road coloring problem, Problem 4.2.5, originated.

Lemma 5.2.5 was needed to generalize the result in [AGW] and H. Furstenberg proved it in 1975. W. Parry used Furstenberg's Lemma to prove Theorem 5.2.6 in 1977 [P3]. His proof focused on "splitting" the matrix F to produce Σ_C and the two maps φ_A and φ_B. His proof is in the spirit of the elementary matrix equations of Section 2.1. It allows a more algebraic approach which has been exploited in related work.

R. Adler and B. Marcus proved Theorem 5.2.7 in 1979 [AM] using the tableau approach presented here. They used an intricate combinatorial construction to fill the tableau in a way which made φ_A and φ_B one-to-one a.e. The argument used here to guarantee the existence of one-to-one a.e. maps φ_A and φ_B relies on J. Ashley's Theorem 4.2.4. In his paper [Ay2] he pointed out that his theorem could be used in this context.

References

[AGW] R. Adler L.W. Goodwyn and B. Weiss, *Equivalence of Topological Markov Shifts,* Israel Journal of Mathematics no. 27 (1977), 49–63.

[AM]   R. Adler and B. Marcus, *Topological Entropy and Equivalence of Dynamical Systems,* Memoirs of the American Mathematical Society no. 219 (1979).

[AW1] R. Adler and B. Weiss, *Entropy, a Complete Metric Invariant for Automorphisms of the Torus,* Proceedings of the National Academy of Sciences, USA no. 57 (1967), 1573–1576.

[AW2] R. Adler and B. Weiss, *Similarity of Automorphisms of the Torus,* Memoirs of the American Mathematical Society no. 98 (1970).

[Ay2] J. Ashley, *Bounded-to-1 Factors of an Aperiodic Shift of Finite Type Are 1-to-1 Almost Everywhere Factors Also,* Ergodic Theory and Dynamical Systems **10** (1990), 615–625.

[Be]   K. Berg, *On the Conjugacy Problem for K-systems,* Ph.D Thesis, University of Minnesota, (1967).

[Ko]   A. Kolmogorov, *A New Metric Invariant for Transient Dynamical Systems,* Academiia Nauk SSSR, Doklady **119** (1958), 861–864. (Russian)

[Or]   D. Ornstein, *Bernoulli Shifts with the Same Entropy are Isomorphic,* Advances in Mathematics no. 4 (1970), 337–352.

[P3]   W. Parry, *A Finitary Classification of Topological Markov Chains and Sofic Systems,* Bulletin of the London Mathematical Society **9** (1977) 86–92.

[Ru]   D.J. Rudolph, *Fundamentals of Measurable Dynamics,* Clarendon Press, 1990.

[Si2] Y. Sinai, *On the Concept of Entropy for a Dynamical System,* Academiia Nauk SSSR, Doklady **124** (1959), 768–771. (Russian)

Chapter 6. Further Topics

In this chapter we introduce some further topics which are closely tied to subshifts of finite type. In the first section we look at sofic systems. The continuous symbolic image of a subshift of finite type need not be a subshift of finite type. It may have an unbounded memory. Sofic systems are the symbolic systems that arise as continuous images of subshifts of finite type. There are three equivalent characterizations of these systems. The characterizations are explained and then we investigate some of the basic dynamical properties of sofic systems. The second section contains a discussion of Markov measures. These are the measures which have a finite memory. We define the measures, compute their measure-theoretic entropy, characterize the measures using conditional entropy and then prove that for a fixed subshift of finite type there is a unique Markov measure whose entropy is greater than the entropy of any other measure on the subshift of finite type. The third section investigates symbolic systems that have a group structure. These are the Markov subgroups. We show that any symbolic system with a group structure is a subshift of finite type. Then we classify them up to topological conjugacy. The fourth section contains a very brief introduction to cellular automata. The point is to see how they fit into the framework we have developed. The final section discusses channel codes as illustrated in Example 1.2.8. We describe a class of codes and develope an algorithm to construct them.

§ 6.1 Sofic Systems

Sofic systems are a generalization of subshifts of finite type. They arise when studying subshifts of finite type because the image of a subshift of finite type under a block map may not be a subshift of finite type. An example illustrates the problem.

Example 6.1.1 Let $\Sigma_A \subseteq \{1,2\}^{\mathbb{Z}}$ be the Golden Mean subshift of finite type defined by the transition matrix $\begin{bmatrix} 1 & 1 \\ 1 & 0 \end{bmatrix}$. Define a two-block map from Σ_A

into $\{0,1\}^{\mathbb{Z}}$ by $\varphi([1,1]) = 1$ and $\varphi([1,2]) = \varphi([2,1]) = 0$. This is illustrated by the graph labelling in Figure 6.1.1. The image of Σ_A is not a subshift of finite type. It is called the *even sofic system* because there must either be no zeros or an even number of zeros between successive occurrences of one. The rule which determines the allowable blocks has an infinite memory.

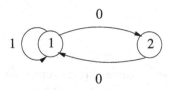

Figure 6.1.1

In automata theory sofic systems are the subshifts, S, whose set of words, $\mathcal{W}(S)$, form a *regular language*.

Sofic systems have many of the dynamical properties of subshifts of finite type but also important differences. There are three different ways of characterizing sofic systems and each has its uses. The first characterization is as a symbolic factor of a subshift of finite type. Another is in terms of follower or predecessor sets and the third is in terms of finite semi-groups. The equivalence of these definitions is explained in Theorem 6.1.2.

Let (X, σ) be an arbitrary subshift. As usual let $\mathcal{W}(X)$ be the words which occur in X. We generalize the predecessor and follower sets defined for subshifts of finite type in Section 2.1 so it is useful in this setting as follows: For $u \in \mathcal{W}(X)$ let $f_X(u) = \{v : uv \in \mathcal{W}(X)\}$ be the *follower set of the word u* and $p_X(u) = \{w : wu \in \mathcal{W}(X)\}$ be the *predecessor set of the word u*. Sofic systems are characterized by having a finite number of distinct follower or predecessor sets, where the sets are thought of solely as subsets of $\mathcal{W}(X)$.

Next let $\mathcal{G}$ be a finite multiplicative semi-group with a zero element. The zero element has the property that $0g = g0 = 0$ for every $g \in \mathcal{G}$. Suppose $\mathcal{A} \subseteq \mathcal{G}$ doesn't contain the zero element and generates the nonzero elements of $\mathcal{G}$. Define a subshift $S \subseteq (\mathcal{A})^{\mathbb{Z}}$ by saying a word $[a_1, \ldots, a_\ell] \in \mathcal{A}^\ell$ is in $\mathcal{W}(S)$ if and only if $a_1 \cdots a_\ell \neq 0$ in $\mathcal{G}$. This is another way of defining sofic systems.

Theorem 6.1.2 *Let $S \subseteq \{1, \ldots, n\}^{\mathbb{Z}}$ be a subshift. Then the following are equivalent:*

(i)   *S is a sofic system,*
(ii) *S is a continuous factor of a subshift of finite type,*
(iii) *S has a finite number of distinct follower sets,*
(iv) *S has a finite number of distinct predecessor sets,*
(v)   *S is defined by a finite semi-group with a zero element and where $\{1, \ldots, n\}$ generates the nonzero elements the semi-group.*

Proof. We will take condition (*ii*) as the definition of a sofic system.

First we show (ii) implies (iii) and (iv). Suppose Σ_A is a subshift of type defined by the transition matrix A and $\varphi : \Sigma_A \to S$ is a one-block factor map. If $u, v \in W(A)$ and they have the same last symbol then $f_A(u) = f_A(v)$. Observe that $f_S(\varphi(u)) = \cup\varphi(f_A(v))$ where the union is taken over all $v \in W(A)$ with $\varphi(v) = \varphi(u)$. Since Σ_A has finitely many follower sets S has finitely many follower sets. The same argument works for predecessor sets.

Next we show (*iii*) or (*iv*) implies (*ii*). Assume there are a finite number of distinct follower sets. We will construct a subshift of finite type and a factor map from it onto S. Define an equivalence relation on $W(S)$ by saying $u \sim v$ if $f_S(u) = f_S(v)$ and the last symbol of u is equal to the last symbol of v. The equivalence classes are the vertices for the transition graph of the subshift of finite type. Define a transition from $[u]$ to $[v]$ if $uj \in [v]$, where j is the last symbol of each word in $[v]$. This is well-defined because if $f_S(\bar{u}) = f_S(u)$ then $f_S(\bar{u}j) = f_S(uj)$. Let this transition graph define a subshift of finite type Σ_A with states $L_A = \{[u] : u \in W(S)\}$. Define a one-block factor map by sending a state $[u]$ to the last symbol of each word in the class. Let $u = [i_0, \ldots i_\ell] \in W(S)$. Consider the state $[i_0] \in L_A$. It maps to i_0 and is followed by the state $[[i_0, i_1]] \in L_A$ which maps to i_1. Following successive transitions in $[i_0, \ldots i_\ell]$ produces a word in Σ_A which maps to u in S. This shows the map is onto.

It is clear that (*v*) implies (*iii*) and (*iv*) since the semi-group is finite and the follower and predecessor sets of a word are determined by the product of the symbols in $\mathcal{G}$.

Next we show (*iii*) and (*iv*) imply (*v*). Given S define an equivalence relation on $W(S)$. Say $u \sim v$ if for every pair of words $w_1, w_2 \in W(S)$

$$w_1 u w_2 \in W(S) \quad \text{if and only if} \quad w_1 v w_2 \in W(S).$$

The elements of the semi-group are the equivalence classes and a zero element. Define multiplication by

$$[u][v] = \begin{cases} 0 & \text{if } uv \notin W(S) \text{ and} \\ [uv] & \text{if } uv \in W(S). \end{cases}$$

To see multiplication is well-defined let $\bar{u} \in [u]$ and $\bar{v} \in [v]$. If $uv \in W(S)$ then $\bar{u}\bar{v} \in W(S)$. So the zero condition is well-defined. If $w_1(uv)w_2 = (w_1 u)v(w_2) \in W(S)$ then $(w_1 u)\bar{v}(w_2) = (w_1)u(\bar{v}w_2) \in W(S)$ and $(w_1)\bar{u}(\bar{v}w_2) = w_1(\bar{u}\bar{v})w_2 \in W(S)$ and multiplication is well-defined. The semi-group is finite since there are finitely many predecessor and follower sets. The words on $\{1, \ldots, n\}$ with nonzero product in this semi-group make up $W(S)$.   □

The subshift of finite type together with the factor map (Σ_A, φ) constructed in the previous proof using follower sets to show (*iii*) implies (*ii*) is called the *full right cover for S*. The subshift of finite type constructed using predecessor sets instead of follower sets which would show (*iv*) implies (*iii*) together with the factor map is called the *full left cover for S*.

A factor map $\varphi : S \to S'$ between sofic systems is *right-resolving* if it is a one-block map and for each state i of S, φ maps the states in $f_S(i)$ in a one-to-one way into $f_{S'}(\varphi(i))$. A factor map $\varphi : S \to S'$ between sofic systems is *left-resolving* if it is a one-block map and for each state i of S, φ maps the states in $p_S(i)$ in a one-to-one way into $p_{S'}(\varphi(i))$. Compare these definitions to the definitions of left and right-resolving maps between subshifts of finite type in Section 4.1. Left and right-closing maps between sofic systems are defined exactly as they are for subshifts of finite type in Section 4.3. Left- and right-resolving or closing factor maps are boundedly finite-to-one.

Observation 6.1.3 *The factor map φ in the full right cover for S is right-resolving.*

Proof. Let $[u]$ be a state for Σ_A and j a state in $f_S(\varphi([u]))$. There is only one candidate for a state in $f_A([u])$ which will map to j and it is $[uj]$.   □

We usually say a subshift is *strictly sofic* if it is a sofic system but not a subshift of finite type. Let X be an arbitrary subshift. A word $m \in \mathcal{W}(X)$ is a *Markov magic word* if for every $u, v \in \mathcal{W}(X)$, $um \in \mathcal{W}(X)$ and $mv \in \mathcal{W}(X)$ implies $umv \in \mathcal{W}(X)$. In a subshift of finite type every word is a Markov magic word. In a strictly sofic system some words are not Markov magic words.

Example 6.1.4 In Example 6.1.1 the factor map φ from the Golden Mean subshift of finite type to the even sofic system is both left and right-resolving. Any word consisting of all 0's is not a Markov magic word for the even sofic system and any word containing a 1 is a Markov magic word.

Observation 6.1.5 *Let S be a sofic system. Then there exist Markov magic words for S.*

Proof. Let $u \in \mathcal{W}(S)$. Let $w \in p_S(u)$ then $f_S(wu) \subseteq f_S(u)$. Since there are finitely many distinct follower sets for S there exist words with minimal follower sets. We mean $m \in \mathcal{W}(S)$ has a minimal follower set in the sense that for all $u \in p_S(m), f_S(um) = f_S(m)$. Such a word is a Markov magic word.   □

The proof of the previous statement works equally well using predecessor sets. A word with a minimal follower set also has a minimal predecessor set.

As for subshifts of finite type a sofic system is said to be *irreducible* if it is transitive and nonwandering. In an irreducible sofic system the doubly transitive points form a dense G_δ (Exercise 4). As for irreducible subshifts of finite type, we differentiate between finite-to-one and infinite-to-one factor maps. For finite-to-one factor maps between irreducible sofic systems define a *magic word* and the *degree* just as for factor maps between subshifts of finite type in Section 4.2. The next theorem shows how to construct a subshift of finite type and a "nice" factor map from the subshift of finite type onto an irreducible sofic system. The proof uses the full right cover.

Theorem 6.1.6 *Let S be an irreducible sofic system. Then there exist an irreducible subshift of finite type Σ_R and a one-to-one a.e. factor map π_R from Σ_R to S.*

Proof. Let S be an irreducible sofic system and (Σ_A, φ) its full right cover. The transition matrix A is probably reducible. Let $w = [w_1, \dots, w_\ell] \in W(S)$ and m be a Markov magic word for S. Then $[w]$ is an arbitrary state in Σ_A. Since S is irreducible there is a word $u \in W(S)$ with $wum \in W(S)$. In the graph of A the path which starts at the state $[w_1]$ and maps to wum passes through the state $[w]$ and ends at the state $[wum] = [m]$. This means A has one maximal component. There are no transitions out of the component and there is a path from every state into this component. Denote this component by R. Then φ restricts to a factor map $\pi_R : \Sigma_R \to S$. To see this map is one-to-one note that every point with m occurring infinitely often in the past has exactly one preimage.   □

This irreducible subshift of finite type together with the factor map (Σ_R, π_R) is called the *right Fischer cover for S*. A similar construction made using the predecessor sets of S produces the *left Fischer cover for S*. It is denoted by (Σ_L, π_L).

Corollary 6.1.7 *Let S be an irreducible sofic system, (Σ_A, φ) its full right cover and (Σ_R, π_R) its right Fischer cover. A word $w \in W(S)$ is a Markov magic word for S if and only if $[w] \in L_R$.*

The topological entropy of a sofic system is defined as it was for a subshift of finite type in Section 1.4. The topological entropy of a sofic system S is

$$h(S, \sigma) = \lim_{\ell \to \infty} \frac{1}{\ell} \log |W(S, \ell)|$$

where $W(S, \ell)$ is the set of words in S of length ℓ and $|\cdot|$ denotes the cardinality of the set. Then we have the following result.

Observation 6.1.8 *Let S be an irreducible sofic system. Then $h(S, \sigma) = \log \lambda$, where λ is the Perron value for the transition matrix of either the left or right Fischer cover.*

The proof is Exercise 5.

If S is an irreducible sofic system then the periodic points are dense in the right Fischer cover for S and so are dense in S. We define the *zeta function* for a sofic system just as we did for a subshift of finite type in Section 1.4. In Observation 1.4.4 we proved the zeta function of a subshift of finite type is the reciprocal of a polynomial. That may not be true for a sofic system (see Exercise 6) but we do have the following theorem about zeta functions for sofic systems. It is stated without proof and references are provided in the notes.

Theorem 6.1.9 *Let S be an irreducible sofic system. Then the zeta function of S is a rational function.*

Next we show the Fischer covers have minimality and uniqueness properties. The next three propositions are stated for the right covers but hold for the left covers as well.

Observation 6.1.10 *Let S be an irreducible sofic system and (Σ_R, π_R) its right Fischer cover. Suppose Σ_B is a subshift of finite type and $\psi : \Sigma_B \to S$ is a right-closing factor map. Then there exists a right-closing factor map $\theta : \Sigma_B \to \Sigma_R$ with $\psi = \pi_R \circ \theta$.*

Proof. Let S, Σ_B, ψ, Σ_R and π_R be as stated. By Exercise 3 we may assume ψ is right-resolving. Fix a word $m \in \mathcal{W}(S)$ which is a Markov magic word for S and is a magic word for ψ. Any word in $\mathcal{W}(S)$ containing m is also a Markov magic word for S and a magic word for ψ. By Corollary 6.1.7 if $mw \in \mathcal{W}(S)$ then $[mw] \in L_R$.

For each $i \in L_B$ choose a word u so $mu\psi(i) \in \mathcal{W}(S)$. Since $mu\psi(i)$ is a magic word for ψ, $\psi(f_B(i)) = f_S(\psi(i))$. Define $\theta : L_B \to L_R$ by $\theta(i) = [mu\psi(i)]$.

We will see that θ maps Σ_B into Σ_R. Let $j \in f_B(i)$. To define $\theta(j)$ we chose a $v \in \mathcal{W}(S)$ with $mv\psi(j) \in \mathcal{W}(S)$ and define $\theta(j) = [mv\psi(j)]$. We need to see $[mu\psi(i)\psi(j)] = [mv\psi(j)]$. But $f_S(mu\psi(i)\varphi(j)) = \varphi(f_B(j)) = f_S(mv\psi(j))$.

Next we show θ is right resolving. Fix an $i \in L_B$ The map ψ is right-resolving so if $j \neq j' \in f_B(i)$ then $\psi(j) \neq \psi(j')$ and $\theta(j) \neq \theta(j')$. We know $\psi(f_B(i)) = f_S(\psi(i))$ so θ maps $f_B(i)$ onto $f_R(\theta(i))$ and θ is right-resolving.

By construction $\pi_R \circ \theta = \psi$.   □

Observation 6.1.11 *Let S be an irreducible sofic system and (Σ_R, π_R) its right Fischer cover. Suppose Σ_B is a subshift of finite type and $\psi : \Sigma_B \to S$ a right-closing factor map. Suppose there is another factor map $\pi : \Sigma_R \to \Sigma_B$ with $\pi_R = \psi \circ \pi$. Then π is a topological conjugacy.*

Proof. By hypothesis we have $\pi_R = \psi \circ \pi$. Apply Proposition 6.1.10 to get $\theta : \Sigma_B \to \Sigma_R$ with $\psi = \pi_R \circ \theta$. Putting these together gives the diagram in Figure 6.1.2 with $\pi_R = \pi_R \circ \theta \circ \pi$.

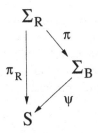

Figure 6.1.2

So $\theta \circ \pi$ is a factor map from Σ_R to itself which preserves $\pi_R^{-1}(x)$ for every $x \in S$. Since π_R is boundedly finite-to-one both π and θ are conjugacies.   □

Theorem 6.1.12 *Let S and S' be irreducible sofic systems. Let (Σ_R, π_R) and $(\Sigma_{R'}, \pi_{R'})$ be their right Fischer covers. Suppose $\varphi : S \to S'$ is a topological conjugacy. Then there is a topological conjugacy $\theta : \Sigma_R \to \Sigma_{R'}$ with $\pi_{R'} \circ \theta = \varphi \circ \pi_R$.*

Proof. We have a right-closing factor map $\varphi \circ \pi_R : \Sigma_R \to S'$ and can apply Proposition 6.1.10 to produce a right-closing $\theta : \Sigma_R \to \Sigma_{R'}$ with $\varphi \circ \pi_R = \pi_{R'} \circ \theta$. Then apply Proposition 6.1.11 to S with $\psi = \varphi^{-1} \circ \pi_{R'}$ and $\Sigma_B = \Sigma_R'$ to conclude θ is a topological conjugacy.   □

In Chapters 2 and 4 we saw that the Jordan form away from zero of the transition matrix for a subshift of finite type contains important dynamical information. We just saw that the right and left Fischer covers for an irreducible sofic system are conjugacy invariants. Next we will associate to an irreducible sofic system another matrix and see that its Jordan form away from zero also contains important dynamical information.

Let S be an irreducible sofic system and (Σ_R, π_R) its right Fischer cover. Recall the Lattice Construction 4.2.7. Carry out the lattice construction for the map π_R.

Core Lattice Construction 6.1.13 Fix a magic word m for π_R. For each word u with $um \in \mathcal{W}(S)$ define a vector $\bar{v}^u \in \mathbb{Z}^{|L_R|}$ by

$$(\bar{v}^u)_j = \begin{cases} 1 & \text{if } j = x_1 \text{ for some } [x_1, \dots, x_{\ell+k}] \in \pi_R^{-1}(uw) \\ 0 & \text{otherwise.} \end{cases}$$

There are a finite number of distinct $\bar{v}^u$ and they generate a sublattice $\mathcal{L}$ of $\mathbb{Z}^{|L_R|}$. As in the Lattice Construction 4.2.7 this lattice is R invariant.

Observation 6.1.14 *The lattice produced by the lattice construction applied to π_R is independent of the choice of magic word.*

Proof. Suppose m and m' are two magic words for π_R. Then the set of generators for $\mathcal{L}$ produced using m' is a subset of the ones produced using m. This follows because if u is a word with $um' \in \mathcal{W}(S)$ then there is a word v with $um'vm \in \mathcal{W}(S)$ and the vectors produced using vm' and $vm'um$ in $\mathbb{Z}^{|L_R|}$ are identical. Similarly the vectors produced using m are a subset of the ones produced using m'. □

Choose a basis for $\mathcal{L}$ and let C be the matrix representing the action of R restricted to $\mathcal{L}$ in the chosen basis. Then C is an integer matrix. Some of its entries may be negative. It is well-defined up to similarity over $\mathbb{Z}$. The matrix C is the right *core matrix for S*. The Jordan form of the core matrix is independent of the choice of basis. We will prove that the Jordan form away from zero of the core matrix is a topological conjugacy invariant and also provides an invariant for the existence of finite-to-one factor maps between irreducible sofic systems.

Lemma 6.1.15 *Let S be an irreducible sofic system and* (Σ_R, π_R) *its right Fischer cover. Let* $S^{[2]}$ *be the two-block presentation of S and* $(\Sigma_{R'}, \pi_{R'})$ *its right Fischer cover. Then R is obtained from R' by a one-step row amalgamation.*

Proof. Let $u = [u_1, \ldots, u_\ell] \in \mathcal{W}(S)$. It corresponds to a two-block word $u' = [(u_1, u_2), \ldots, (u_{\ell-1}, u_\ell)]$ in $\mathcal{W}(S^{[2]})$. Let $[u'] \in L_{R'}$. By Corollary 6.1.7 u' is a Markov magic word for $S^{[2]}$ and so the corresponding word $u \in \mathcal{W}(S)$ is a Markov magic word for S. Define a map $\pi : L_{R'} \to L_R$ by $\pi([u']) = [u]$. Where u' is the two-block description of u. This is well-defined since when $v' \in [u']$, $v \in [u]$. It is onto since all $[u] \in L_R$ are equivalence classes of Markov magic words and so contain words of arbitrarily long length.

Suppose $[u'] \neq [v']$ and $\pi([u']) = \pi([v'])$. Then $f_{S'}(u') = f_{S'}(v')$, the final symbols $(u_{\ell-1}, u_\ell)$ and (v_{k-1}, v_k) in $L_{S[2]}$ are different but $u_\ell = v_k$. This means $f_{R'}([u']) = f_{R'}([v'])$. It also shows that if $[w'] \in p_{R'}([u'])$ then $[w'] \notin p_{R'}([u'])$ since if the word $w'(u_{\ell-1}, u_\ell) \in \mathcal{W}(S^{[2]})$ then the word $w'(v_{k-1}, v_k) \notin \mathcal{W}(S^{[2]})$.

So π amalgamates symbols in $L_{R'}$ with identical followers and disjoint predecessors. It is the description of a one-step row amalgamation from Section 2.1. □

Let R, R' and $\pi : L_{R'} \to L_R$ be as described in Lemma 6.1.15 and its proof. Then π determines an onto map $\bar{\pi} : \mathbb{Z}^{L_{R'}} \to \mathbb{Z}^{L_R}$ by linearity and the diagram in Figure 6.1.3 commutes.

Let U be the $|L_{R'}| \times |L_R|$, $\{0, 1\}$ matrix defined by

$$U_{[u'][v]} = \begin{cases} 1 & \text{if } \pi([u']) = [v] \text{ and} \\ 0 & \text{otherwise.} \end{cases}$$

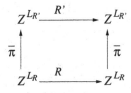

Figure 6.1.3

For $[u] \in L_R$, $Ue_{[u]} \in \mathbb{Z}^{L_{R'}}$ is the sum over the $e_{[u']}$ with $\bar{\pi}(e_{[u']}) = e_{[u]}$. Let V be the $|L_R| \times |L_{R'}|$, $\{0,1\}$ matrix with row $[u]$ the same as a $[u']$ row of R' for any $[u']$ with $\pi([u']) = [u]$. Note $\bar{\pi} \circ R' = V$ as a map from $\mathbb{Z}^{L_{R'}}$ to $\mathbb{Z}^{L_R}$.

Corollary 6.1.16 *Let R', R, U and V be as stated. Then $U : \mathbb{Z}^{L_R} \to \mathbb{Z}^{L_{R'}}$, $V : \mathbb{Z}^{L_{R'}} \to \mathbb{Z}^{L_R}$ with $R' = UV$ and $VU = R$.*

Proof. This follows from Lemma 6.1.15 and the discussion in Section 2.5 about describing one-step column or row amalgamations by matrix equations.   □

Now we apply the observations in Lemma 6.1.15 and Corollary 6.1.16 to the core matrices of S and $S^{[2]}$.

Lemma 6.1.17 *Let S be an irreducible sofic system and and $S^{[2]}$ be the two-block presentation of S. The right core matrix for S and the right core matrix for $S^{[2]}$ have the same Jordan form away from zero.*

Proof. Let $\mathcal{L}'$ be the lattice for $S^{[2]}$ and $\mathcal{L}$ the lattice for S. First we observe that the map $\bar{\pi} : \mathbb{Z}^{L_{R'}} \to \mathbb{Z}^{L_R}$ takes $\mathcal{L}'$ onto $\mathcal{L}$. Let $m' \in \mathcal{W}(S^{[2]})$ be a magic word for $\pi_{R'}$ so its corresponding word $m \in \mathcal{W}(S)$ is a magic word for π_R. If $u'm' \in \mathcal{W}(S^{[2]})$ and um is its corresponding word in $\mathcal{W}(S)$ then $\bar{\pi}(\bar{v}^{u'}) = \bar{v}^u$.

Next observe that for a generator $\bar{v}^u \in \mathcal{L}$, $U\bar{v}^u = \Sigma\bar{v}^{u'}$ where the sum is over the $\bar{v}^{u'} \in \mathcal{L}'$ with $\bar{\pi}(\bar{v}^{u'}) = \bar{v}^u$. So $U(\mathcal{L}) \subseteq \mathcal{L}'$. We have the commuting diagram in Figure 6.1.4.

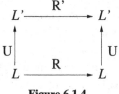

Figure 6.1.4

Next observe that for a generator $\bar{v}^{u'} \in \mathcal{L}'$, $V\bar{v}^{u'} = \bar{\pi}R'\bar{v}^{u'}$. This completes the diagram in Figure 6.1.5.

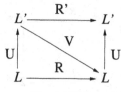

Figure 6.1.5

The reasoning used in Observation 2.2.5 shows that R' restricted to $\mathcal{L}'$ and R restricted to $\mathcal{L}$ have the same Jordan form away from zero.   □

Now we generalize Theorem 4.2.8 to the core matrices of sofic systems. Let $\varphi : S' \to S$ be a one-block finite-to-one factor map between irreducible sofic systems. Then φ may have a *diamond* in the sense of Theorem 4.1.7 (see Exercise 8). But if $m' \in \mathcal{W}(S')$ is a Markov magic word for S', $\varphi(m') = m$ and $mum \in \mathcal{W}(S)$ then $\varphi^{-1}(mum)$ cannot contain a diamond.

Theorem 6.1.18 *Suppose* $\varphi : S' \to S$ *is a finite-to-one factor map between irreducible sofic systems. Then the Jordan form away from zero of the core matrix of S is a principal submatrix of the Jordan form away from zero of the core matrix for S'.*

Proof. By Lemma 6.1.17 we may assume that φ is a one-block, d-to-one a.e. factor map. Let $(\Sigma_{R'}, \pi_{R'})$ and (Σ_R, π_R) be the right Fischer covers for S' and S. We need to carefully choose a magic word for φ in S so that its preimages in S' all contain subwords which are Markov magic words. We can construct a word $m \in \mathcal{W}(S)$ of length ℓ with three properties. First, it is a magic word for φ. Second, it is a Markov magic word for S. Third, there are d Markov magic words $m^1, \dots, m^d \in \mathcal{W}(S')$ of length k and an integer $t \geq 0$ with $k + t - 1 \leq \ell$ so that for every $[i_1, \dots, i_\ell] \in \varphi^{-1}(m)$ there is a j with $[i_t, \dots, i_{t+k-1}] = m^j$. See Figure 6.1.6.

Let $\mathcal{L}'$ and $\mathcal{L}$ be the lattices for S' and S. We use m and $m^1, \dots, m^d$ to construct an R' invariant sublattice $\mathcal{L}_\varphi$ of $\mathcal{L}'$ and an onto linear map from $\mathcal{L}_\varphi$ to $\mathcal{L}$ which commutes with R' and R. The construction is a slightly modified version of the Lattice Construction 4.2.7. For each word u with $um \in \mathcal{W}(S)$ and each $j = 1, \dots d$. Define a vector $\bar{v}^{(u,j)}$. To do this consider $\varphi^{-1}(um)$. The blocks $\varphi^{-1}(um)$ fall into d classes as shown in Figure 6.1.7.

Each block $w \in \varphi^{-1}(um)$ has one of the m^j lying over m and beginning over the tth coordinate of m. The vector $\bar{v}^{(u,j)}$ is the sum of the vectors in $\mathcal{L}'$ which are produced by $\pi_{R'}(w)$ when w has m^j lying correctly over m. These are well defined vectors in $\mathcal{L}'$ by Observation 6.1.14. Let $\mathcal{L}_\varphi$ be the sublattice generated by all possible $\bar{v}^{(u,j)}$.

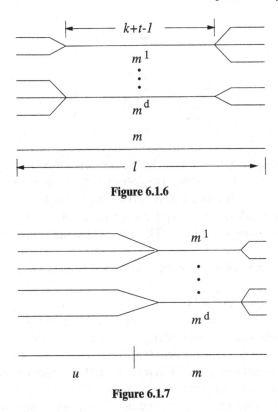

Figure 6.1.6

Figure 6.1.7

First we check that $\mathcal{L}_\varphi$ is R' invariant. This is exactly as in Construction 4.2.7. Next define a map $\theta : \mathcal{L}_\varphi \to \mathcal{L}$ which sends the vector $\bar{v}^{(u,j)} \in \mathcal{L}_\varphi$ to the vector $\bar{v}^u \in \mathcal{L}$. We check to see that this map is well defined and linear just as in Construction 4.2.7.

Finally we extend θ to the rational spans of $\mathcal{L}_\varphi$ and $\mathcal{L}$ to conclude that the Jordan form away from zero of R restricted to $\mathcal{L}$ is a principal submatrix of the Jordan form away from zero of R' restricted to $\mathcal{L}'$ just as we did for Theorem 4.2.7.   □

Corollary 6.1.19 *If S and S' are topologically conjugate irreducible sofic systems then the Jordan forms away from zero of their core matrices are the same.*

Exercises

1. Let S be the even sofic system of Example 6.1.1
   a. What are the follower and predecessor sets for S?
   b. Construct the full left and right covers for S.
   c. Produce a semi-group that describes S.

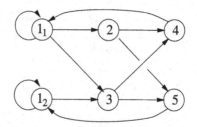

2. Let S be the sofic system defined by the vertex labelling in the graph below.
    Prove that S is strictly sofic. Describe the Markov magic words for S.
    Describe the words that are not Markov magic words.

3. Prove that a right-closing map between irreducible sofic systems can be
    recoded to be right-resolving. Hint: Consider the proof of Proposition
    4.3.3.

4. Prove that the doubly transitive points form a dense G_δ in an irreducible
    sofic system.

5. Prove Observation 6.1.8.

6. Compute the zeta function for the even sofic system using the zeta function
    of the Golden Mean subshift of finite type, matrix A of Example 1.4.5, and
    the product formula for the zeta function in Exercise 3 of Section 1.4.

7. Consider the sofic system of Exercise 2. Find the right Fischer cover, use
    Construction 6.1.13 to produce the lattice for (Σ_R, π_R) and use this to
    compute the core matrix for S and the Jordan form away from zero for
    the core matrix.

8. Give an example of a finite-to-one map between irreducible sofic systems
    with a diamond.

9. Let S be an irreducible sofic system. We define the *Krieger cover* for S.
    Denote the left-infinite rays of S by

    $$W(S, (-\infty, 0]) = \{x(-\infty, 0] : x \in S\}.$$

    Similarly, denote the right-infinite rays in S by $W(S, [1, \infty))$. Define the
    follower set of a left-infinite ray $x(-\infty, 0]$ by

    $$F_S(x(-\infty, 0]) = \{y[1, \infty) \in W(S, [1, \infty)) : x(-\infty, 0]y[1, \infty) \in S\}.$$

    This is the infinite ray version of the usual follower set. Since S is sofic
    there are a finite number of distinct follower sets. Define an equivalence
    relation on left-infinite rays as we did to construct the full cover. Two left-
    infinite rays are equivalent if they have the same follower sets and the same
    symbol in the zeroth coordinate. The equivalence classes are the vertices
    for the transition graph of a subshift of finite type. Define a transition
    from $[x(-\infty, 0]]$ to $[y(-\infty, 0]]$ if $x(-\infty, 0]y_0 \in [y(-\infty, 0]]$. Show this is

well-defined. Let this transition graph define a subshift of finite type Σ_K. Define a one-block factor map to S by sending a state $[x(-\infty, 0]]$ to x_0. Show this is a one-to-one a.e. right-resolving factor map. Show $\Sigma_F \subseteq \Sigma_K \subseteq \Sigma_A$, where Σ_F is the right Fischer cover and Σ_A is the full right cover. Mimic the proof of Theorem 6.1.12 to show the entire Krieger cover is a conjugacy invariant.

10. Compute the right Krieger cover for the sofic system in Exercise 2.
11. Recall that a map is open if the image of every open set is open. Let X be any closed, shift-invariant subset of $\{1, \ldots, n\}^{\mathbb{Z}}$. Show X is a subshift of finite type if and only if σ restricted to X is open (a k-step subshift of finite type for some k).

§6.2 Markov Measures and the Maximal Measure

To begin the study of measures we describe the Borel σ-algebra of measurable sets for a subshift of finite type. Let Σ_A be an irreducible subshift of finite type. Let $C(\Sigma_A)$ denote the collection of cylinder sets in Σ_A. At the beginning of Section 1 we observed that $C(\Sigma_A)$ is a basis for the topology of Σ_A. Recall that an *algebra* of sets is a collection $\mathcal{A}$ of subsets of a space if (i) the complement of A is in $\mathcal{A}$ whenever A is and (ii) any finite union of sets in $\mathcal{A}$ is in $\mathcal{A}$. A σ-*algebra* of sets is a collection $\mathcal{B}$ of subsets of a space if (i) the complement of A is in $\mathcal{B}$ whenever A is and (ii) any countable union of sets in $\mathcal{B}$ is in $\mathcal{B}$. Let $\mathcal{A}(\Sigma_A)$ denote the algebra of open-closed sets in Σ_A. It is the smallest algebra containing $C(\Sigma_A)$. If A is in $\mathcal{A}(\Sigma_A)$ then A is a finite union of cylinder sets. That is, there is a t, ℓ and $w_1, \ldots, w_k \in W(A, \ell)$ with $A = \cup[w_i]_t$. Let $\mathcal{B}(\Sigma_A)$ denote the Borel σ-algebra of sets in Σ_A. It is the smallest σ-algebra containing $C(\Sigma_A)$ or $\mathcal{A}(\Sigma_A)$.

A *Borel probability measure* on Σ_A is a measure defined for each set in $\mathcal{B}(\Sigma_A)$ and with the measure of Σ_A equal to one. We are interested in the shift-invariant, Borel probability measures on Σ_A. Denote the collection of such measures by $\mathcal{M}(\Sigma_A)$. We define a metric on $\mathcal{M}(\Sigma_A)$ as follows. For each ℓ define

$$\varrho_\ell(\mu, \nu) = \max\{|\mu([w]) - \nu([w])| : w \in W(A, \ell)\}.$$

Then define

$$d(\mu, \nu) = \Sigma_{\ell=1}^{\infty} \frac{\varrho_\ell(\mu, \nu)}{2^\ell}.$$

The metric d makes $\mathcal{M}(\Sigma_A)$ a compact metric space (Exercise 1). It gives $\mathcal{M}(\Sigma_A)$ the *weak** topology.

Now we describe Markov measures. A square nonnegative matrix is *stochastic* if the sum of the entries in each row is one. A stochastic matrix has one as an eigenvalue. A row vector p is a *stationary probability vector for* P if it is nonnegative, its entries sum to one and $pP = p$. Observe that when P is irreducible the Perron-Frobenius Theorem guarantees the existence of a unique, strictly positive, stationary probability vector for P.

Let P be a stochastic matrix and p a stationary probability vector for P. Define a $\{0, 1\}$ transition matrix P^0 by $P_{ij}^0 = 0$ if $P_{ij} = 0$ and $P_{ij}^0 = 1$ if $P_{ij} > 0$. The transition matrix P^0 defines a subshift of finite type Σ_{P^0} and we use the pair (p, P) to define a shift-invariant, probability measure $\mu_{(p,P)}$ on it.

First define $\mu_{(p,P)}$ on $C(\Sigma_{P^0})$. For $i \in L_{P^0}$ define $\mu_{(p,P)}([i]_t) = p_i$. For a word $[i_0, \ldots, i_\ell] \in W(P^0)$ define

$$\mu_{(p,P)}([i_0, \ldots, i_\ell]_t) = p_{i_0} P_{i_0 i_1} \cdots P_{i_{\ell-1} i_\ell}.$$

Think of p_i as the probability of choosing i and P_{ij} as the transition probability of going from state i to state j. Note that the value of $\mu_{(p,P)}$ on a cylinder set is independent of t.

Lemma 6.2.1 *For* $\mu_{(p,P)}$ *as defined on* $C(\Sigma_A)$

$$\mu_{(p,P)}([i_0, \ldots, i_\ell]_t) = \Sigma \mu_{(p,P)}([i_0, \ldots, i_\ell, j]_t)$$

where the sum is over all $j \in f(i_\ell)$ *and*

$$\mu_{(p,P)}([i_0, \ldots, i_\ell]_t) = \Sigma \mu_{(p,P)}([i, i_0, \ldots, i_\ell]_{t-1})$$

where the sum is over all $i \in p(i_0)$.

Proof.
    The first equality holds because P is stochastic. The second because p is a stationary probability vector for P.   □

A *measure* μ on an algebra $\mathcal{A}$ must have two properties. The first is that $\mu(\phi) = 0$ and the second is that if $\{A_i\}$ is a sequence of disjoint sets in $\mathcal{A}$ whose union is also in $\mathcal{A}$ then

$$\mu(\cup A_i) = \Sigma \mu(A_i).$$

Lemma 6.2.2 $\mu_{(p,P)}$ *on the cylinder sets* $C(\Sigma_{P^0})$ *induces a measure on the algebra* $\mathcal{A}(\Sigma_{P^0})$.

Proof. First define $\mu_{(p,P)}(\phi) = 0$. Then use the fact that every set in $\mathcal{A}(\Sigma_{P^0})$ is a finite union of cylinder sets to define $\mu_{(p,P)}$ on each open-closed set. Since each element of $\mathcal{A}(\Sigma_{P^0})$ is open-closed, a countable union of sets from $\mathcal{A}(\Sigma_{P^0})$ which is also in the set is a finite union of cylinder sets. Lemma 6.2.1 shows $\mu_{(p,P)}$ is well-defined.   □

Lemma 6.2.3 *The measure* $\mu_{(p,P)}$ *on the algebra* $\mathcal{A}(\Sigma_{P0})$ *induces a shift-invariant, Borel probability measure on* Σ_{P0}.

Proof. The measure $\mu_{(p,P)}$ on the algebra $\mathcal{A}(\Sigma_{P0})$ extends to a measure on the σ-algebra $\mathcal{B}(\Sigma_{P0})$ by the Carathéodory Extension Theorem. It is shift-invariant because the definition on $\mathcal{C}(\Sigma_{P0})$ did not depend on t. It is a probability measure because p is a probability vector. □

The measure $\mu_{(p,P)}$ is the *one-step Markov measure* defined by the pair (p, P).

Given a measure $\mu \in \mathcal{M}(\Sigma_A)$ the *support of* μ is the smallest closed set with measure one. Suppose $\mu_{(p,P)}$ is a one-step Markov measure on Σ_{P0} defined by the pair (p, P). If $i \in L_{P0}$ and $p_i = 0$ then the $\mu_{(p,P)}$ measure of any cylinder set containing i is zero. Conversely, if every p_i is positive then the $\mu_{(p,P)}$ measure of every cylinder set is positive and the support of $\mu_{(p,P)}$ is Σ_{P0}. If the stationary probability vector has some entries which are zero we can discard the entries of p and the corresponding rows and columns of P to get a new pair (p', P') where every entry of p' is positive.

Example 6.2.4 Let $P = \begin{bmatrix} 1/2 & 1/2 \\ 0 & 1 \end{bmatrix}$. Then P is a stochastic matrix and the only stationary probability vector for P is $(0, 1)$. The first row and column of P should be ignored.

Lemma 6.2.5 *Let* P *be a stochastic matrix and* p *a strictly positive, stationary probability vector for* P. *Then there is a permutation of the indices of* P *which puts* P *in block diagonal form with each diagonal block stochastic and irreducible.*

Proof. By Theorem 1.3.10 we may put P in the form

$$\begin{bmatrix} P_1 & * & * & \cdots & * \\ 0 & P_2 & * & \cdots & * \\ 0 & 0 & P_3 & \cdots & * \\ \vdots & \vdots & \vdots & \ddots & \vdots \\ 0 & 0 & 0 & \cdots & P_m \end{bmatrix}$$

where each P_i is either an irreducible matrix or the one by one matrix $[0]$. Note each P_i has every row with sum one or less so the Perron value for each is smaller than or equal to 1 (Exercise 2). Let $p = (p_1, \ldots, p_m)$ be the stationary probability vector for P in this form with the entries of p grouped so the vector p_i corresponds to P_i. Then

$$(p_1, \ldots, p_m) = pP = (p_1 P_1, *, \ldots, *).$$

So P_1 has Perron value 1 and p_1 is its Perron eigenvector. Since the rows of P_1 sum to one or less the Perron-Frobenius Theorem 1.3.5 shows it is stochastic. If not we could increase the value of an entry without increasing the Perron value. Consequently, all the $*$'s in the first row of P are zero. Continue this reasoning to show all the P_i's are stochastic and all the $*$'s in P are zero.   □

Let A be an irreducible, $\{0,1\}$ transition matrix. Suppose P is a stochastic matrix indexed by a subset L_{P^0} of L_A and for $i,j \in L_{P^0}, P_{ij} > 0$ implies $A_{ij} = 1$. When this condition holds we say P is compatible with A. Further suppose p is a strictly positive, stationary probability vector for P. Then the pair (p,P) defines a one-step Markov measure $\mu_{(p,P)}$ on Σ_A with support $\Sigma_{P^0} \subseteq \Sigma_A$.

A measure $\mu \in \mathcal{M}(\Sigma_A)$ is ergodic if for $U \in \mathcal{B}(\Sigma_A)$, $\sigma(U) = U$ implies $\mu(U)$ is zero or one. A measure $\mu \in \mathcal{M}(\Sigma_A)$ is mixing if for all $U, V \in \mathcal{B}(\Sigma_A)$, $\lim_{n\to\infty} \mu(\sigma^n(U) \cap V) = \mu(U) \cdot \mu(V)$.

Theorem 6.2.6 *Let A be an irreducible, $\{0,1\}$ transition matrix. Suppose P is a stochastic matrix compatible with A and p is a strictly positive, stationary probability vector for P. Let $\mu_{(p,P)}$ be the one-step Markov measure defined by the pair (p,P). Then*

(i) *the measure $\mu_{(p,P)}$ is ergodic if and only if P is irreducible and*
(ii) *the measure $\mu_{(p,P)}$ is mixing if and only if P is irreducible and aperiodic.*

Proof. Suppose the matrix P is reducible. By Lemma 6.2.5 P can be put in block diagonal form with each diagonal block an irreducible, stochastic matrix. Then for each i, $\Sigma_{P_i^0} \subseteq \Sigma_A$ is a proper subshift of finite type which is shift-invariant and has positive measure. This means $\mu_{(p,P)}$ is not ergodic.

Next suppose P is irreducible with period ϱ. Note that $\Sigma_A - \Sigma_{P^0}$ is a shift invariant set but it has $\mu_{(p,P)}$ measure zero. So, we need only consider Σ_{P^0}. Recall from Section 1.3 the periodic subsets $\mathcal{C}_0, \ldots, \mathcal{C}_{\varrho-1}$ of L_{P^0}. Suppose $[i_0, \ldots, i_\ell]$ and $[j_0, \ldots, j_k]$ are in $\mathcal{W}(A)$ with $i_\ell \in \mathcal{C}_0$ and $j_0 \in \mathcal{C}_r$. Observe that

$$\mu_{(p,P)}(\sigma^n([i_0, \ldots, i_\ell]_0) \cap [j_0, \ldots, j_k]_0)$$
$$= p_{i_0} P_{i_0 i_1} \cdots P_{i_{\ell-1} i_\ell} (P^{n-\ell})_{i_\ell j_0} P_{j_0 j_1} \cdots P_{j_{\ell-1} j_\ell}$$

when $n \geq \ell$. Then $(P^n)_{i_\ell j_0} = 0$ when $n \neq r \mod \varrho$ and by Theorem 1.3.8

$$\lim_{n\to\infty} (P^{n\varrho+r})_{i_\ell j_0} = P_{j_0}.$$

So no nonempty cylinder set is invariant under the shift and

$$\overline{\lim} \, \mu_{(p,P)}(\sigma^n([i_0, \ldots, i_\ell]_0) \cap [j_0, \ldots, j_k]_0))$$
$$= \mu_{(p,P)}([i_0, \ldots, i_\ell]_0) \times \mu_{(p,P)}([j_0, \ldots, j_k]_0).$$

Apply this observation about cylinder sets and Lemma 6.2.1 to each open-closed set $A \in \mathcal{A}(\Sigma_{P0})$ to conclude that

$$\overline{\lim} \, \mu_{(p,P)}(\sigma^n(A) \cap A)$$
$$\leq \mu_{(p,P)}(A) \times \mu_{(p,P)}(A).$$

The only open-closed sets invariant under σ are the empty set and Σ_{P0}.

Next we extend this observation to all of $\mathcal{B}(\Sigma_{P0})$ by an approximation argument. Let $B \in \mathcal{B}(\Sigma_{P0})$. Then for every $\varepsilon > 0$ there is a set $A \in \mathcal{A}(\Sigma_{P0})$ with $\mu_{(p,P)}(A \bigtriangleup B) < \varepsilon$. Here $A \bigtriangleup B$ is the symmetric difference, $(A - B) \cup (B - A)$, of A and B. With A and B as stated we see

$$|\mu_{(p,P)}(\sigma^n(B) \cap B) - \mu_{(p,P)}(\sigma^n(A) \cap A)| < 2\varepsilon.$$

We know the values for $\mu_{(p,P)}(\sigma^n(A) \cap A)$ are different from $\mu_{(p,P)}(A)$ unless A is the empty set or Σ_{P0}. This means the values for $\mu_{(p,P)}(\sigma^n(B) \cap B)$ are different from $\mu_{(p,P)}(B)$ unless B has measure zero or one. This finishes the proof of (i).

If P is reducible, then $\mu_{(p,P)}$ is not ergodic and so not mixing. Suppose P is irreducible with period $\varrho > 1$. As in the second part of the proof of (i) we see that for $i, j \in L_{P0}$ the sequence $\mu_{(p,P)}(\sigma^n([i]_0) \cap [j]_0)$ doesn't converge. So $\mu_{(p,P)}$ is not mixing.

Suppose P is irreducible and aperiodic. We proceed as in the second part of the proof of (i). Now $\varrho = 1$ and $r = 0$. This shows

$$\lim_{n \to \infty} \mu_{(p,P)}(\sigma^n([i_0, \dots, i_\ell]_0) \cap [j_0, \dots, j_k]_0)$$
$$= \mu_{(p,P)}([i_0, \dots, i_\ell]_0) \times \mu_{(p,P)}([j_0, \dots, j_k]_0).$$

For any $[i_0, \dots, i_\ell]$ and $[j_0, \dots, j_k]$ in $\mathcal{C}(\Sigma_{P0})$. Then continue as before to see that this is true for any two sets in $\mathcal{A}(\Sigma_{P0})$ and use an approximation argument to see that it holds for any sets in $\mathcal{B}(\Sigma_{P0})$. □

Let A be an irreducible, $\{0, 1\}$ transition matrix. We have seen that if P is a stochastic matrix compatible with A and p is a strictly positive, stationary probability vector for P then the pair (p, P) defines a one-step Markov measure on Σ_A. Similarly we can define an *n-step Markov measure* on Σ_A. Let P be a stochastic matrix compatible with $A^{[n]}$, the transition matrix for the *n*-block presentation of A. Let p be a strictly positive, stationary probability vector for P. Then the pair (p, P) defines a one-step Markov measure on $\Sigma_{A^{[n]}}$ and an *n*-step Markov measure on Σ_A. These are the *Markov measures on* Σ_A. A Markov measure is intrinsically characterized by the following observation.

Observation 6.2.7 *The measure $\mu \in \mathcal{M}(\Sigma_A)$ is a Markov measure on Σ_A if and only if there is an $n \geq 0$ so that*

$$\frac{\mu([i_0, \dots, i_n]_0)}{\mu([i_0, \dots, i_{n-1}]_0)} = \frac{\mu([i_{-k}, \dots, i_n]_{-k})}{\mu([i_{-k}, \dots, i_{n-1}]_{-k})}$$

for every $[i_{-k}, \dots, i_n] \in \mathcal{W}(A, n+k+1)$ and $k \geq 0$.

Proof. Exercise 4

Let $\mu \in \mathcal{M}(\Sigma_A)$. Define the *n-step Markov approximation to μ* as follows. Define the matrix P by

$$P_{[i_0, \dots, i_{n-1}][i_1, \dots, i_n]} = \frac{\mu([i_0, \dots, i_n]_0)}{\mu([i_0, \dots, i_{n-1}]_0)}$$

and the vector p by

$$p_{[i_0, \dots, i_{n-1}]} = \mu([i_0, \dots, i_{n-1}]).$$

The proof that this defines a Markov measure on Σ_A is left as Exercise 5. Next we have two observations about Markov approximations.

Observation 6.2.8 *If $\mu \in \mathcal{M}(\Sigma_A)$ and $\mu_{(p,P)}$ is the n-step Markov approximation to μ, then the support of μ is contained in the support of $\mu_{(p,P)}$.*

Proof. Exercise 6

Observation 6.2.9 *Let Σ_A be an irreducible subshift of finite type. The ergodic Markov measures with support equal to Σ_A are dense in $\mathcal{M}(\Sigma_A)$.*

Proof. Exercise 8

In the next part of the discussion we will compute the measure-theoretic entropy of a Markov measure. Then for a fixed irreducible subshift of finite type we will construct a one-step Markov measure whose measure-theoretic entropy is equal to the topological entropy of the subshift of finite type. The final part of the discussion is devoted to showing that every other shift invariant, Borel probability measure has a measure-theoretic entropy strictly less than the topological entropy of the subshift of finite type.

We begin with a brief review of measure-theoretic entropy. Let Σ_A be an irreducible subshift of finite type, $\mu \in \mathcal{M}(\Sigma_A)$ and $\mathcal{C} = \{C_i\}$ a measurable partition of Σ_A. Define the *entropy of the partition $\mathcal{C}$* to be

$$H_\mu(\mathcal{C}) = -\sum_i \mu(C_i) \log \mu(C_i),$$

using the convention that $0 \log 0 = 0$.

The entropy of a partition is a measure of uncertainty. If the space is viewed as a dart board partitioned by the pieces of C then the entropy of the partition measures the chance of predicting in which piece a dart thrown will land. The higher the entropy the less the chance of predicting.

Define the *time n-wedge of C* to be

$$V_{t=0}^{n}\sigma^{-t}(C) = \{C_{i_0} \cap \sigma^{-1}(C_{i_1}) \cap \cdots \cap \sigma^{-n}(C_{i_n})\}.$$

It is also a measurable partition of Σ_A. The entropy of the time n-wedge of C now measures the chances of predicting where $n+1$ successive darts will land. This means the entropy increases with n unless there is no uncertainty.

Let the *entropy of the shift with respect to C* be

$$h_\mu(\Sigma_A, C) = \lim_{n \to \infty} \frac{1}{n+1} H_\mu(V_{t=0}^{n}\sigma^{-t}(C)).$$

Here we are taking the long-term average of the uncertainty and the limit.

Finally define the *entropy of the shift with respect to μ* or simply the *entropy of μ* to be

$$h_\mu(\Sigma_A) = \sup_C h_\mu(\Sigma_A, C),$$

where the supremum is over all measurable partitions of Σ_A.

From the ergodic theory point of view the entropy of the shift with respect to a measure is a measure of the uncertainty or randomness of the shift acting on the measure space (Σ_A, μ). From the information theory point of view one thinks of incoming signals where the probability of receiving a given block at any time is given by the measure μ. Then the entropy measures the amount of information being transmitted in a unit of time.

The generator theorem makes it possible to compute the entropy of a measure. It shows that when C is a partition where the Borel σ-algebra is the smallest σ-algebra containing $\cup_n V_{t=-n}^{n} \sigma^t(C)$ then $h_\mu(\Sigma_A) = h_\mu(\Sigma_A, C)$. This theorem allows us to use the time zero partition $C = \{[i]_0 : i \in L_A\}$ to examine the entropy of a measure. For the rest of this section C will always denote the time zero partition of the subshift of finite type.

We can use the time zero partition of a subshift of finite type and apply the definition directly to compute the entropy of the shift with respect a Markov measure.

Observation 6.2.10 *Let Σ_A be an irreducible subshift of finite type and $\mu_{(p,P)}$ be the one-step Markov measure on Σ_A defined by the pair (p, P). Then the entropy of the shift with respect to $\mu_{(p,P)}$ is*

$$h_{\mu_{(p,P)}} = -\sum p_i P_{ij} \log P_{ij}$$

where the sum is over the $[i,j] \in \mathcal{W}(A, 2)$.

Proof. Let C be the time zero cylinder set partition for Σ_A.
First compute $H_{\mu_{(p,P)}}(\vee_{t=0}^2 \sigma^{-t}(C))$.

$$H_{\mu_{(p,P)}}(C \vee \sigma^{-1}(C) \vee \sigma^{-2}(C))$$

$$= -\sum p_{i_0} P_{i_0 i_1} P_{i_1 i_2} \log p_{i_0} P_{i_0 i_1} P_{i_1 i_2}$$

$$= -\sum p_{i_0} P_{i_0 i_1} P_{i_1 i_2} (\log p_{i_0} + \log P_{i_0 i_1} + \log P_{i_1 i_2})$$

$$= -\sum p_{i_0} \log p_{i_0} - \sum p_{i_0} P_{i_0 i_1} \log P_{i_0 i_1} - \sum p_{i_1} P_{i_1 i_2} \log P_{i_1 i_2}$$

$$= -\sum p_i \log p_i - 2 \sum p_i P_{ij} \log P_{ij}.$$

Then observe that this generalizes to

$$H_{\mu_{(p,P)}}(\vee_{t=0}^n \sigma^{-t}(C))$$

$$= -\sum p_i \log p_i - n \sum p_i P_{ij} \log P_{ij}.$$

Then

$$\lim_{n \to \infty} \frac{1}{n+1} H_{\mu_{(p,P)}}(\vee_{t=0}^n \sigma^{-t}(C))$$

$$= -\sum p_i P_{ij} \log P_{ij}. \qquad \square$$

Let Σ_A be an irreducible subshift of finite type. We construct a special one-step Markov measure for Σ_A. The entropy of the shift with respect to this measure will be equal to the topological entropy of Σ_A. Suppose A is a $\{0,1\}$ transition matrix. Let λ denote the Perron value of A, ℓ the left Perron eigenvector for A and r the right Perron eigenvector for A. Normalize the vectors so $\sum \ell_i r_i = 1$. Define

$$p_i = \ell_i r_i \qquad P_{ij} = A_{ij} \frac{r_j}{\lambda r_i}.$$

A simple computation will show P is a stochastic matrix and p is a strictly positive, stationary probability vector for P (Exercise 11). For reasons which will become obvious this measure is called the *measure of maximal entropy for* Σ_A or simply the *maximal measure for* Σ_A.

Observation 6.2.11 *Let Σ_A be an irreducible subshift of finite type and let $\mu_{(p,P)}$ be the maximal measure for Σ_A. Then the entropy of the shift with respect $\mu_{(p,P)}$ is equal to the topological entropy of Σ_A.*

Proof. Simply use Observation 6.2.10 to compute the entropy. $\qquad \square$

Most of the rest of this section is devoted to showing that if Σ_A is an irreducible subshift of finite type then every measure in Σ_A, except the maximal measure, has entropy strictly less than the topological entropy of Σ_A. This is done in three steps. First we show that the entropy of the shift with respect to any invariant, Borel probability measure is less than or equal to the topological entropy of Σ_A. That is Observation 6.2.13. Next we see that the maximal measure is the only one-step Markov measure whose entropy is equal to the topological entropy. This is shown by Observation 6.2.15. The last step is to show that every other measure has entropy strictly less than its one-step approximation - unless it is already a one-step Markov measure. This is done by Corollary 6.2.19.

Let Σ_A be an irreducible subshift of finite type and $\mu \in \mathcal{M}(\Sigma_A)$. We will show $h_\mu(\Sigma_A) \leq h(\Sigma_A)$. To do this we will rely on the fact that the function $\phi : [0, \infty) \to \mathbb{R}$ defined by

$$\phi(x) = \begin{cases} 0 & \text{if } x = 0 \\ -x \log x & \text{if } x \neq 0. \end{cases}$$

is strictly concave down.

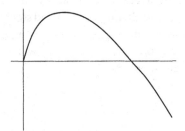

Figure 6.2.1

Lemma 6.2.12 *Let $\phi : [0, 1) \to \mathbb{R}$ be as above. Suppose $x_j \in [0, \infty)$ and $t_j \geq 0$ with $\sum t_j = 1$. Then*

$$\sum t_j \phi(x_j) \leq \phi\left(\sum t_j x_j\right)$$

with equality if and only if all the x_j's corresponding to nonzero t_j are equal.

Proof. The second derivative of ϕ is $\phi''(x) = -1/x$ which is strictly negative on $(0, \infty)$ so ϕ is strictly concave down. This means that the statement is true for two nonzero t_j. It follows for more than two nonzero t_j by induction.   □

Observation 6.2.13 *Let Σ_A be an irreducible subshift of finite type and $\mu \in \mathcal{M}(\Sigma_A)$. Then the entropy of the shift with respect to μ is less than or equal to the topological entropy of Σ_A.*

Proof. We have

$$H_\mu(\vee_{t=0}^n \sigma^{-t}(\mathcal{C})) = \sum \phi(\mu([i_0, \dots, i_n]))$$

where the sum is over all $[i_0, \dots, i_n] \in \mathcal{W}(A, n+1)$. Apply Lemma 6.2.12 to $H_\mu(\vee_{t=0}^n \sigma^{-t}(\mathcal{C}))$ with $x_{[i]} = \mu([i_0, \dots, i_n])$ and $t_{[i]} = \frac{1}{|\mathcal{W}(A, n+1)|}$ for all $[i] \in \mathcal{W}(A, n+1)$. This shows that

$$H_\mu(\vee_{t=0}^n \sigma^{-t}(\mathcal{C})) \le \log |\mathcal{W}(A, n+1)|.$$

The conclusion follows from this.   □

To examine the one-step Markov measures we give an alternate description of the measures. Then we give two proofs that the maximal measure is the only one-step Markov measure with entropy equal to the topological entropy. The first proof is combinatorial and the second uses elementary calculus. Both will be used in the second part of this book when we are dealing with countable state Markov shifts. Let A be an irreducible, $\{0, 1\}$ transition matrix, G_A its directed graph and Σ_A the irreducible subshift of finite type described by A. Consider (p, P) a stationary probability vector and stochastic matrix describing a one-step Markov measure on Σ_A. Then $p_i P_{ij} = x_{ij}$ can be thought of as a weight on the edge from i to j in G_A. The set of weights on G_A coming from (p, P) satisfies three conditions. They are:

(i) $x_{ij} \ge 0$,
(ii) $\sum_{(i,j)} x_{ij} = 1$ and
(iii) for fixed j, $\sum_i x_{ij} = \sum_k x_{jk}$.

Conversely, a set of weights on the edges of G_A which satisfy the three conditions determine a stochastic matrix P and a stationary probability vector p for P. They are

$$p_j = \sum_i x_{ij} = \sum_k x_{jk} \qquad P_{ij} = x_{ij} / \sum_k x_{ik} = x_{ij}/p_i.$$

We say that a set of weights on the edges of G_A which satisfies these three conditions is a *Markov weight set for* G_A. Denote the set of Markov weight sets on G_A by MW_A. If (x_{ij}) and (y_{ij}) are Markov weight sets then for $0 \le t \le 1$, $(tx_{ij} + (1-t)y_{ij})$ is also a Markov weight set. Say a Markov weight set is in the interior of MW_A when $x_{ij} > 0$ for each (i, j) with $A_{ij} = 1$. These statements are summarized by the following lemma.

Lemma 6.2.14 *Let A be an irreducible, $\{0, 1\}$ transition matrix and G_A its directed graph. The Markov weight sets on G_A and the one-step Markov measures on Σ_A are in one-to-one correspondence. The collection of Markov weight sets*

MW_A is a convex subset of $\mathbb{R}^{E_A}$, where E_A is the set of edges of G_A. The one-step Markov measures with support equal to Σ_A are defined by the weight sets in the interior of MW_A. These correspond to the edge sets with all $x_{ij} > 0$.

Observation 6.2.15 Let A be an irreducible, $\{0, 1\}$ transition matrix. There is a unique, one-step Markov measure whose entropy is equal to the topological entropy of Σ_A and every other one-step Markov measure has entropy strictly less than the topological entropy of Σ_A.

Proof. The maximal measure exists by Observation 6.2.11.

Suppose there are two one-step Markov measures $\mu_{(p,P)}$ and $\mu_{(q,Q)}$ with entropy equal to the topological entropy of Σ_A. By Observation 6.2.13 and the Perron-Frobenius Theorem both measures must have support equal to Σ_A. Let $\mu_{(p,P)}$ be defined by the Markov edge set (x_{ij}) and $\mu_{(q,Q)}$ by the Markov edge set (y_{ij}). Then for $0 \leq t \leq 1$, $(tx_{ij} + (1-t)y_{ij})$ defines a Markov edge set for G_A. Fix $0 < t < 1$ and let $\mu_{(r,R)}$ be the one-step Markov measure defined by $(tx_{ij} + (1-t)y_{ij})$. Then r and R are specified by the formulas

$$r_i = tp_i + (1-t)q_i \quad \text{and}$$

$$R_{ij} = \frac{tp_i P_{ij} + (1-t)q_i Q_{ij}}{tp_i + (1-t)q_i} = t_i P_{ij} + (1-t_i)Q_{ij},$$

where $t_i = tp_i/r_i$. Since both $\mu_{(p,P)}$ and $\mu_{(q,Q)}$ have full support $0 < t_i < 1$ for each i. This is crucial. Now compute the entropy of the measure $\mu_{(r,R)}$. We see

$$\begin{aligned} h_{\mu_{(r,R)}}(\Sigma_A) &= -\sum r_i R_{ij} \log R_{ij} \\ &= -\sum r_i (t_i P_{ij} + (1-t_i)Q_{ij}) \log(t_i P_{ij} + (1-t_i)Q_{ij}) \\ &\geq -\sum r_i (t_i P_{ij} \log P_{ij} + (1-t_i)Q_{ij} \log Q_{ij}) \\ &= -\sum (tp_i P_{ij} \log P_{ij} + (1-t)q_i Q_{ij} \log Q_{ij}) \\ &= th_{\mu_{(p,P)}}(\Sigma_A) + (1-t)h_{\mu_{(q,Q)}}(\Sigma_A). \end{aligned}$$

The inequality comes from applying Lemma 6.2.12 to each pair (i, j). For each i, both t_i and $(1-t_i)$ are nonzero so we can conclude that there is equality if and only if $P_{ij} = Q_{ij}$ for every pair (i, j). □

For the second proof of Observation 6.2.15 we compute the entropy of a one-step Markov measure in terms of its Markov edge set.

Lemma 6.2.16 Let A be an irreducible, $\{0, 1\}$ transition matrix, G_A its directed graph and $x = (x_{ij})$ a Markov weight set on G_A. It defines a one-step Markov *measure μ on Σ_A with entropy*

$$h_\mu(\Sigma_A) = h(x) = -\sum_{ij} x_{ij} \log x_{ij} + \sum_i (\sum_k x_{ik}) \log(\sum_k x_{ik}).$$

Proof. This is a computation using Observation 6.2.10   □

At this point we have a real valued function h defined on the positive quadrant of R^{E_A} and we want to maximize it subject to the edge set constraints. These constraints are given by the equations

$$g_j(x) = \sum_i x_{ij} - \sum_k x_{jk} = 0 \quad \text{for all } j \quad \text{and} \quad g(x) = 1 - \sum_{ij} x_{ij} = 0.$$

This can be done using Lagrange multipliers. To do so consider the function

$$F(x, \kappa, \eta) = h(x) + \sum_s \kappa_s g_s(x) + \eta g(x).$$

Then compute

$$\frac{\partial F}{\partial x_{ij}} = -(\log x_{ij} + 1) + (\log(\sum_k x_{ik}) + 1) + \kappa_i(\delta_{ij} - 1) + \kappa_j(1 - \delta_{ji}) - \eta,$$

$$\frac{\partial F}{\partial \kappa_s} = \sum_i x_{ij} - \sum_k x_{jk}, \quad \text{and} \quad \frac{\partial F}{\partial \eta} = 1 - \sum_{ij} x_{ij}.$$

Solve

$$\frac{\partial F}{\partial x_{ij}} = 0 \quad \text{to get} \quad \log \frac{x_{ij}}{\sum_k x_{ik}} = \kappa_j - \kappa_i - \eta, \quad \text{or}$$

$$\frac{x_{ij}}{\sum_k x_{ik}} = \frac{e^{\kappa_j}}{e^\eta e^{\kappa_i}}$$

summing both sides over j gives

$$e^\eta e^{\kappa_i} = \sum_j e^{\kappa_j}.$$

For every (i,j) we consider $A_{ij} = 1$ so the previous equation becomes a right eigenvector equation for A. Let $\lambda = e^\eta$ and r be the vector with $r_k = e^{\kappa_k}$ so $Ar = \lambda r$. Moreover, all the terms r_k are positive so the Perron-Frobenius Theorem shows λ is the Perron value of A and r is the right Perron eigenvector for A.

Proof. Second Proof of Observation 6.2.15 The maximal measure exists by Observation 6.2.11.

Any measure whose entropy is equal to the topological entropy must have support equal to Σ_A by Observation 6.2.13 and the Perron-Frobenius Theorem. This means that any one-step Markov measure with entropy equal to the topological entropy must have its Markov weight set in the interior of MW_A. The previous argument using Lagrange multipliers shows that there is one maximal value for the entropy function in the interior of MW_A. It shows that the maximal measure is unique among the one-step Markov measures and produces a formula for the measure.   □

The final step is to show that if a measure has entropy equal to the topological entropy then it must be a one-step Markov measure. The proof once again depends on the convexity of the function ϕ and Lemma 6.2.12. To simplify the following computations we define the conditional entropy. Let Σ_A be an irreducible subshift of finite type and $\mu \in \mathcal{M}(\Sigma_A)$. Let $\mathcal{D}$ and $\mathcal{E}$ be two finite partitions of Σ_A. Then the *entropy of $\mathcal{D}$ given $\mathcal{E}$* is

$$H_\mu(\mathcal{D}|\mathcal{E}) = -\sum_{i,j} \mu(D_i \cap E_j) \log \frac{\mu(D_i \cap E_j)}{\mu(E_j)}.$$

The idea of the conditional entropy is to consider each E_j as a measure space with μ normalized on it. Let the partition $\mathcal{D}$ define a partition of E_j and take the entropy of $\mathcal{D}$ on E_j. This gives

$$-\sum_i \frac{\mu(D_i \cap E_j)}{\mu(E_j)} \log \frac{\mu(D_i \cap E_j)}{\mu(E_j)}.$$

Now average these entropies taking into account the measures of the E_j.

Let $\mathcal{C}$ be the usual time zero partition. We are only interested in the entropy of $\mathcal{C}$ given $\vee_{t=1}^n \sigma^t(\mathcal{C})$. This is the entropy of $\mathcal{C}$ given the past to time n. It is

$$H_\mu(\mathcal{C}| \vee_{t=1}^n \sigma^t(\mathcal{C})) = -\sum \mu([i_{-n}, \ldots, i_0]) \log \frac{\mu([i_{-n}, \ldots, i_0])}{\mu([i_{-n}, \ldots, i_{-1}])}.$$

The idea of a conditional entropy is related to the idea of a Markov approximation to a measure. In particular observe that if $\mu_{(p,P)}$ is a one-step Markov measure then for $n \geq 1$,

$$H_{\mu_{(p,P)}}(\mathcal{C}| \vee_{t=1}^n \sigma^t(\mathcal{C})) = -\sum p_i P_{ij} \log P_{ij}.$$

We will use several properties of conditional entropy.

Lemma 6.2.17 *Let Σ_A be an irreducible subshift of finite type, $\mu \in \mathcal{M}(\Sigma_A)$ and C the time zero cylinder set partition. Then*

(i)

$$H_\mu(\vee_{t=0}^n \sigma^t(C)) = H_\mu(C) + \sum_{k=1}^n H_\mu(C| \vee_{t=1}^k \sigma^t(C)).$$

(ii)

$$H_\mu(C| \vee_{t=1}^n \sigma^t(C)) \leq H_\mu(C| \vee_{i=1}^{n-1} \sigma^t(C))$$

with equality if and only if

$$\frac{\mu([i_{-n}, \ldots, i_0])}{\mu([i_{-n}, \ldots, i_{-1}])} = \frac{\mu([i_{-n+1}, \ldots, i_0])}{\mu([i_{-n+1}, \ldots, i_{-1}])},$$

for all $[i_{-n}, \ldots, i_0]$.

(iii)

$$h_\mu(\Sigma_A) = \lim_{n \to \infty} H_\mu(C| \vee_{t=1}^n \sigma^t(C)) = \inf H_\mu(C| \vee_{t=1}^n \sigma^t(C)).$$

Proof. To prove (i) simply compute

$$H_\mu(C \vee \sigma(C))$$

$$= -\sum \mu([i_{-1}, i_0]) \log \mu([i_{-1}, i_0])$$

$$= -\sum \mu([i_{-1}, i_0]) \left[(\log \mu([i_0]) + \log \frac{\mu([i_{-1}, i_0])}{\mu([i_0])} \right]$$

$$= H_\mu(C) + H_\mu(C|\sigma(C)).$$

The general statement follows the same reasoning.

The proof of (ii) is also a computation and relies on Lemma 6.2.12

$$H_\mu(C| \vee_{t=1}^n \sigma^t(C))$$

$$= -\sum \mu([i_{-n}, \ldots, i_0]) \log \frac{\mu([i_{-n}, \ldots, i_0])}{\mu([i_{-n}, \ldots, i_{-1}])}$$

$$= - \sum_{[i_{-n+1}, \ldots, i_0]} \mu([i_{-n+1}, \ldots, i_{-1}]) \left[\sum_{i_n} \frac{\mu([i_{-n}, \ldots, i_{-1}])}{\mu([i_{-n+1}, \ldots, i_{-1}])} \right.$$

$$\left. \times \frac{\mu([i_{-n}, \ldots, i_0])}{\mu([i_{-n}, \ldots, i_{-1}])} \log \frac{\mu([i_{-n}, \ldots, i_0])}{\mu([i_{-n}, \ldots, i_{-1}])} \right]$$

$$\leq - \sum_{[i_{-n+1}, \ldots, i_0]} \mu([i_{-n+1}, \ldots, i_0]) \log \frac{\mu([i_{-n+1}, \ldots, i_0])}{\mu([i_{-n+1}, \ldots, i_{-1}])}$$

$$= H_\mu(C| \vee_{i=1}^{n-1} \sigma^i(C)).$$

The inequality follows by Lemma 6.2.12 taking for a fixed $[i_{-n+1}, \ldots i_0]$

$$t_n = \frac{\mu([i_{-n}, \ldots, i_{-1}])}{\mu([i_{-n+1}, \ldots, i_{-1}])}, \quad x_n = \frac{\mu([i_{-n}, \ldots, i_0])}{\mu([i_{-n}, \ldots, i_{-1}])}$$

and

$$\sum t_n x_n = \frac{\mu([i_{-n+1}, \ldots, i_0])}{\mu([i_{-n+1}, \ldots, i_{-1}])}.$$

There is equality if and only if

$$\frac{\mu([i_{-n}, \ldots, i_0])}{\mu([i_{-n}, \ldots, i_{-1}])}$$

is constant over all i_{-n}.

The proof of (iii) uses (ii) to see that the conditional entropies are nonincreasing and then (i) to see that

$$\lim_{n \to \infty} \frac{1}{n+1} H_\mu(\vee_{t=0}^n \sigma^{-t}(\mathcal{C})) = \lim_{n \to \infty} H_\mu(\mathcal{C} \mid \vee_{t=1}^n \sigma^t(\mathcal{C}))$$
$$= \inf H_\mu(\mathcal{C} \mid \vee_{t=1}^n \sigma^t(\mathcal{C}))$$

$\square$

Corollary 6.2.18 *Let Σ_A be an irreducible subshift of finite type, $\mu \in \mathcal{M}(\Sigma_A)$ and $\mathcal{C}$ the time zero cylinder set partition. Then μ is an n-step Markov measure if and only if*

$$H_\mu(\mathcal{C} \mid \vee_{t=1}^n \sigma^t(\mathcal{C})) = H_\mu(\mathcal{C} \mid \vee_{t=1}^m \sigma^t(\mathcal{C})),$$

for all $m \geq n$

Proof. We have already made this computation for a one-step Markov measure. The same type of computation shows that the equalities hold for an n-step Markov measure.

If the equalities hold then use Lemma 6.2.17 (ii) to see that

$$\frac{\mu([i_{-m}, \ldots, i_0])}{\mu([i_{-m}, \ldots, i_{-1}])} = \frac{\mu([i_{-n}, \ldots, i_0])}{\mu([i_{-n}, \ldots, i_{-1}])},$$

for all $m \geq n$. Then apply Observation 6.2.7 to conclude that μ is an n-step Markov measure. $\square$

Corollary 6.2.19 *Let Σ_A be an irreducible subshift of finite type, $\mu \in \mathcal{M}(\Sigma_A)$ and $\mathcal{C}$ the time zero cylinder set partition. Let $\mu_{(p,P)}$ be the one-step Markov approximation to μ. Then $h_\mu(\Sigma_A) \leq h_{\mu_{(p,P)}}(\Sigma_A)$ with equality if and only if $\mu = \mu_{(p,P)}$.*

Proof. This follows from Lemma 6.2.17 (*ii*) and Corollary 6.2.18.

Lemma 6.2.17 shows that conditional entropy is closely related to the idea of a Markov measure. It shows that an n-step Markov approximation is a randomization of the measure it approximates. We put these results together for the following theorem.

Theorem 6.2.20 *Let Σ_A be an irreducible subshift of finite type. Then there is a one-step Markov measure whose entropy is equal to the topological entropy of Σ_A. Every other shift-invariant, Borel probability measure has entropy strictly less than the topological entropy of Σ_A*

Proof. By Observation 6.2.11 there is a one-step Markov measure with entropy equal to the topological entropy of Σ_A. By Corollary 6.2.19 we know that for any shift-invariant, Borel probability measure there is a one-step Markov measure whose entropy is greater than or equal to the entropy of the original measure. By Observation 6.2.15 every one-step Markov measure, except the maximal measure, has entropy strictly less than the topological entropy.   □

We end this section with a theorem describing a very nice property of the maximal measure.

Earlier in this section we defined ergodicity of a measure. To prove the next theorem we use a well-known, alternate characterization of ergodicity. It states that a measure on a subshift of finite type is ergodic for the shift if the only measurable, shift-invariant functions are the constant functions. A function f is shift-invariant if $f = f \circ \sigma$ almost everywhere. The equivalence of the two definitions of ergodic is easily seen by considering the characteristic functions of the Borel sets.

Theorem 6.2.21 *Let Σ_A and Σ_B be irreducible subshifts of finite type with the same topological entropy. Suppose $\varphi : \Sigma_A \to \Sigma_B$ is a factor map. Then φ takes the maximal measure of Σ_A to the maximal measure of Σ_B.*

Proof. Let Σ_A, Σ_B and φ be as stated. Then A and B have the same Perron value. Call it λ. Let μ_A denote the maximal measure for Σ_A and μ_B denote the maximal measure for Σ_B. Using the formula for μ_A we can find find constants c_1 and c_2 so that for each n and $[i_1, \ldots , i_n] \in \mathcal{W}(A, n)$

$$c_1(1/\lambda)^n \leq \mu_A([i_1, \ldots , i_n]) \leq c_2(1/\lambda)^n.$$

The same is true for μ_B. Consequently, we can find constants m_1 and m_2 so that for each n, $[i_1, \ldots , i_n] \in \mathcal{W}(A, n)$ and $[j_1, \ldots , j_n] \in \mathcal{W}(B, n)$

$$m_1\mu_A([i_1, \ldots , i_n]) \leq \mu_B([j_1, \ldots , j_n]) \leq m_2\mu_A([i_1, \ldots , i_n]).$$

Since φ is a factor map it pushes μ_A down to a shift-invariant, Borel probability measure μ on Σ_B defined by $\mu(E) = \mu_A(\varphi^{-1}(E))$ for each $E \in \mathcal{B}(\Sigma_B)$.

By Lemma 4.1.5 we may assume that φ is a one-block map and by Theorem 4.1.7 we know that φ is uniformly bounded-to-one and has no diamonds. This means that there is an M so that for each n and $[j_1, \dots, j_n] \in \mathcal{W}(B, n)$ there is at least one and no more than M words in $\mathcal{W}(A, n)$ which map to $[j_1, \dots, j_n]$. This in turn means that

$$m_1 \mu([j_1, \dots, j_n]) \leq \mu_B([j_1, \dots, j_n]) \leq M \cdot m_2 \mu([j_1, \dots, j_n]).$$

So μ and μ_B are absolutely continuous with respect to each other. By the Radon-Nikodym Theorem there is a nonnegative measurable function $[d\mu/d\mu_B]$ with

$$\mu(E) = \int_E \left[\frac{d\mu}{d\mu_B}\right] d\mu_B$$

for each $E \in \mathcal{B}(\Sigma_B)$. Moreover, since μ and μ_B are both shift-invariant measures we see $[d\mu/d\mu_B]$ is a shift-invariant function. That is, $[d\mu/d\mu_B] = [d\mu/d\mu_B] \circ \sigma$ almost everywhere with respect to μ_B. We know μ_B is ergodic by Theorem 6.2.6 since Σ_B is irreducible and μ_B has full support. By the definition of ergodic discussed just before this theorem we conclude $[d\mu/d\mu_B]$ is a constant function. Since μ and μ_B are probability measures $[d\mu/d\mu_B] \equiv 1$ and $\mu = \mu_B$.   □

Example 6.2.22 To see that Theorem 6.2.21 is false when the factor map is not finite-to-one let $\varphi : \{0, 1\} \to \{0, 1, 2\}$ be defined by $\varphi(0) = 0$ and $\varphi(1) = \varphi(2) = 1$. The proof is Exercise 14.

Exercises

1. Let $\mathcal{M}(\Sigma_A)$ denote the set of shift-invariant, Borel probability measures on Σ_A. For $\mu, \nu \in \mathcal{M}(\Sigma_A)$ and for each ℓ, set

$$\varrho_\ell(\mu, \nu) = \max\{|\mu([w]) - \nu([w])| : w \in \mathcal{W}(A, \ell)\}.$$

Then define

$$d(\mu, \nu) = \Sigma_{\ell=1}^\infty \frac{\varrho_\ell(\mu, \nu)}{2^\ell}.$$

Show that d is a metric on $\mathcal{M}(\Sigma_A)$ and $\mathcal{M}(\Sigma_A)$ is compact in the topology defined by this metric.

2. Let A be a square, nonnegative, irreducible matrix with the sum of the entries in each row less than or equal to one. Show that the Perron value for A is less than or equal to one.

3. Let
$$P = \begin{bmatrix} 1/2 & 0 & 1/2 & 0 & 0 \\ 0 & 1/4 & 0 & 3/4 & 0 \\ 1/8 & 0 & 0 & 0 & 7/8 \\ 0 & 1 & 0 & 0 & 0 \\ 1/2 & 0 & 1/2 & 0 & 0 \end{bmatrix}.$$
Find two strictly positive, stationary probability vectors for P. Then find a permutation of the indices which puts P in block diagonal form.

4. Prove Observation 6.2.7.

5. Let $\mu \in \mathcal{M}(\Sigma_A)$. Define the pair (p, P) by
$$p_{[i_0,\dots,i_{n-1}]} = \mu([i_0,\dots,i_{n-1}])$$
and
$$P_{[i_0,\dots,i_{n-1}][i_1,\dots,i_n]} = \frac{\mu([i_0,\dots,i_n]_0)}{\mu([i_0,\dots,i_{n-1}]_0)}.$$
Show that P is stochastic and p is a stationary probability vector for it.

6. Prove Observation 6.2.8.

7. Let $A = \begin{bmatrix} 1 & 1 \\ 1 & 0 \end{bmatrix}$. Give an example of an ergodic, shift-invariant probability measure on Σ_A that is not a Markov measure. Compute the pair (p, P) which defines the 2-step Markov approximation to your measure.

8. Prove Observation 6.2.9.

9. Suppose (X, μ) is a nonatomic probability space. Consider all partitions of X with m elements and their entropies. Let $\mathcal{D}$ be one of these partitions and suppose it maximizes the entropy over all such partitions. Show $\mu(C_i) = 1/m$ for all i.

10. Let $P = \begin{bmatrix} 1/2 & 1/2 \\ 1/2 & 1/2 \end{bmatrix}$ and $Q = \begin{bmatrix} 1/4 & 3/4 \\ 3/4 & 1/4 \end{bmatrix}$ define measures $\mu_{(p,P)}$ and $\mu_{(q,Q)}$ on $\{1,2\}^{\mathbb{Z}}$. Compute their entropies.

11. Show that the p_i and P_{ij} defined for the maximal measure just before Observation 6.2.11 define a stochastic matrix and its stationaary probability vector. Carry out the computation for the proof of Observation 6.2.11.

12. Let $A = \begin{bmatrix} 1 & 1 \\ 1 & 0 \end{bmatrix}$. Compute the pair (p, P) which defines the maximal measure for Σ_A.

13. Compute the Markov weight sets for the measures in Problem 10.

14. Make the computation to show the factor map φ of Example 6.2.22 does not send the maximal measure of the three-shift to the maximal measure of the two-shift.

§ 6.3 Markov Subgroups

In this section we will describe the structure of subshifts with a group structure. If G is a finite group then $G^{\mathbb{Z}}$ with coordinate by coordinate addition is a compact, totally disconnected group. The shift is a continuous, group automorphism. If $X \subseteq G^{\mathbb{Z}}$ is a closed, shift-invariant subgroup then we say (X, σ) is a *Markov subgroup*. We will see that any Markov subgroup is a subshift of finite type and this allows us to describe and classify the dynamics of all Markov subgroups.

Example 6.3.1 Begin with the finite group $\mathbb{Z}/4\mathbb{Z} \oplus \mathbb{Z}/2\mathbb{Z}$ and form the full shift on the group. It is a group with coordinate by coordinate addition. Define a subshift of finite type, which is also a subgroup, $X \subseteq (\mathbb{Z}/4\mathbb{Z} \oplus \mathbb{Z}/2\mathbb{Z})^{\mathbb{Z}}$ by the transitions given by the diagram in Figure 6.3.1. The follower sets are:

$$f(0,0) = f(2,0) = f(1,1) = f(3,1) = \{(0,0), (2,0), (1,0), (3,0)\}$$
$$f(1,0) = f(3,0) = f(0,1) = f(2,1) = \{(1,1), (3,1), (2,1), (0,1)\}.$$

One must check the rule is closed under addition (Exercise 1).

Figure 6.3.1

Example 6.3.2 Let G be any finite group. Consider the full-shift $G^{\mathbb{Z}}$ with coordinate by coordinate addition. Let $H \subseteq G^k$ be a subgroup of the product of G with itself k times. The subgroup H is the set of allowable words of length k. Define $X_H \subseteq G^{\mathbb{Z}}$ to be the set $X_H = \{x : [x_i, \dots, x_{i+k-1}] \in H \text{ for all } i \in \mathbb{Z}\}$. This defines a $(k-1)$-step subshift of finite type. It is also a subgroup of $G^{\mathbb{Z}}$.

The goal of this section is to describe and classify all closed, Markov subgroups.

If X and $\bar{X}$ are compact topological groups and T and $\bar{T}$ are continuous automorphisms of X and $\bar{X}$ respectively, we say (X, T) and $(\bar{X}, \bar{T})$ are *isomorphic* if there is a continuous group isomorphism $\varphi : X \to \bar{X}$ so that $\bar{T} \circ \varphi = \varphi \circ T$. We say (X, T) and $(\bar{X}, \bar{T})$ are *conjugate* if there is a homeomorphism $\varphi : X \to \bar{X}$. We will prove the following theorem using a sequence of lemmas.

Theorem 6.3.3 *If G is a finite group and X is a closed, shift-invariant subgroup of $G^{\mathbb{Z}}$ then (X, σ) is conjugate to one of the following.*

(i)   *An automorphism of a finite group with the discrete topology. This happens if and only if $h(X, \sigma) = 0$.*

(ii)   *A full m-shift. This happens if and only if $h(X, \sigma) > 0$ and σ is transitive. Then $h(X, \sigma) = \log m$.*

(iii) *The direct product of an automorphism of a finite group with a full m-shift. Then $h(X, \sigma) = \log m$.*

Suppose G is a finite group and X is a Markov subgroup of $G^{\mathbb{Z}}$. For $w \in \mathcal{W}(X, k)$ we will use $f(w)$ to be the follower set of w in $\mathcal{W}(X, 1)$. That is, $\{g \in G : wg \in \mathcal{W}(X, k+1)\}$. We will use $p(w)$ as the predecessor set in the same way. Let e be the identity element of G and $[e^k] \in \mathcal{W}(X; k)$ the word of k e's.

The next lemma contains a crucial observation about shift spaces with a group structure.

Lemma 6.3.4 *Let G be a finite group and $X \subseteq G^{\mathbb{Z}}$ a Markov subgroup. Then*

(i)   *the words of length k in X form a subgroup of G^k;*

(ii)   *the follower set of the word of k e's is a normal subgroup of $G = \mathcal{W}(X, 1)$;*

(iii) *the predecessor set of the word of k e's is a normal subgroup of $G = \mathcal{W}(X, 1)$;*

(iv)   *the follower set of each $[g_0, \dots, g_{k-1}] \in \mathcal{W}(X, k)$ is a coset of $f([e^k])$;*

(iv)   *the predecessor set of each $[g_0, \dots, g_{k-1}] \in \mathcal{W}(X, k)$ is a coset of $p([e^k])$.*

Proof. (i) $\mathcal{W}(X, k)$ is clearly a subgroup of G^k because X is a subgroup of $G^{\mathbb{Z}}$ with coordinate by coordinate addition from G.

(ii) Let $[e^k]$ denote the word of k e's in $\mathcal{W}(X, k)$ and $f([e^k])$ denote its follower set as a subset of $\mathcal{W}(X, 1)$. Suppose $g, g' \in f([e^k])$. In $\mathcal{W}(X, k+1)$

$$[e, \dots, e, g][e, \dots, e, g'] = [e, \dots, e, gg'] \text{ and}$$
$$[e, \dots, e, g]^{-1} = [e, \dots, e, g^{-1}],$$

so $f([e^k])$ is a subgroup. If $h \in G$, there is a $[h_0, \ldots, h_{k-1}] \in \mathcal{W}(X, k)$ with $h \in f([h_0, \ldots, h_{k-1}])$ and

$$[h_0, \ldots, h_{k-1}, h][e, \ldots, e, g][h_0, \ldots, h_{k-1}, h]^{-1}$$
$$= [e, \ldots, e, hgh^{-1}].$$

This means $f([e^k])$ is normal in G.

(iv) If $[g_0, \ldots, g_{k-1}] \in \mathcal{W}(X, k)$ and $g \in f([g_0, \ldots, g_{k-1}])$ then $g \cdot f([e^k]) \subseteq f([g_0, \ldots, g_{k-1}])$. Conversely, if $h \in f([g_0, \ldots, g_{k-1}])$, then

$$[g_0, \ldots, g_{k-1}, g]^{-1}[g_0, \ldots, g_{k-1}, h] = [e, \ldots, e, g^{-1}h],$$

so $f([g_0, \ldots, g_{k-1}]) \subseteq g \cdot f([e^k])$.   □

(iii) and (v) The proof for predecessor sets is the same as for follower sets.

Lemma 6.3.5 *Let $\bar{G}$ be a finite group and $\bar{X}$ a closed, shift invariant subgroup of $\bar{G}^{\mathbb{Z}}$. Then $\bar{X}$ is isomorphic to a one-step Markov subgroup of $G^{\mathbb{Z}}$ for some finite group G.*

Proof. Suppose G is a finite group and $\bar{X}$ is a closed, shift invariant subgroup of $\bar{G}^{\mathbb{Z}}$. Then the follower sets of the words

$$G = \mathcal{W}(X, 1) \supseteq f([e]) \supseteq f([e^2]) \supseteq \cdots \supseteq f([e^n]) \supseteq \cdots$$

is a descending chain of normal subgroups. Since $\bar{G}$ is finite there is an N such that $f([e^n]) = f([e^N])$ for all $n \geq N$. Since the other follower sets are cosets of this,

$$f([g_0, \ldots, g_{N-1}]) = f([g_{-j}, \ldots, g_{N-1}]) \quad \text{for all} \quad j \geq 0$$

with $[g_{-j}, \ldots, g_{N-1}] \in \mathcal{W}(X, N + j)$. The N-block presentation of X has a one-step transition rule. Setting $X = X^{[N]}$ and $G = \mathcal{W}(X, N)$ completes the proof.   □

Lemma 6.3.6 *Let G be a finite group and X a Markov subgroup of $G^{\mathbb{Z}}$. Then:*

(i) *$f : G \to G/f(e)$ is an onto, continuous, group homomorphism with kernel $p(e)$;*

(ii) *$p : G \to G/p(e)$ is an onto, continuous, group homomorphism with kernel $f(e)$.*

Proof. The fact that $f : G \to G/f(e)$ is a well-defined, continuous, onto group homomorphism follows immediately from Lemma 6.3.4. To see the kernel of the map f is $p(e)$ observe that $g \in p(e)$ if and only if $f(g)$ contains e. But then $f(g) = f(e)$. The same is true when the roles of $p(e)$ and $f(e)$ are reversed.   □

A one-step Markov subgroup of $G^{\mathbb{Z}}$ is defined by a subgroup $H \subseteq G^2$. In terms of predecessor and follower sets $H = \{(p(g), g) : g \in G\} = \{(g, f(g)) : g \in G\}$.

Proof of Theorem 6.3.3. By Lemma 6.3.5 we may assume that (X, σ) is a one-step Markov subgroup of $G^{\mathbb{Z}}$ for some finite group G. By Lemma 6.3.6 the map $f : G \to G/f(e)$ is a homomorphism with kernel $p(e)$. When G is finite this implies that $|G| = |G/f(e)| \times |p(e)|$, where $|\cdot|$ denotes the cardinality of a set. For a finite group G and normal subgroup $f(e)$ we also have $|G| = |G/f(e)| \times |f(e)|$. Together these imply $|f(e)| = |p(e)|$.

If $|f(e)| = |p(e)| = 1$, define an automorphism $\tau : G \to G$ by $\tau(g) = f(g)$. This is case (i).

If $|f(e)| = |p(e)| > 1$ then $|f(g)| = |p(g)| = m$ for some m and all $g \in G$. This implies that $h(X, \sigma) = \log m$. By Lemma 6.3.4 we know that $p(e)$ and $f(e)$ are normal subgroups of G. Let $K = p(e) \cap f(e)$. It is a normal subgroup of G. If $g \in K$ then $p(g) = p(e)$ and $f(g) = f(e)$ and so $K^{\mathbb{Z}}$ is a normal subgroup of X.

There are two cases, either $K = \{e\}$ or K is larger. First we take the case with $K \neq \{e\}$. We can write $G \simeq G/K \times K$, where the multiplication in $G/K \times K$ is obtained by viewing G as an extension of K by G/K. We are not interested in the algebra here - only the fact that we have a set isomorphism between G and $G/K \times K$. This is the reason we lose the isomorphism and end up with only a conjugacy. Now we have (X, σ) is conjugate to $(Z, \sigma) \times (K^{\mathbb{Z}}, \sigma)$, where Z is a Markov subgroup of $(G/K)^{\mathbb{Z}}$. The conjugacy is obtained by the correspondence on the symbol level between G and $G/K \times K$ and by observing that $(G^{\mathbb{Z}}, \sigma)$ is conjugate to $((G/K \times K)^{\mathbb{Z}}, \sigma)$ which is conjugate to $((G/K)^{\mathbb{Z}}, \sigma) \times (K^{\mathbb{Z}}, \sigma)$. The Markov subgroup Z is contained in $(G/K)^{\mathbb{Z}}$. Now examine Z. If $f(e) = \{e\}$ in Z stop. If $f(e) \neq \{e\}$ then by construction the subgroup K in Z is $\{e\}$.

Now consider the case with $K = \{e\}$. We define a Markov subgroup Y of $(G/f(e))^{\mathbb{Z}}$ which is isomorphic to (X, σ). To do this define a one-block map from X into $(G/f(e))^{\mathbb{Z}}$ by g goes to $g \cdot f(e)$. The image of X is a Markov subgroup with the transition rule which stipulates $h \cdot f(e)$ can follow $g \cdot f(e)$ if and only if there is a g' in $g \cdot f(e)$ with $f(g') = h \cdot f(e)$. Denote the image of X by Y. The map from X to Y is invertible by a two-block map because $K = \{e\}$. If $h \cdot f(e)$ can be followed by $g \cdot f(e)$, the $g' \in g \cdot f(e)$ with $f(g') = h \cdot f(e)$ is unique. The map is an isomorphism.

To wind up the proof, we work by induction. Start with (X, σ) such that $f(e) \neq \{e\}$ and use the first of the two constructions above to get a conjugate system $(Z, \sigma) \times (K_1^{\mathbb{Z}}, \sigma)$. Either $f(e) = \{e\}$ in Z and we are done or in Z the subgroup $K = \{e\}$. In this case apply the second of the two constructions above to get (X_1, σ). We continue the process alternating the two constructions.

Finally we reach the case where X_k is a Markov subgroup of $G_k^{\mathbb{Z}}$ and $f(e) = e$. The result is that (X, σ) is conjugate to $(X_k, \sigma) \times (K_k^{\mathbb{Z}}, \sigma) \times \cdots \times (K_1^{\mathbb{Z}}, \sigma)$. This in turn is conjugate to $(G_k, \tau) \times (K^{\mathbb{Z}}, \sigma)$ with $K = K_k \times \cdots \times K_1$.   □

Exercises

1. Show that the transition rule defining the subshift of finite type in Example 6.3.1 is closed under addition
2. Compute the subgroup described in Example 6.3.2 for the Markov subgroup in Example 6.3.1.
3. Carry out the reductions used in the proof of Theorem 6.3.3 for the Markov subgroup in Example 6.3.1.
4. Show that the Markov subgroup in Example 6.3.1 is not isomorphic to any Markov group of the form $G^{\mathbb{Z}}$. Hint: Figure out the possibilities for such a G and then examine the order as group elements of the periodic points.

§6.4 Cellular Automata

This section is included to very briefly point out the relationship between cellular automata and the symbolic systems we have studied. Some of the standard questions about cellular automata will be stated in terms of symbolic dynamics.

A one-dimensional *cellular automaton* is a continuous, shift-commuting map from a subshift of finite type to itself. The most common definition of a one-dimensional cellular automaton is as a block map from a full shift to itself. But, Theorem 1.4.9 states that a map from one symbolic system to another is continuous and shift commuting if and only if it is a block map. And, there is no particular reason to restrict our attention to full shifts. An n-dimensional cellular automaton is a continuous shift-commuting map from an n-dimensional subshift of finite type to itself. An n-dimensional subshift of finite type is a subshift of $\{1, \ldots, m\}^{\mathbb{Z}^n}$ defined by specifying an allowable list of n-dimensional cubes and saying the subshift is the set of points where every appropriately sized subblock is an allowable cube. It is a straightforward generalization of the usual subshift of finite type.

Cellular automata are used as models in almost every imaginable field and there are many different types of questions asked concerning them. An important question is to find a cellular automaton "rule" which models a particular system. In the symbolic dynamics setting a cellular automaton is a dynamical system. We ask the usual questions about the dynamical behavior of classes of cellular automata or about individual cellular automata. Common questions are about entropy, mixing properties, periodic points, nonwandering points, invariant sets and the eventual image. Answering many of these questions is impossible in general. It has been shown that the many natural questions are undecidable.

We can apply some of the techniques we have developed to one-dimensional cellular automata. A cellular automaton is *reversible* if it is a homeomorphism. Theorem 2.1.14 explains how to decompose a reversible cellular automaton into a composition of elementary conjugacies. If a cellular automaton isn't a homeomorphism it is either onto or not. If the cellular automaton is onto then by Corollary 4.1.8 it is finite-to-one and by Theorem 6.2.21 it preserves the measure of maximal entropy. If a cellular automaton is not onto then it is uncountable-to-one on some point and its image is a sofic system by Theorem 6.1.2. By the same reasoning its image after any finite number of iterations is a sofic system. If φ is a cellular automaton acting on Σ_A then $\cap \varphi^n(\Sigma_A)$ is its *eventual image*. There are examples of cellular automata whose eventual images are not sofic systems.

6.4.1 Let $\varphi : \{0,1\}^{\mathbb{Z}} \to \{0,1\}^{\mathbb{Z}}$ be defined by $(\varphi(x))_i = x_i + x_{i+1} \mod 2$ as in Examples 1.4.7 and 4.1.2. This is the simplest nontrivial cellular automaton. It is onto, it is not forwardly expansive and the periodic points are dense (Exercise). But, it is not known how to describe the closed, shift and φ invariant subsets. It is also not known which shift invariant measures are also φ invariant.

This example illustrates that even in the simplest cellular automata understanding the dynamics can be very difficult.

§ 6.5 Channel Codes

Channel codes are used to send and store data efficiently. Often the physical constraints of a storage or transmission system impose constraints on the strings of 0's and 1's that can be efficiently stored or transmitted. Information is thought of as unconstrained strings of 0's and 1's and a code is a way of transforming the unconstrained strings of 0's and 1's into the constrained strings of symbols. In this section we will discuss an application of symbolic dynamics to the construction of channel codes. These codes are used in magnetic tape machines, magnetic disks and optical data storage disks. The method is applicable to the construction of run-length limited codes. These codes and the physical reasons behind their use are described in Example 1.2.8.

We first describe the problem in terms of symbolic dynamics and then return to the physical consequences. Let Σ_A be an irreducible subshift of finite type with entropy $\log N$, where N is an integer. Theorem 4.2.5 shows that we can apply a sequence of state-splittings to Σ_A to produce a conjugate subshift of finite type whose transition matrix has row sum N. Then we can label the edges of the new transition graph to produce a right-resolving factor map onto the full N-shift. This allows us to "code" between the N-shift and Σ_A.

Example 6.5.1 Consider the matrix

$$A = \begin{bmatrix} 1 & 1 & 1 \\ 1 & 0 & 0 \\ 1 & 0 & 0 \end{bmatrix}$$

from Exercise 4 of Section 4.2. Label the states a, b and c. Split the state a into two states to get the matrix

$$C = \begin{bmatrix} 1 & 1 & 0 & 0 \\ 0 & 0 & 1 & 1 \\ 1 & 1 & 0 & 0 \\ 1 & 1 & 0 & 0 \end{bmatrix}.$$

There is a one-block conjugacy from Σ_C to Σ_A with a two block inverse given by

$$[a, a] \rightarrow a_1 \quad [a, b], [a, c] \rightarrow a_2 \quad [b, x] \rightarrow b \quad [c, x] \rightarrow c_2.$$

Consider the transition graph of C and an edge labelling as in Figure 6.5.1 which produces a right-resolving factor map onto $\{0, 1\}^{\mathbb{Z}}$.

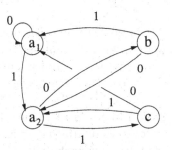

Figure 6.5.1

Fix a "start state", say a_1, in G_A. Now we "code" from finite strings in $\{0, 1\}^{\mathbb{Z}}$ to strings in Σ_A. Take a string from $\{0, 1\}^{\mathbb{Z}}$, say 01101001. Start at a_1 and follow the directions giving the path $a_1 a_1 a_2 c a_1 a_2 b a_2 c$ in G_C. This maps to *aaacaabac* in Σ_A. We have encoded an arbitrary string of eight 0's and 1's to a string of nine Σ_A constrained a's, b's and c's. The encoding mechanism; that is, the labelled graph of C is a *finite state automaton*. We decode the string with a two-block code simply by following the maps. First go from Σ_A to Σ_C and then to $\{0, 1\}^{\mathbb{Z}}$. It gives

$$aaacaabac \;\;\rightarrow\;\; a_1a_1a_2ca_1a_2ba_2c \;\;\rightarrow 01101001.$$

Note that the decoder is a block code while the encoder is dependent on the finite-state automaton. This has important consequences.

This example contains all the ideas needed to solve the coding problem. First we generalize Theorem 4.2.5.

Lemma 6.5.2 *Suppose A is an irreducible transition matrix with Perron value greater than or equal to a positive integer N. Then there is a positive integral vector v with $Av \geq Nv$.*

Proof. Let λ be the Perron value of A and $\lambda \geq N$. If $\lambda = N$ then, as we have already remarked in the proof of Theorem 4.2.5, there is such a v. Suppose $\lambda > N$. Let r be the right Perron eigenvector for A so $Ar = \lambda r > Nr$. The entries in the vectors Ar and Nr both change continuously as the entries of r vary. This means there is a vector w with positive rational entries arbitrarily close to r which satisfies $Aw > Nw$. Then we can multiply w by a positive integer to produce the vector v. □

Lemma 6.5.3 *Suppose A is an irreducible transition matrix with Perron value greater than or equal to a positive integer N. Then there exists a subshift of finite type Σ_C conjugate to Σ_A whose transition graph has every row sum greater than or equal to N.*

Proof. Use Lemma 6.5.2 to produce the positive integral vector v with $Av \geq Nv$. Now repeat the state splitting argument in the proof of Theorem 4.2.5 reducing the entries of v until arriving at a transition matrix C and the vector v' of all ones satisfying $Cv' \geq Nv'$. Then C is the desired matrix. □

Now we use Lemma 6.5.3 to construct channel codes. We assume that the data to be transmitted consists of arbitrary strings of 0's and 1's and the constrained strings consist of strings of symbols described by a subshift of finite type. The code will encode data blocks of length p into constrained blocks of length q and we say the code has *rate* p/q.

The code must be implemented by a device. We say a code is *practical* if the encoder is described by a finite state automaton and the decoder is described by a block code. The reason we require the decoder to be described by a block code is that errors may occur in storing or transmitting the constrained string. A finite state automaton may have an infinite memory. If a two strings differ by one bit and are coded by a finite state automaton then the encoded strings may differ in arbitrarily many places. This means if one error occurs in storing or transmitting the constrained string and the decoder is described by an arbitrary finite state automaton, the data string which was encoded and the data string obtained after decoding may differ in arbitrarily many places. It is best to avoid that situation. If the decoder is described by a block code it will not happen. If one error occurs in storing or transmitting the constrained string, the data string which was encoded and the data string obtained after decoding can differ in only a bounded number of places.

Putting these considerations together we arrive at the next theorem.

Theorem 6.5.4 *Suppose data consists of arbitrary strings of 0's and 1's, and a channel constraint is described by a transition matrix A with Perron value λ. Then there is a practical channel code of rate p/q for every $p/q \leq \log_2 \lambda$.*

Proof. Suppose A, λ and p/q are as described. So $2^p \leq \lambda^q$ and A^q is the transition matrix for σ^q acting on Σ_A. The symbols for this subshift of finite type are $W(A, q)$, the blocks of length q of Σ_A. The transition matrix A^q has Perron value λ^q. Apply Lemma 6.5.3 to find a subshift of finite type Σ_C conjugate to Σ_{A^q} and with each row sum in C greater than or equal to 2^p. Then the transition graph for C contains an irreducible subgraph whose transition matrix $\hat{C}$ has each row sum equal to 2^p. An edge coloring for the transition graph of $\hat{C}$ produces a right-resolving factor map from $\Sigma_{\hat{C}}$ onto the full-shift on 2^p symbols. As was illustrated in Example 6.5.1 this factor map describes the desired code.   □

We end with an example which produces a run-length limited code.

Example 6.5.5 The $(1, 3)$ Run-length limited sequences have at least one zero and no more than three zeros between successive occurrences of ones. The constraints are described by the graph

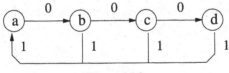

Figure 6.5.2

in Figure 6.5.2 which has transition matrix

$$A = \begin{bmatrix} 0 & 1 & 0 & 0 \\ 1 & 0 & 1 & 0 \\ 1 & 0 & 0 & 1 \\ 1 & 0 & 0 & 0 \end{bmatrix}.$$

There is a conjugacy between Σ_A and the shift acting on the $(1, 3)$ constrained sequences. It can be described by a one-block map from Σ_A to the sequences which sends the symbol a to 1 and the symbols b, c and d to 0. The inverse can be described by a three block map with memory two (Exercise 1).

The matrix A has characteristic polynomial $x^4 - x^2 - x - 1$. The Perron value of A is greater than the square root of two. By Theorem 6.5.4 we can code at rate 1/2. The first step is to compute

$$A^2 = \begin{bmatrix} 1 & 0 & 1 & 0 \\ 1 & 1 & 0 & 1 \\ 1 & 1 & 0 & 1 \\ 0 & 1 & 0 & 0 \end{bmatrix}.$$

The subshift of finite type Σ_{A^2} is canonically conjugate to the shift squared acting on Σ_A so there is a conjugacy between Σ_{A^2} and the shift squared acting on the $(1, 3)$ constrained sequences. This will show up in the decoder described below. The symbols for A^2 are still labelled a, b, c and d. Let

$$v = \begin{bmatrix} 1 \\ 2 \\ 1 \\ 1 \end{bmatrix}.$$

Then $A^2v \geq 2v$. We split the state labelled b in the graph of A^2 into two states by partitioning its followers into $\{b\}$ and $\{a, d\}$. This results in the transition matrix

$$C = \begin{bmatrix} 1 & 0 & 0 & 1 & 0 \\ 0 & 1 & 1 & 0 & 0 \\ 1 & 0 & 0 & 0 & 1 \\ 1 & 1 & 1 & 0 & 0 \\ 0 & 1 & 1 & 0 & 0 \end{bmatrix}.$$

Every row has sum greater than or equal to two. Label the edges of the graph for C by 0's and 1's discarding the edge from c to a. Let $\hat{C}$ denote the transition matrix after discarding the edge. This produces the labelled transition graph in Figure 6.5.3.

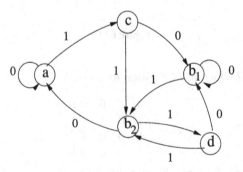

Figure 6.5.3

The code is now defined. It needs only to be unraveled. The map from $\Sigma_{\hat{C}}$ into the $(1,3)$ constrained sequences is a one-block conjugacy described by

$$a \rightarrow 10 \quad b_1 \rightarrow 01 \quad b_2 \rightarrow 00 \quad c \rightarrow 01 \quad d \rightarrow 01.$$

The encoder is described by the labelled graph in Figure 6.5.3. In table form it is.

		0	1
a	$\rightarrow$	$10/a$	$10/c$
b_1	$\rightarrow$	$01/b_1$	$01/b_2$
b_2	$\rightarrow$	$00/a$	$00/d$
c	$\rightarrow$	$01/b_1$	$01/b_2$
d	$\rightarrow$	$01/b_1$	$01/b_2$

The encoder works by starting in state a and looking at a data string. It looks at the first data bit, say 1, then it outputs the two bits 10 of constrained information and goes to state c. Then it looks at the second data bit, say 0, and outputs the two bits 01 of constrained information and goes to state b_1. It continues until the data string is encoded. In practice the states b_1, c and d can be combined to one state.

The decoder is described by the conjugacy which goes from the $(1,3)$ constrained data strings in the image of $\Sigma_{\hat{C}}$ to $\Sigma_{\hat{C}}$ followed by the right-resolving factor map from $\Sigma_{\hat{C}}$ onto the two-shift. This composition is given by the decoder table.

aa	$.101x$	ac	$.100x$
$b_1 b_1$	$.0101$	$b_1 b_2$	$.0100$
$b_2 a$	$.001x$	$b_2 d$	$.000x$
cb_2	$.0101$	cb_2	$.0100$
db_1	$.0101$	db_2	$.0100$
	0		1

In the table x means either 0 or 1 and the first column of constrained bits decodes to 0 while the constrained bits in the second column decode to 1. In practice the rows corresponding to pairs of symbols beginning with c and d are redundant.

The encoder and decoder together describe the code. Both can be read off from the graphs in Figures 6.5.2 and 6.5.3.

Exercises

1. Show that the map described in Example 6.5.5 from the subshift of finite type Σ_A to the $(1,3)$ constrained sequences is a conjugacy by producing the inverse map.

2. Show that the labelled graph in Figure 6.5.3 produces the encoder and decoder described at the end of Example 6.5.5.

3. Use the method followed in Example 6.5.5 to produce a $(2,7)$ run-length limited code with rate 1/2.

Notes

Section 6.1

Sofic systems were introduced into dynamics by B. Weiss [We] in 1973. There were two motivations for the work. One was to produce examples of dynamical systems with a unique invariant probability measure with maximal entropy. See Section 6.2 for a discussion. The other motivation was to characterize the images of subshifts of finite type under block maps. He used the word *sofic* to describe the finite nature of the description used for a sofic system. In that paper Weiss proved Theorem 6.1.2 where he gave three characterizations of sofic systems; namely: a sofic system is the image of a subshift of finite type under a block map, a sofic system has a finite number of distinct follower and predecessor sets and a sofic system can be defined by a finite semi-group.

The set of words $\mathcal{W}(S)$ for a sofic system is a *regular language*. A regular language is a set of words which can be recognized by a *finite state automaton*. A finite state automaton is defined by a directed graph with the edges labelled by inputs and outputs. Examples of finite state automata can be found in Section 6.5. Finite state automata and regular languages have been studied since the 1940's by people interested in language and communication theory. See [HU] for an introduction to languages and automata.

In 1975 [CP2], [CP3] E. Coven and M. Paul began an investigation of sofic systems. They showed every sofic system is the boundedly finite-to-one image of a subshift of finite type with the same entropy, Observation 6.1.8. They also characterized the sofic systems with a unique invariant probability measure with maximal entropy. At about the same time R. Fischer [Fi1], [Fi2] also studied sofic systems. He was interested in the existence of maximal measures and used graphs to investigate the problem. He pointed out the connection with language and automata theory, and used the *minimal Shannon graph* to describe a sofic system. Using this description he produced The *Fischer covers* and proved Theorem 6.1.6.

Theorem 6.1.9 states that the zeta function of a sofic system is a rational function. This is a special case of a theorem due to A. Manning [Mg] in 1971 where he showed that all Axiom A diffeomorphisms have a rational zeta function. D. Fried [Fd] in 1987 defined a class of dynamical systems called *finitely presented dynamical systems*. They are the dynamical systems which are expansive images of subshifts of finite type. This class includes the sofic systems, Axiom A diffeomorphisms and psuedo-Anosov diffeomorphisms. Fried proved Theorem 6.1.9 in this general setting.

W. Krieger in the early 1980's [Kg6], [Kg7] showed that the Fischer covers of a sofic system are invariants of topological conjugacy, Theorem 6.1.12. He actually proved more than just this – see Exercise 9. The proof given here using

Observations 6.1.10 and 6.1.11 is due to M. Boyle, B. Kitchens and B. Marcus [BKM].

The core matrix for a sofic system was defined by M. Nasu in 1985 [Nu3]. He defined the core matrix and proved that the Jordan form away from zero of the core matrix is a conjugacy invariant, Corollary 6.1.19. Then he generalized Theorem 4.2.8 to show that the Jordan form away from zero provides an invariant for finite-to-one factor maps between irreducible sofic systems. The construction used here relies on the Lattice Construction of Chapter 4 and is different from the one Nasu used. It is the approach used by S. Williams [Wm] to give an alternate construction of the core matrix and a proof of Theorem 6.1.18.

Section 6.2

Markov measures have been studied in probability theory and statistical mechanics since the turn of the century. There they are called finite state Markov chains. For an introduction to and a history of finite state Markov chains see Feller [Fe].

The concept of entropy is derived from statistical mechanics. It was first used in the mid-ninetheenth century . The function defining the entropy of a partition as a measure of its ability to help predict the outcome of an event also dates to the mid-nineteenth century [Kh]. C. Shannon defined the entropy of a Markov measure in his 1948 paper [S], basing his definition on the notion of conditional entropy. His idea was to find a measure of the information content of an information source. In 1958 A. Kolmogorov [Ko] defined the entropy of a measure preserving transformation as an invariant for measure-theoretic isomorphisms. In 1959 Y. Sinai [Si2] simplified Kolmogorov's work and proved the generator theorem. For a general introduction to entropy theory see [Pe] or [Wa].

C. Shannon in his 1948 paper [S] studied subshifts of finite type and Markov measures as models of communication channels. He used directed graphs to represent subshifts of finite type and applied the Perron-Frobenius Theorem to compute the *capacity* of the channel. We call the capacity the topological entropy of the subshift of finite type. Then he considered a Markov measure on a subshift of finite type as a statistical model of the output of an information source which is to be transmitted through a channel with a fixed capacity. He defined the entropy of the Markov measure as a measure of the information content of the source. He did this using conditional entropies on the graph representing the source. Next he proved his noiseless coding theorem. This can be stated as follows: Let an information channel be represented by a directed graph, so that it corresponds to a subshift of finite type Σ_A. It has a capacity which is to us the topological entropy $h(\Sigma_A)$. If a source will transmit through this channel it can be represented by a Markov measure on the graph of A. If the measure is $\mu_{(p,P)}$ it is defined by the pair (p, P). The source has an entropy. To us this is the measure-theoretic entropy $h_{\mu_{(p,P)}}(\Sigma_A)$. Then using this source and this channel it is possible to transmit through the channel at any *rate* up to $h_{\mu_{(p,P)}}(\Sigma_A)/h(\Sigma_A)$. Rate means that the bit streams of length ℓ emitted by the source can be coded into strings of length *rate* $\times$ ℓ of bits acceptable to the channel. In Section 6.5 we precisely formulated and proved a version of this theorem. Also in Shannon's paper is a formula for the maximal measure and he showed that its entropy is greater than the entropy of any other Markov measure.

W. Parry approached these questions independently of Shannon and from a very different perspective. In his 1964 paper [P1] he initiated the study of finite state Markov shifts from an ergodic theory viewpoint. He considered the collection of all shift-invariant probability measures on a subshift of finite type. Then he used the formulation of entropy for an invariant measure given by Kolmogorov [Ko]. He defined the maximal measure and proved its entropy is greater than that of any other invariant, Borel probability measure, Theorem 6.2.20. He defined the absolute entropy of a subshift of finite type to be the entropy of the maximal measure and showed it is the growth rate of the number of words allowed in a subshift of finite type. That is our definition of topological entropy. He also characterized n-step Markov measures in terms of conditional entropy, Corollary 6.2.18.

Theorem 6.2.20 is the first instance of the *Variational Principle*. The Variational Principle can be stated as follows. Let X be a compact metric space and f a homeomorphism from X to itself. Let $\mathcal{M}(X)$ denote the set of all f-invariant, Borel probability measures on X. The Variational Principle states that the topological entropy of f is equal to the supremum of the measure-theoretic entropies of the measures in $\mathcal{M}(X)$ with respect to f. There are examples where no measure realizes the supremum and examples where more than one measure realizes the supremum. The systems where exactly one measure realizes the supremum are called *intrinsically ergodic*. The general statement of the Variational Principle was proved in 1970. For a thorough treatment of this subject see [DGS] or [Wa].

Theorem 6.2.21 was proved by E. Coven and M. Paul in 1974 [CP1].

Section 6.3

Group automorphisms have been studied in ergodic theory since P. Halmos examined toral automorphisms for ergodicity in 1943 [Hs]. Toral automorphisms motivated the construction of Markov partitions and the use of symbolic dynamics as is illustrated by Example 1.2.7. Theorem 6.3.3 classifies one-dimensional Markov subgroups up to topological conjugacy. It was proven by B.Kitchens in 1987 and the approach used here follows [K2]. The dynamics of higher dimensional Markov subgroups are much more interesting than the dynamics in one dimension. There are many accesible and open problems concerning two-dimensional Markov subgroups. For an introduction see [Sd].

Section 6.4

J. von Neumann is generally considered to have introduced cellular automata in his 1966 paper "Theory of Self-reproducing Automata". In recent years cellular automata have been used in the theory of computation, physics, biology, mathematics and many other disciplines as simple models of complex systems which evolve over time. There is a huge literature on different aspects of cellular automata. The conference proceedings [Gu] and [FTW] contain good introductions to many of these ideas.

Section 6.5

Coding for information transfer and storage has been around since the telegraph was invented. The run-length limited codes described in this section have been used in magnetic storage devices since the late 1960's. Today essentially every magnetic and optical storage device uses some type of channel code. Run-length limited codes were originally found by a variety of add-hoc methods on a case by case basis. P. Franaszek [Fz] discovered a general method to construct such codes. The treatment here is from the 1983 paper by R. Adler, D. Coppersmith and M. Hassner [ACH]. They made the connection with symbolic dynamics, put the problem in the mathematical context presented here and gave the algorithm to prove Theorem 6.5.4.

References

[ACH]   R. Adler, D. Coppersmith and M. Hassner, *Algorithms for Sliding Block Codes*, IEEE Transactions on Information Theory no. 29 (1983), 5–22.

[BKM]   M. Boyle, B. Kitchens and B. Marcus, *A Note on Minimal Covers for Sofic Systems*, Proceedings of the American Mathematical Society **95** (1985), 403–411.

[CP1]   E. Coven and M. Paul, *Endomorphisms of Irreducible Subshifts of Finite Type*, Mathematical Systems Theory **8** (1974), 167–175.

[CP2]   E. Coven and M. Paul, *Sofic Systema*, Israel Journal of Mathematics **20** (1975) 165–177.

[CP3]   E. Coven and M. Paul, *Finite Procedures for Sofic Systems*, Monatshefte für Mathematik **83** (1977), 265–278.

[DGS] M. Denker, C. Grillenberger and K. Sigmund, *Ergodic Theory on Compact Spaces* Lecture Notes in Mathematics, 527, Springer-Verlag, (1976)

FTW] D. Farmer, T. Toffoli and S. Wolfram, editors, *Cellular Automata*, North-Holland, 1984.

Fe] W. Feller, *An Introduction to Probability Theory and Its Applications*, 3rd. edition, Wiley, 1968.

[Fi1] R. Fischer, *Sofic Systems and Graphs*, Monatshefte für Mathematik **80** 1975, 179–186.

[Fi2] R. Fischer, *Graphs and Symbolic Dynamics*, Proceedings of the Colloquia Mathematical Society János Bolyai, Information Theory **16** (1975) 229–244.

[Fz] P. Franaszek, *Sequence State Methods for Run-Length Limited Codes*, IBM Journal of Research and Development **14** (1970), 376–383.

[Fd] D. Fried, *Finitely Presented Dynamical Systems*, Ergodic Theory and Dynamical Systems **7** (1987), 489–507.

[Gu] H. Gutowitz, editor, *Cellular Automata: Theory and Experiment*, MIT Press, 1991.

[Hs] P. Halmos, *On Automorphisms of Compact Groups*, Bulletin of the American Mathematical Society **49** (1943), 619–624.

[HU] J.E Hopcroft and J.D. Ullman, *Frmal Languages and Their Relation to Automata*, Addison-Wesley Publishing Co., 1969.

[Kh] A. Khinchin, *Mathematical Foundations of Statistical Mechanics*, Dover Publications, 1949.

[K2] B. Kitchens, *Expansive Dynamics on Zer Dimensional Groups*, Ergodic Theory and Dynamical Systems **7** (1987), 249–261.

[Ko] A. Kolmogorov, *A New Metric Invariant for Transient Dynamical Systems*, Academiia Nauk SSSR, Doklady **119** (1958), 861–864. (Russian)

[Kg6] W. Krieger, *On Sofic Systems I*, Israel Journal of Mathematics **48** (1984), 305–330.

[Kg7] W. Krieger, *On Sofic Systems II*, Israel Journal of Mathematics **60** (1987), 167–176.

[Mg] A. Manning, *Axiom A Diffeomorphisms have Rational Zeta Functions*, Bulletin of the London Mathematical Society **3** (1971), 215–220.

[Nu3] M. Nasu, *An Invariant for Bounded-to-one Factor Maps Between Transitive Sofic Subshifts*, Ergodic Theory and Dynamical Systems **5** (1985), 85–105.

[Or] D. Ornstein, *Bernoulli Shifts with the Same Entropy are Isomorphic*, Advances in Mathematics no. 4 (1970), 337–352.

[P1] W. Parry, *Intrinsic Markov Chains*, Transactions of the American Mathematical Society **112** (1964), 55–66.

[Pe] K. Petersen, *Ergodic Theory*, Cambridge Press, 1983.

[Sd] K. Schmidt, *Dynamical Systems of Algebraic Origin*, Birkhäuser, 1995.

[S] C. Shannon, *A Mathematical Theory of Communication*, Bell System Technical Journal (1948), 379–473 and 623–656, reprinted in *The Mathematical Theory of Communication* by C. Shannon and W. Weaver, University of Illinois Press, 1963.

[Si2] Y. Sinai, *On the Concept of Entropy for a Dynamical System*, Academiia Nauk SSSR, Doklady **124** (1959), 768–771. (Russian)

[vN] J. von Neumann, *Theory of Self-reproducing Automata* (A.W. Burks, eds.), Univerity of Illinois Press, 1966.

[Wa]   P. Walters, *An Introduction to Ergodic Theory* GTM, Springer-Verlag, 1981.

[We]   B. Weiss, *Subshifts of Finite Type and Sofic Systems,* Monatshefte für Mathematik **77** (1973), 462–474.

[Wm]   S. Williams, *Lattice Invariants for Sofic Shifts,* Ergodic Theory and Dynamical Systems **11** (1991), 787–801.

Chapter 7.   Countable State Markov Shifts

This chapter is concerned with countable state Markov shifts. The first problem is to extend the Perron-Frobenius theory to nonnegative, countably infinite matrices. There are several difficulties. Countable matrices are classified by recurrence properties. There are three classes: positive recurrent, null recurrent and transient. The corresponding version of the Perron-Frobenius Theorem is successively weakened for each class. The necessary matrix theory is treated in the first section. The treatment is complete and there are a number of examples.

The second section begins the investigation of some of the basic dynamical properties of countable state Markov shifts. When the Markov shift has a countably infinite state space the Markov shift is noncompact. This makes it difficult to define topological entropy. Several possible definitions of topological entropy are discussed. No one definition has all the desired properties. After looking at several possible formulations for topological entropy we turn to the Variational Principle. There is no problem defining the measure-theoretic entropy for a shift-invariant, Borel probability measure. We prove the logarithm of the Perron value is equal to the supremum over the measure-theoretic entropies just as for subshifts of finite type. We end the chapter by proving that an irreducible, countable state Markov shift has a maximal measure if and only if the transition matrix is positive recurrent. In which case, the measure is unique and has the form of the maximal measure for a subshift of finite type.

§ 7.1 Perron-Frobenius Theory

We say a matrix is a *countable* matrix if its rows and columns are indexed by the same set and the set is either finite or countably infinite. In this section we extend the Perron-Frobenius theory of Section 1.3 to countable, real, nonnegative matrices. Let T denote a countable, real, nonnegative matrix. As in Section 1.3 associate with it a directed graph with labelled edges. The vertices of the graph are the indices for the rows and columns of the matrix. If $T_{ij} > 0$

there is an edge from vertex i to vertex j labelled T_{ij}. If $T_{ij} = 0$ there is no edge from vertex i to vertex j. Denote the graph by G_T. The *weight* of a path $[i_0, \ldots, i_\ell]$ in the graph G_T is the product $T_{i_0 i_1} \cdots T_{i_{\ell-1} i_\ell}$.

We extend the definition of irreducible to such matrices by saying such a matrix is *irreducible* if for every pair of vertices i and j there is an $\ell > 0$ with $(T^\ell)_{ij} > 0$. Equivalently, the matrix T is irreducible when the graph G_T is strongly connected. We also extend the notion of the *period* of a matrix to these matrices. Fix an index i and let $p(i) = \gcd\{\ell : (T^\ell)_{ii} > 0\}$. This is the *period of the index i*. When T is irreducible the period of every index is the same and is called the *period of T*. When T is irreducible the period is also the greatest common divisor of the lengths of the closed, directed paths in G_T. As before, a matrix with period one is said to be *aperiodic*.

We will parallel the development of Section 1.3 by first examining irreducible and aperiodic matrices and then extending the results to arbitrary irreducible matrices.

Lemma 7.1.1 *Let T be a countable, real, nonnegative, irreducible and aperiodic matrix. For a fixed state i:*

(i)   *there exists a k so that $(T^n)_{ii} > 0$ for all $n \geq k$;*

(ii) $(T^{n+m})_{ii} \geq (T^n)_{ii} \cdot (T^m)_{ii}$;

(iii) $\lim_{n \to \infty} \sqrt[n]{(T^n)_{ii}}$ *exists and equals* $\sup_n \sqrt[n]{(T^n)_{ii}}$;

(iv) *if $\lambda = \lim \sqrt[n]{(T^n)_{ii}}$ then $(T^n)_{ii}/\lambda^n \leq 1$ for all n.*

Proof. (i) If T is as supposed its graph will have a finite subgraph containing i that is irreducible and aperiodic. The statement holds for the finite subgraph so it holds for T. Statement (ii) is clear just as it is for finite matrices. To prove (iii) first suppose $\limsup \sqrt[n]{(T^n)_{ii}}$ is finite. Call the $\limsup \lambda$. For $\varepsilon > 0$ there is a q with $\sqrt[q]{(T^q)_{ii}} > \lambda - \varepsilon$. Now use (i) to find a k so that $(T^r)_{ii} > 0$ for all $r \geq kq$. For any $n \geq kp$ we write $n = pq + r$ with $kp \leq r < (k+1)p$. Then

$$\sqrt[n]{(T^n)_{ii}} \geq \sqrt[\frac{p}{pq+r}]{(T^q)_{ii}} \cdot \sqrt[n]{(T)_{ii}^r} \geq (1 - \varepsilon)^2$$

for all sufficiently large n. This means that the $\liminf$ of the sequence is equal to the $\limsup$ and so λ is the limit. To finish the proof when λ is finite suppose, there is a q with $\sqrt[q]{(T^q)_{ii}} = \lambda^* > \lambda$. Then for every $n \geq 1$ $\sqrt[nq]{(T^{nq})_{ii}} \geq \lambda^*$ which is a contradiction. The same type argument works if the $\limsup$ is infinite (Exercise 1). Statement (iv) follows immediately from (iii).   □

We define λ to be the *Perron value* of T. For the rest of this section we assume λ is finite. Note that the irreducibility of T implies $\lambda = \lim_{n \to \infty} \sqrt[n]{(T^n)_{ij}}$

for any pair of states i and j, and that this limit gives the usual Perron value when T is a finite, nonnegative, irreducible and aperiodic matrix.

We will make use of several generating functions. For any pair of states i and j, define

$$t_{ij}(0) = \delta_{ij}, \quad t_{ij}(1) = T_{ij}, \quad \text{and} \quad t_{ij}(n) = (T^n)_{ij}.$$

The first type of generating functions are defined by

$$T_{ij}(z) = \sum_{n=0}^{\infty} t_{ij}(n)z^n.$$

For the next type of generating functions we define coefficients inductively. Let

$$\ell_{ij}(0) = 0, \quad \ell_{ij}(1) = T_{ij} \quad \text{and} \quad \ell_{ij}(n+1) = \sum_{r \neq i} \ell_{ir}(n)t_{rj}.$$

The coefficient $\ell_{ij}(n)$ is the sum of the weights of the paths that go from i to j in n steps without returning to i at any time prior to n. Define the functions

$$L_{ij}(z) = \sum_{n=1}^{\infty} \ell_{ij}(n)z^n.$$

Similarly, define coefficients

$$r_{ij}(0) = 0, \quad r_{ij}(1) = T_{ij} \quad \text{and} \quad r_{ij}(n+1) = \sum_{r \neq j} t_{ir}r_{rj}(n).$$

The coefficient $r_{ij}(n)$ represents the sum of the weights of the paths that go from i to j in n steps without hitting j at any time prior to n. Define the functions

$$R_{ij}(z) = \sum_{n=1}^{\infty} r_{ij}(n)z^n.$$

Observe that the radius of convergence of every $T_{ij}(z)$ is $1/\lambda$.

Next we make two crucial definitions. The matrix T is *recurrent* if $T_{ii}(1/\lambda) = \infty$. The matrix T is *transient* if $T_{ii}(1/\lambda) < \infty$. Lemma 7.1.8 will show these definitions do not depend on the choice of i. There is a further refinement of the recurrent matrices. The matrix T is *positive recurrent* if

$$\sum_{n=1}^{\infty} n\ell_{ii}(n)/\lambda^n < \infty$$

and *null recurrent* if

$$\sum_{n=1}^{\infty} n\ell_{ii}(n)/\lambda^n = \infty.$$

Lemma 7.1.13 will show this definition is also independent of the choice of i. Theorem 7.1.3 (d) and Remark 7.1.5 show a finite, nonnegative, irreducible and aperiodic matrix is positive recurrent.

The terminology comes from probability theory. Recall from Section 6.2 that a nonnegative matrix is *stochastic* if every row has sum one. We use the same definition in this setting. If $T = P$ is a stochastic matrix we may think of it as defining a probabilistic walk on the graph of T. The probability of going from state i to state j in one step is P_{ij}. The probability of going from state i to state j in n steps is $(P^n)_{ij}$ and the probability of going from state i to state j in n steps without returning to i in the meantime is $\ell_{ij}(n)$. Then

$$T_{ii}(1) = \sum_{n=0}^{\infty} t_{ii}(n) = \sum_{n=0}^{\infty} (P^n)_{ii}$$

is the expected number of returns to i of a walk beginning at i.

$$L_{ii}(1) = \sum_{n=1}^{\infty} \ell_{ii}(n)$$

is the probability that a walk, which begins at i, returns to i.

$$L'_{ii}(1) = \sum_{n=1}^{\infty} n\ell_{ii}(n)$$

is the expected return time to i. The matrix P is transient if the expected number of returns to i of a walk beginning at i is finite and recurrent if it is infinite. We will see that this is the same as saying that the matrix is transient if the probability of returning to i is less than one and recurrent if the probability of returning to i is one. A recurrent matrix P is null recurrent if the expected time of return to i of a walk beginning at i is infinite and positive recurrent if it is finite.

Remark 7.1.2 The terminology here is slightly nonstandard. Transient, null recurrent, and positive recurrent are traditionally used when talking about stochastic matrices where $T_{ii}(z)$, $L_{ii}(z)$ and $L'_{ii}(z)$ are always evaluated at one to determine recurrence properties. D. Vere-Jones used the terms R-transitive, R-null recurrent, and R-positive recurrent where $T_{ii}(z)$, $L_{ii}(z)$, and $L'_{ii}(z)$ are evaluated at $R = 1/\lambda$ to determine recurrence properties. The confusion may come about when T is stochastic and $T_{ii}(1) < \infty$. When T is stochastic and $T_{ii}(1) < \infty$ we will say that T is *probabilisticly transitive*. In this case, λ may be equal to one or it may be less than one. The matrix T may be transient or recurrent in the sense that we have defined. This is illustrated in Example 7.1.31 (iii). When T is stochastic and $T_{ii}(1) = \infty$, $\lambda = 1$ and there is no ambiguity.

Roughly speaking the three classes are based on how well the usual Perron-Frobenius Theorem can be extended. The differences will also be mirrored in the dynamics of the Markov shifts they define. Next we state the two most important theorems about recurrent matrices. The proofs will be carried out in a sequence of lemmas and remarks. It requires a good deal of work. After stating the theorems for recurrent matrices the weakened versions for transient matrices is explained in Remark 7.1.5.

Generalized Perron-Frobenius Theorem 7.1.3 Suppose T is a countable, nonnegative matrix. Further, suppose it is irreducible, aperiodic and recurrent. Then there exists a Perron value $\lambda > 0$ (assumed to be finite) such that:

(a) $\lambda = \lim\limits_{n\to\infty} \sqrt[n]{(T^n)_{ij}}$ for any pair of indices i and j, so that $1/\lambda$ is the radius

of convergence of the power series $T_{ij}(z) = \sum\limits_{n=0}^{\infty} t_{ij}(n)z^n$;

(b) λ has strictly positive left and right eigenvectors;

(c) the eigenvectors are unique up to constant multiples;

(d) let ℓ, r be the left and right eigenvectors for λ then $\ell \cdot r < \infty$ if and only if T is positive recurrent;

(e) if $0 \le S \le T$ and β is the Perron value for S then $\beta \le \lambda$, if S is recurrent then there is equality if and only if $S = T$;

(f) $\lim\limits_{n\to\infty} T^n/\lambda^n = 0$ if T is null recurrent, and $\lim\limits_{n\to\infty} T^n/\lambda^n = r\ell$, normalized so that $\ell r = 1$, if T is positive recurrent.

Finite Approximation Theorem 7.1.4 Let T be a countable, nonfinite, nonnegative matrix. Assume T is irreducible, aperiodic, and recurrent.

(a) If λ is the Perron value for T, $\lambda = \sup\{\lambda(A) : A$ is a finite, irreducible and aperiodic submatrix of T, $\lambda(A)$ is A's Perron value $\}$.

(b) Let ℓ and r be the left and right eigenvectors of T for the Perron value λ, normalized so that for some index i, $\ell_i = r_i = 1$. Let $\{A_n\}$ be an increasing family of finite irreducible, aperiodic submatrices of T that converge to T. Let $\ell^{(n)}, r^{(n)}$ be the left and right eigenvectors corresponding to their Perron values, normalized so that $r_i^{(n)} = \ell_i^{(n)} = 1$. Then $\lim \ell_j^{(n)} = \ell_j$ and $\lim r_j^{(n)} = r_j$ for all j in the index set.

The statements in Theorem 7.1.3 are proved in the following collection of lemmas and remarks. The content of each will be found as follows. Statement (a) is the definition of the Perron value for such matrices. Statements (b) and (c) are explained in Remarks 7.1.12 and 7.1.17. Statement (d) is Lemma 7.1.14. Lemma 7.1.23 is statement (e) and statement (f) follows from Lemmas 7.1.19

through 7.1.22. In Theorem 7.1.4 statement (a) is Lemma 7.1.33 and statement (b) is Lemma 7.1.34.

Remark 7.1.5 For a transient matrix Theorem 7.1.3 breaks down as follows. Statement (a) is still the definition of the Perron value but statements (b) and (c) may not be true. The situation is discussed in Remark 7.1.12. When the matrix is transitive there do not exist left and right eigenvectors whose inner product is finite. This is shown in Lemma 7.1.16. The equality part of statement (e) is false and is demonstrated by Lemma 7.1.23. For a transitive matrix $\lim_{n\to\infty} T^n/\lambda^n = 0$ as is shown by Lemma 7.1.19. Statement (a) of Theorem 7.1.4 is Lemma 7.1.33 but statement (b) does not apply.

We begin the discussion with several combinatorial lemmas.

Lemma 7.1.6 *Suppose T is a countable, nonnegative matrix. Further, suppose it is irreducible and aperiodic.*

(i) $L_{ii}(z) = R_{ii}(z)$.

(ii)   *Let $\ell_{ii}^{(k)}(n)$ be n-th coefficient of $(L_{ii}(z))^k = L_{ii}^{(k)}(z)$ and $r_{ii}^{(k)}(n)$ be the same for $R_{ii}(z)$. Each represents the sum of the weights of the paths of length n from i to i with exactly $k-1$ i's in between.*

(iii) $t_{ij}(n) = \displaystyle\sum_{s=1}^{n} t_{ii}(n-s)\ell_{ij}(s) = \sum_{s=0}^{n-1} r_{ij}(n-s)t_{jj}(s)$.

(iv) $T_{ii}(z) = 1/(1 - L_{ii}(z)) = 1/(1 - R_{ii}(z))$, for $|z| < 1/\lambda$.

(v) $L_{ii}(1/\lambda) = R_{ii}(1/\lambda) \leq 1$.

(vi)   *for $i \neq j$, $T_{ij}(z) = T_{ii}(z)L_{ij}(z) = R_{ij}(z)T_{jj}(z)$, for $|z| < 1/\lambda$.*

(vii) $T_{ij}(z) = z\displaystyle\sum_{r} T_{ir}(z)t_{rj} + \delta_{ij}$,

$T_{ij}(z) = z\displaystyle\sum_{r} t_{ir}T_{rj}(z) + \delta_{ij}$, for $|z| < 1/\lambda$.

(viii) $L_{ij}(z) = z\displaystyle\sum_{r} L_{ir}(z)t_{rj} + z \cdot t_{ij}(1 - L_{ii}(z))$,

$R_{ij}(z) = z\displaystyle\sum_{r} t_{ir}R_{rj}(z) + z \cdot t_{ij}(1 - R_{ii}(z))$, for $|z| < 1/\lambda$.

Proof. (i) The definitions of $\ell_{ii}(n)$ and $r_{ii}(n)$ show that $\ell_{ii}(n) = r_{ii}(n)$.
(ii) By the formula for coefficients of products of power series we have that

$$\ell_{ii}^{(2)}(1) = \ell_{ii}^{(2)}(0) = 0 \text{ and } \ell_{ii}^{(2)}(n) = \sum_{s=1}^{n-1} \ell_{ii}(n-s)\ell_{ii}(s) \text{ for } n \geq 2.$$

Then by induction

$$\ell_{ii}^{(k)}(n) = 0 \text{ for } n < k \text{ and } \ell_{ii}^{(k)}(n) = \sum_{s=1}^{n-1} \ell_{ii}^{(k-1)}(n-s)\ell_{ii}(s) \text{ for } n \geq k.$$

(iii) The first equality follows from grouping paths by the last occurrence of i and the second from grouping paths by the first occurrence of j.

(iv) The 0th coefficient of $T_{ii}(z)L_{ii}(z)$ is zero and for $n \geq 1$ the nth coefficient is $\sum_{s=1}^{n} t_{ii}(n-s)\ell_{ii}(s)$. Now apply (iii) to see that this is $t_{ii}(n)$ which means that $T_{ii}(z) = 1 + T_{ii}(z)L_{ii}(z)$.

(v) This follows from (iv) which implies that $|L_{ii}(z)| < 1$ for $|z| < 1/\lambda$.

(vi) This follows from the same argument as (iv) but taking into account that $t_{ij}(0) = \delta_{ij}$.

(vii)

$$T_{ij}(z) = \delta_{ij} + \sum_{n=1}^{\infty} t_{ij}(n)z^n$$

$$= \delta_{ij} + \sum_{n=1}^{\infty} \left(\sum_{r} t_{ir}(n-1)t_{rj} \right) z^n$$

$$= \sum_{r} \left(\sum_{n=1}^{\infty} t_{ir}(n-1)z^{n-1} \right) t_{rj}z + \delta_{ij}$$

(viii)

$$L_{ij}(z) = \sum_{n=1}^{\infty} \ell_{ij}(n)z^n$$

$$= t_{ij}z + \sum_{n=2}^{\infty} \ell_{ij}(n)z^n$$

$$= t_{ij}z + \sum_{n=2}^{\infty} \left(\sum_{r \neq i} \ell_{ir}(n-1)t_{rj} \right) z^n$$

$$= \sum_{r \neq i} \left(\sum_{n=2}^{\infty} \ell_{ir}(n-1)z^{n-1} \right) t_{rj}z + t_{ij}z$$

$$= z \sum_{r \neq i} L_{ir}(z)t_{rj} + t_{ij}z$$

$$R_{ij}(z) = t_{ij}z + \sum_{n=2}^{\infty} r_{ij}(n)z^n$$

$$= t_{ij}z + \sum_{n=2}^{\infty} \left(\sum_{r \neq j} t_{ir}r_{rj}(n-1) \right) z^n$$

$$= z \sum_{r \neq j} t_{ir}R_{rj}(z) + t_{ij}z. \qquad \square$$

Lemma 7.1.7 *Suppose T is a countable, nonnegative matrix. Further, suppose it is irreducible and aperiodic. Then $L_{ij}(1/\lambda)$ and $R_{ij}(1/\lambda)$ are finite for all i and j.*

Proof. This follows from Lemma 7.1.6 (v), (viii) and the irreducibility of T.

$$L_{ii}(z) = z \sum_r L_{ir}(z)t_{ri} + z \cdot t_{ii}(1 - L_{ii}(z)), \text{ for } |z| < 1/\lambda$$

and so

$$1/\lambda \sum_r L_{ir}(1/\lambda)t_{ri} + (1/\lambda)t_{ii}(1 - L_{ii}(1/\lambda)) = L_{ii}(1/\lambda) \le 1.$$

This means $L_{ij}(1/\lambda)$ is finite for all j with $t_{ji} \neq 0$. Then repeat the argument on these $L_{ij}(z)$ to see that $L_{ij'}(1/\lambda)$ is finite for all j' where there exists a j with $t_{j'j}t_{ji} \neq 0$. By the irreducibility of T, every $L_{ij}(1/\lambda)$ is finite.   □

Lemma 7.1.8 *Suppose T is a countable, nonnegative, irreducible and aperiodic matrix. Then*

(i)   *either every $T_{ij}(1/\lambda)$ is finite or every $T_{ij}(1/\lambda)$ is infinite, and*
(ii)   *either every $L_{ii}(1/\lambda)$ is less than one or every $L_{ii}(1/\lambda)$ is equal to one.*

Proof. From Lemma 7.1.6 (vii) we have

$$T_{ij}(z) = z \sum_r T_{ir}(z)t_{rj} + \delta_{ij}.$$

Using the irreducibility of T and applying the same reasoning as in the proof of Lemma 7.1.7 to this equation we see that either every $T_{ij}(1/\lambda)$ is finite or every one is infinite.
(iii) This follows by combining (i) and Lemma 7.1.7 (iv).   □

Next we define a collection of vectors. For each i define a row vector

$$\ell^{(i)} = (L_{i1}(1/\lambda), L_{i2}(1/\lambda), \dots, L_{ij}(1/\lambda), \dots).$$

For each j define a column vector

$$r^{(j)} = \begin{bmatrix} R_{1j}(1/\lambda) \\ R_{2j}(1/\lambda) \\ \vdots \\ R_{ij}(1/\lambda) \\ \vdots \end{bmatrix}.$$

These are well defined by Lemma 7.1.7.

Lemma 7.1.9 *Let T be a countable, nonnegative, irreducible and aperiodic matrix.*

(i)   *If T is recurrent*

$$\ell^{(i)}T = \lambda\ell^{(i)} \text{ for all } i, \text{ and}$$

$$Tr^{(j)} = \lambda r^{(j)} \text{ for all } j.$$

(ii) *If T is transient*

$$\ell^{(i)}T \leq \lambda\ell^{(i)} \text{ for each } i, \text{ with inequality occurring in some entry, and}$$

$$Tr^{(j)} \leq \lambda r^{(j)} \text{ for each } j, \text{ with inequality occurring in some entry.}$$

Proof. (i) This is a result of combining Lemmas 7.1.6 (viii) and 7.1.8 (ii). By Lemma 7.1.6 (viii) we see

$$L_{ij}(1/\lambda) = 1/\lambda \sum_r L_{ir}(1/\lambda)t_{rj} + (1/\lambda)t_{ij}(1 - L_{ii}(1/\lambda)).$$

Lemma 7.1.8 (ii) states that when T is recurrent $L_{ii}(1/\lambda) = 1$ for all i and this gives the desired equality. The equations for $r^{(j)}$ follow similarly.
(ii) Here, we use Lemma 7.1.6 (viii) again but now $L_{ii}(1/\lambda) < 1$. There is inequality for those $L_{sj}(1/\lambda)$ for which $t_{sj} > 0$.   □

Lemma 7.1.10 *If $x \geq 0$ is a nonzero row vector satisfying $xT \leq \lambda x$ or $y \geq 0$ is a nonzero column vector satisfying $Ty \leq \lambda y$ then:*

(i)   *x (or y) is strictly positive;*
(ii)   *if x is normalized so that $x_i = 1$ then $x_r \geq L_{ir}(1/\lambda)$, for all r;*
(iii) *if y is normalized so that $y_j = 1$ then $y_s \geq R_{sj}(1/\lambda)$, for all s.*

Proof. (i) This is a consequence of irreducibility. For each i and j there is an n so that $(T^n)_{ij} > 0$.
(ii) Assume $x_i = 1$, then $x_i \geq L_{ii}(1/\lambda)$ by Lemma 7.1.6 (v). For $r \neq 1$ we want to see that

$$x_r \geq L_{ir}(1/\lambda) = \sum_{n=1}^{\infty} \ell_{ir}(n)/\lambda^n.$$

We will show by induction that

$$x_r \geq \sum_{n=1}^{m} \ell_{ir}(n)/\lambda^n, \text{ for all } m.$$

For $m = 1$,

$$x_r \geq 1/\lambda \left(\sum_s x_s t_{sr} \right) \text{ (by assumption)} \geq (1/\lambda) \cdot x_i t_{ir} = (1/\lambda)\ell_{ir}(1)$$

since $x_i = 1$ and $t_{ir} = \ell_{ir}(1)$. Assume it is true up to m. Then

$$x_r \geq 1/\lambda \sum_s x_s t_{sr} = 1/\lambda \left(\sum_{s \neq i} x_s t_{sr} + t_{ir} \right)$$

$$\geq 1/\lambda \left(\sum_{s \neq i} \left(\sum_{n=1}^{m} \ell_{is}(n)/\lambda^n \right) t_{sr} + t_{ir} \right)$$

$$= 1/\lambda \left(\sum_{n=1}^{m} \left(\sum_{s \neq i} \ell_{is}(n) t_{sr} \right) 1/\lambda^n + t_{ir} \right)$$

$$= 1/\lambda \left(\sum_{n=1}^{m} \ell_{ir}(n+1)/\lambda^n + \ell_{ir}(1) \right) = \sum_{n=1}^{m+1} \ell_{ir}(n)/\lambda^n.$$

(iii) This is the same argument as (ii).   □

Lemma 7.1.11 *If T is recurrent, let $\ell = \ell^{(1)}$ and $r = r^{(1)}$ so $\ell T = \lambda \ell$ and $Tr = \lambda r$.*

(i)   *If $x \geq 0$ is a nonzero row vector satisfying $xT \leq \lambda x$ (in particular every $\ell^{(i)}$), then x is a multiple of ℓ.*

(ii) *If $y \geq 0$ is a nonzero column vector satisfying $Ty \leq \lambda y$ (in particular every $r^{(j)}$), then y is a multiple of r.*

Proof. Lemma 7.1.9 shows that $\ell T = \lambda \ell$. Suppose $x \geq 0, x \neq 0$ and $xT \leq \lambda x$. Normalize x so that $x_1 = 1$. Then $(x - \ell) \geq 0$ by Lemma 7.1.10 (ii) and satisfies $(x - \ell)T \leq \lambda(x - \ell)$. But $(x - \ell)_1 = 0$, and so by lemma 7.1.10 (i), $(x - \ell) = 0$. The same holds for $y \geq 0$ satisfying $Ty \leq \lambda y$.   □

Remark 7.1.12 Lemmas 7.1.9, 7.1.10 and 7.1.11 say that when T is recurrent there exist unique strictly positive row and column vectors, ℓ and r, with $\ell T = \lambda \ell$ and $Tr = \lambda r$. Moreover, for recurrent T there does not exist a nonnegative (nonzero) row vector x with $xT \leq \lambda x$ and inequality occurring in some entry. Neither does there exist such a column vector y. When T is transient there exist strictly positive row and column vectors x, y with $xT \leq \lambda x$, $Ty \leq \lambda y$ and inequality occurring. A note of caution is that for transient T there may exist strictly positive x and y with $xT = \lambda x$ and $Ty = \lambda y$. This is demonstrated in Example 7.1.29.

Lemma 7.1.13 *Either $L'_{ij}(1/\lambda) = \sum_{n=1}^{\infty} n\ell_{ij}(n)/\lambda^{n-1}$ is finite for all i and j or it is infinite for all i and j.*

Proof. From Lemma 7.1.6 (viii) we can compute that

$$L'_{ij}(z) = L_{ij}(z) + \sum_{r \neq i} L'_{ir}(z) t_{ij}, \text{ for } |z| < 1/\lambda.$$

By Lemma 7.1.7 $L_{ij}(1/\lambda)$ is finite for all i and j. Then by the irreducibility of T and the reasoning used in the proofs of Lemmas 7.1.7 and 7.1.8 we see that either every $L'_{ij}(1/\lambda)$ is finite or every one is infinite.   □

For each i, define $\mu(i) = (1/\lambda)L'_{ii}(1/\lambda) = \sum_{n=1}^{\infty} n\ell_{ii}(n)/\lambda^n$. It is the *mean recurrence weight*. By definition a recurrent matrix is positive recurrent if every mean recurrent weight is finite and null recurrent if each is infinite.

The terminology once again comes from probability theory. When T is stochastic and recurrent, μ is called the *mean recurrence measure*. When T is stochastic it defines a probabilistic walk on the graph of T and when T is also recurrent $\mu(i)$ represents the expected return time to i of a walk beginning at i. When T is stochastic and transient, the vector μ no longer represents the expected return times. For transient T, $\mu(i)$ may be either finite or infinite, see Example 7.1.31 (ii) and (iii).

Lemma 7.1.14 *Suppose T is recurrent. Then T is positive recurrent if and only if there exists a pair of nonzero vectors $x, y \geq 0$ satisfying $xT = \lambda x$, $Ty = \lambda y$ with $x \cdot y$ finite.*

Proof. Suppose T is recurrent and there are $x \geq 0$ and $y \geq 0$, nonzero satisfying $xT = \lambda x$, $Ty = \lambda y$. By Lemma 7.1.11, x is a constant multiple of ℓ and y is a constant multiple of r. We need only consider $\ell \cdot r$ since multiplication by constants will not change whether the dot product is finite or infinite. So

$$x \cdot y = \sum_r L_{1r}(1/\lambda)R_{r1}(1/\lambda).$$

$$\sum_r L_{1r}(z)R_{r1}(z) = \sum_r \sum_{n=2}^{\infty} \left(\sum_{s=1}^{n-1} \ell_{1r}(n-s)r_{r1}(s) \right) z^n$$

$$= \sum_{n=2}^{\infty} \left(\sum_{s=1}^{n-1} \sum_r \ell_{1r}(n-s)r_{r1}(s) \right) z^n$$

$$= \sum_{n=2}^{\infty} \left(\sum_{s=1}^{n-1} (\ell_{11}(n) + \ell_{11}(n-s)\ell_{11}(s)) \right) z^n$$

$$= \sum_{n=2}^{\infty} \left((n-1)\ell_{11}(n) + \ell_{11}^{(2)}(n) \right) z^n$$

$$= \sum_{n=2}^{\infty} (n-1)\ell_{11}(n)z^n + \sum_{n=2}^{\infty} \ell_{11}^{(2)}(n)z^n$$

$$= z\, L'_{11}(z) - L_{11}(z) + (L_{11}(z))^2.$$

Consequently, $x \cdot y = (1/\lambda)L'_{11}(1/\lambda) = \mu(1)$ when T is recurrent.   □

The previous proofs have all been combinatorial. The next ones will be very different. They are much more analytic. Lemma 7.1.15 contains two observations that will be used in the next proofs.

Lemma 7.1.15

(i)   *Suppose that* $0 \le a_i$ *and* $\sum a_i = a < \infty$,

$$0 \le b_i^{(n)} \le B \text{ and } \lim b_i^{(n)} = b_i \text{, then } \lim_{n \to \infty} \Sigma a_i b_i^{(n)} = \Sigma a_i b_i.$$

(ii) *Suppose that* $0 \le a_i$ *and* $\sum a_i = \infty$

$$0 \le b_i^{(n)}, \lim_n b_i^{(n)} = b \text{ and } \Sigma a_i b_i^{(n)} \le B \text{ for all } n,$$

then $b = 0$.

Proof. (i) We have that $0 \le \Sigma a_i b_i^{(n)}$ and $\Sigma a_i b_i \le aB$, and since they are nonnegative term series they converge. Say $\Sigma a_i b_i^{(n)} = \ell(n)$ and $\Sigma a_i b_i = \ell$. Choose M so that

$$\ell - \sum_{i=1}^{M} a_i b_i < \varepsilon/3$$

and so that

$$\sum_{i=M+1}^{\infty} a_i b_i^{(n)} \le B \sum_{i=M+1}^{\infty} a_i < \varepsilon/3.$$

Then choose N so that for $n \ge N$ and $1 \le i \le M$, $|b_i - b_i^{(n)}| < \varepsilon/3a$. Then for $n \ge N$

$$|\ell - \ell(n)| \le |\ell - \sum_{i=1}^{M} a_i b_i| + \sum_{i=1}^{M} a_i |b_i - b_i^{(n)}| + \Sigma a_i b_i^{(n)} < \varepsilon.$$

(ii) Suppose $b \ne 0$. Choose M so that

$$\sum_{i=0}^{M} a_s > \frac{2B}{b}.$$

Then choose N so that for $n \ge N$ and $1 \le i \le M$, $b_i^{(n)} > b/2$. For $n \ge N$

$$\sum_{i=0}^{\infty} a_i b_i^{(n)} \ge \sum_{i=0}^{M} a_i b_i^{(n)} > \sum_{i=0}^{M} a_i b/2 > B,$$

a contradiction.   □

Lemma 7.1.16 *If T is transient there does not exist a pair of nonzero vectors $x, y \ge 0$ satisfying $xT = \lambda x$, $Ty = \lambda y$ with $x \cdot y$ finite.*

Proof. Suppose such a pair exists. By Lemma 7.1.10 they are strictly positive. Since

$$\lambda^n y_i = \sum_j (T^n)_{ij} y_j,$$

$$(T^n)_{ij} y_j / \lambda^n y_i \leq 1 \text{ for all } n.$$

Next we see that

$$x_j y_j = (xT^n/\lambda^n)_j y_j = \left(\sum_i x_i (T^n)_{ij}/\lambda^n \right) y_j$$

$$= \sum_i (x_i y_i) \left[(T^n)_{ij} y_j / \lambda y_i \right].$$

Now apply Lemma 7.1.15 (i) with $a_i = x_i y_i$, $b_i^{(n)} = (T^n)_{ij} y_j / \lambda^n y_i \leq 1$ by the first observation, and $b_i = 0$ since $\sum_n (T^n)_{ij}/\lambda^n$ converges. This says $x_j y_j = 0$, a contradiction. □

Remark 7.1.17 Lemmas 7.1.14 and 7.1.16 together say that a matrix T is positive recurrent if and only if there exists a pair of nonzero vectors $x, y \geq 0$ with $xT = \lambda x$, $Ty = \lambda y$ and $x \cdot y$ finite.

The next theorem is a special case of the Renewal Theorem of probability. The notes at the end of the chapter contain notes and references for this theorem.

Theorem 7.1.18 *If T is a countable, nonnegative, irreducible and aperiodic matrix with a finite Perron value λ, then:*

(i) $\lim_{n \to \infty} (T^n)_{ii}/\lambda^n = 0$ *for all i when T is transient;*

(ii) $\lim_{n \to \infty} (T^n)_{ii}/\lambda^n = 1/\mu(i)$ *for all i when T is recurrent.*

Proof. Case (i) is clear since $\sum_{n=0}^{\infty} (T^n)_{ii}/\lambda^n$ converges. The proof of case (ii) will be divided into 7 steps. Fix i and use the notation $t_{ii}(n), \ell_{ii}(n), \ell_{ii}^{(k)}(n)$ as previously established.

Step 1.
There exists a k so that for $m \geq k$, $t_{ii}(m) > 0$ and for each such m there is an $m_1 \leq m$ so that $\ell_{ii}^{(m_1)}(m) > 0$. Lemma 7.1.1 (i) tells us that such a k exists. Then if $t_{ii}(m) = (T^m)_{ii} > 0$, choose an arbitrary path from i to i of length m and let m_1 be the number of times it returns to i.

Step 2.

Let $L = \limsup t_{ii}(n)/\lambda^n$. Then there is a subsequence $\{n^*\}$ of $\mathbb{N}$ so that for $s = 0, 1, 2, \ldots$

$$\lim_{n^*} \frac{t_{ii}(n^* - s)}{\lambda^{(n^* - s)}} = L.$$

To prove this we will construct a sequence $\{n'\}$ so that $\lim_{n'} t_{ii}(n' - m)/\lambda^{(n' - m)} = L$ for all $m \geq k$. and then take $\{n^*\}$ to be $\{n' - k\}$. Begin with a subsequence $\{n'\}$ with $\lim_{n'} t_{ii}(n')/\lambda^{n'} = L$. We will prove that for all $m \geq k$, $\lim_{n'} t_{ii}(n' - m)/\lambda^{(n' - m)} = L$. Using step 1, fix $m \geq k$ and choose $m_1 \leq m$ so that $\ell_{ii}^{(m_1)}(m) > 0$. Let

$$L(m) = \liminf_{n'} \frac{t_{ii}(n' - m)}{\lambda^{(n' - m)}}.$$

Using a diagonal argument choose a subsequence $\{n''\}$ of $\{n'\}$ so that

$$\lim_{n''} \frac{t_{ii}(n'' - m)}{\lambda^{(n'' - m)}} = L(m) \quad \text{and} \quad \lim_{n''} \frac{t_{ii}(n'' - s)}{\lambda^{(n'' - s)}} = L(s)$$

for each $s \geq k, s \neq m$ and some $L(s) \leq L$.

Now observe that

$$t_{ii}(r) = \left(\sum_{k < m_1} \ell_{ii}^{(k)}(r) \right) + \left(\sum_{k = m_1}^{r} \ell_{ii}^{(k)}(r) \right)$$

$$= \left(\sum_{k < m_1} \ell_{ii}^{(k)}(r) \right) + \left(\sum_{k = m_1}^{r} \ell_{ii}^{(m_1)}(s) t_{ii}(r - s) \right).$$

The first equality follows from Lemma 7.1.6 (ii) and grouping the paths from i to i of length r by the number of occurrences of i. The second equality is similar to Lemma 7.1.6 (iii) but the paths are grouped according to the m_1st occurrence of i. Use this equality and take the limit along $\{n''\}$.

$$L = \lim_{n''} t_{ii}(n'')/\lambda^{n''}$$

$$= \lim_{n''} \sum_{k < m_1} \ell_{ii}^{(k)}(n'')/\lambda^{n''}$$

$$+ \lim_{n''} \sum_{s = m_1}^{n''} \left(\ell_{ii}^{(m_1)}(s)/\lambda^s \right) \left(t_{ii}(n'' - s)/\lambda^{(n'' - s)} \right).$$

Since T is recurrent $\Sigma \ell_{ii}^{(k)}(n)/\lambda^n = 1$, for all k. This means that the first sum goes to zero as n'' goes to infinity. We will apply Lemma 7.1.15 (i) to the second sum. Take $a_s = \ell_{ii}^{(m_1)}(s)/\lambda^s$. As we have just observed, $\Sigma \ell_{ii}^{(m_1)}(s)/\lambda^s = 1$. Take $b_s^{(n'')} = t_{ii}(n'' - s)/\lambda^{(n'' - s)}$, which is less than or equal to one by Lemma 7.1.1.

Then $b_s = L(s)$, and

$$L = \lim_{n''} \sum_{s=m_1}^{n''} \left(\ell_{ii}^{(m_1)}(s)/\lambda^s\right)\left(t_{ii}(n'' - s)/\lambda^{(n''-s)}\right)$$

$$= \sum_{s=m_1} \left(\ell_{ii}^{(m_1)}(s)/\lambda^s\right) L(s).$$

But $L \geq L(s)$ for each s and

$$\sum_{s=m_1}^{\infty} \ell_{ii}^{(m_1)}(s)/\lambda^s = 1.$$

This means $L(s) = L$ for each s with $\ell_{ii}^{(m_1)}(s) \neq 0$. In particular, $L(m) = L$ and the sequence $\{n'\}$ has the desired property.

Step 3.

$$\sum_{s=0}^{n} \left[t_{ii}(n - s)/\lambda^{n-s}\left(\sum_{j>s} \ell_{ii}(j)/\lambda^j\right)\right] = 1 \text{ for all } n.$$

To see this we use the fact that

$$\sum_{n=1}^{\infty} \ell_{ii}(n)/\lambda^n = 1,$$

which means that the above expression is equal to

$$\sum_{s=0}^{n} \left[t_{ii}(n - s)/\lambda^{n-s}\left(1 - \sum_{j\leq s} \ell_{ii}(j)/\lambda^j\right)\right].$$

We prove this is equal to one by induction. For $n = 0$ we have

$$t_{ii}(0)/\lambda^0 \left(1 - \ell_{ii}(0)/\lambda^0\right) = 1.$$

Assume it is true for n and examine

$$\sum_{s=0}^{n+1} \left[\frac{t_{ii}(n + 1 - s)}{\lambda^{n+1-s}}\left(1 - \sum_{j\leq s} \frac{\ell_{ii}(j)}{\lambda^j}\right)\right]$$

$$= \sum_{s=0}^{n} \left[\frac{t_{ii}(n - s)}{\lambda^{n-s}}\left(1 - \sum_{j\leq s} \frac{\ell_{ii}(j)}{\lambda^j}\right)\right]$$

$$+ \left[\frac{t_{ii}(n + 1 - 0)}{\lambda^{n+1-0}} - \frac{t_{ii}(n + 1 - 1)}{\lambda^{n+1-1}} \cdot \frac{\ell_{ii}(1)}{\lambda^1}\right.$$

$$\left. - \frac{t_{ii}(n + 1 - 2)}{\lambda^{n+1-2}} \cdot \frac{\ell_{ii}(2)}{\lambda^2} - \cdots - \frac{t_{ii}(0)}{\lambda^0} \cdot \frac{\ell_{ii}(n + 1)}{\lambda^{(n+1)}}\right].$$

The first term is one by the induction hypothesis. The second term is

$$\frac{t_{ii}(n+1) - \sum_{j=1}^{n+1} t_{ii}(n+1-j)\ell_{ii}(j)}{\lambda^{n+1}},$$

and the numerator is zero by Lemma 7.1.6 (iii).

Step 4.
Observe that

$$\mu(i) = \sum_{s=1}^{\infty} s\ell_{ii}(s)/\lambda^s = \sum_{s=0}^{\infty} \sum_{j>s} \ell_{ii}(j)/\lambda^j.$$

Step 5.
When T is null recurrent we apply Lemma 7.1.15 (ii), using the sequence $\{n^*\}$, in the equation in step 3. Take

$$a_s = \sum_{j>s} \ell_{ii}(j)/\lambda^j, \quad b_s^{(n^*)} = t_{ii}(n^*-s)/\lambda^{(n^*-s)}$$

and $b = L$. The conclusion is that

$$\lim_{n^*} t(n^*-s)/\lambda^{(n^*-s)} = L = \limsup\, t_{ii}(n)/\lambda^n = 0 = 1/\mu(i).$$

Step 6.
When T is positive recurrent we apply Lemma 7.1.15 (i), using $\{n^*\}$, to the equation in step 3. We again take

$$a_s = \sum_{j>s} \ell_{ii}(j)/\lambda^j, \quad b_s^{(n^*)} = t_{ii}(n^*-s)/\lambda^{(n^*-s)}$$

and $b_s = L$. We have that $\sum_s a_s = \mu(i) < \infty$, by step 4, and that $b_s^{(n^*)} \leq 1$ by Lemma 7.1.1 (iv). Then

$$1 = \lim_{n^*} \sum_{s=0}^{n^*} \left(\frac{t_{ii}(n^*-s)}{\lambda^{(n^*-s)}} \right) \left(\sum_{j>s} \ell_{ii}(j)/\lambda^j \right)$$

$$= \sum_{s=0}^{\infty} L \left(\sum_{j>s} \ell_{ii}(j)/\lambda^j \right) = L \cdot \mu(i).$$

This means $\limsup\, t_{ii}(n)/\lambda^n = 1/\mu(i)$.

Step 7.
Finally we reverse the argument taking

$$L = \liminf\, t_{ii}(n)/\lambda^n \qquad \text{to get}$$

$$= \lim_{n} t_{ii}(n)/\lambda^n = 1/\mu(i). \qquad \qquad \square$$

Lemma 7.1.19 *If T is a nonnegative, irreducible and aperiodic, countable matrix with a finite Perron value λ then:*

(i) $\lim\limits_{n\to\infty} (T^n)_{ij}/\lambda^n = 0$ *for all pairs i and j when T is transient;*

(ii) $\lim\limits_{n\to\infty} (T^n)_{ij}/\lambda^n = L_{ij}(1/\lambda)/\mu(i) = R_{ij}(1/\lambda)/\mu(j)$ *for every pair i and j when T is recurrent.*

Proof. By Lemma 7.1.6 (iii)

$$t_{ij}(n)/\lambda^n = \sum_{s=1}^{n} \frac{t_{ii}(n-s)}{\lambda^{n-s}} \cdot \frac{\ell_{ij}(s)}{\lambda^s}.$$

Use Lemma 7.1.15 (i) with $a_s = \ell_{ij}(s)/\lambda^s$, $b_s^{(n)} = t_{ii}(n-s)/\lambda^{n-s} \leq 1$ to see that

$$\lim_{n\to\infty} (T^n)_{ij}/\lambda^n = \sum_{s=1}^{\infty} \ell_{ij}(s)/\lambda^s \left(\lim_{n\to\infty} \frac{t_{ii}(n-s)}{\lambda^{n-s}} \right).$$

By Theorem 7.1.18 this is zero when T is transient and $L_{ij}(1/\lambda)/\mu(i)$ when T is recurrent. The second equality follows similarly but by using the second equality, $t_{ij}(n) = \sum_{s=0}^{n} r_{ij}(n-s)t_{ij}(s)$, of Lemma 7.1.6 (iii). $\qquad\square$

Lemma 7.1.20

(i)   *If T is recurrent, then*

(a) $L_{ij}(1/\lambda) = L_{ik}(1/\lambda)L_{kj}(1/\lambda), \qquad 1 = L_{ik}(1/\lambda)L_{ki}(1/\lambda),$

and

(b) $R_{ij}(1/\lambda) = R_{ik}(1/\lambda)R_{kj}(1/\lambda), \qquad 1 = R_{ij}(1/\lambda)R_{ji}(1/\lambda).$

(ii) *If T is positive recurrent, then*

$$L_{ij}(1/\lambda) = \frac{\mu(i)}{\mu(j)} R_{ij}(1/\lambda).$$

Proof. (i) Recall the vectors $\ell^{(i)}$ defined before Lemma 7.1.9. By Lemma 7.1.9 these are eigenvectors for T and by Lemma 7.1.11 each is a multiple of any other. Since T is recurrent $L_{ii}(1/\lambda) = 1$ for each i. Consider $\ell^{(i)}$ and $\ell^{(k)}$. We see that $L_{ik}(1/\lambda)\ell^{(k)} = \ell^{(i)}$, or $L_{ik}(1/\lambda)L_{kj}(1/\lambda) = L_{ij}(1/\lambda)$ for each j. The second of the first pair of equalities, is just the entry where $j = i$. The second pair of equalities follows by considering the column vectors $r^{(j)}$, also defined before Lemma 7.1.9, and applying the same reasoning. Statement (ii) is a repeat of Lemma 7.1.19 (ii). $\qquad\square$

Lemma 7.1.21 *If T is positive recurrent, then $\ell^{(i)} \cdot r^{(i)} = \mu(i)$.*

Proof. Since $r^{(i)}$ is a right eigenvector for T

$$\frac{(T^n)_{ki}}{\lambda^n} \cdot \frac{R_{ij}(1/\lambda)}{R_{kj}(1/\lambda)} \leq 1 \text{ for all } i, j, k.$$

Since $\ell^{(i)}$ is a left eigenvalue for T

$$L_{ij}(1/\lambda) = \sum_k L_{ik}(1/\lambda)(T^n)_{kj}/\lambda^n.$$

Now

$$1 = L_{ii}(1/\lambda)\frac{R_{ii}(1/\lambda)}{R_{kk}(1/\lambda)} = \sum_k L_{ik}(1/\lambda)\frac{(T^n)_{ki}}{\lambda^n} \cdot \frac{R_{ii}(1/\lambda)}{R_{kk}(1/\lambda)}$$

$$= \sum_k L_{ik}(1/\lambda)R_{ki}(1/\lambda)\left[\frac{(T^n)_{ki}}{\lambda^n} \cdot \frac{R_{ik}(1/\lambda)}{R_{kk}(1/\lambda)}\right] \text{ by Lemma 7.1.11 (i).}$$

We will apply Lemma 7.1.15 (i) to this equation. Let $a_k = L_{ik}(1/\lambda)R_{ki}(1/\lambda)$. The sum, $\sum_k a_k$, is finite by Lemma 7.1.21. Let

$$b_k^{(n)} = \left[\frac{(T^n)_{ki}}{\lambda^n} \cdot \frac{R_{ik}(1/\lambda)}{R_{kk}(1/\lambda)}\right] \leq 1, \quad \text{then}$$

$$\lim_{n \to \infty} b_k^{(n)} = \frac{R_{ki}(1/\lambda)}{\mu(i)} \cdot \frac{R_{ik}(1/\lambda)}{R_{kk}(1/\lambda)} = \frac{1}{\mu(i)}$$

by Lemmas 7.1.19 (ii) and 7.1.20 (i). We now have that

$$1 = \lim_{n \to \infty} \sum_k a_k b_k^{(n)}$$

$$= \sum_k L_{ik}(1/\lambda)R_{ki}(1/\lambda)\left[\frac{(T^n)_{ki}}{\lambda^n} \cdot \frac{R_{ik}(1/\lambda)}{R_{kk}(1/\lambda)}\right]$$

$$= \frac{1}{\mu(i)} \sum_k L_{ik}(1/\lambda)R_{ki}(1/\lambda). \square$$

Lemma 7.1.22 *If T is positive recurrent*

$$\sum_i \frac{1}{\mu(i)} = 1.$$

Proof.

$$\mu(j) = \sum_i L_{ji}(1/\lambda)R_{ij}(1/\lambda) \text{ by Lemma 7.1.21}$$

$$= \sum_i \frac{\mu(j)}{\mu(i)}R_{ji}(1/\lambda)R_{ij}(1/\lambda) \text{ by Lemma 7.1.20 (ii)}$$

$$= \mu(j) \; \Sigma\frac{1}{\mu(i)} \text{ by Lemma 7.1.20 (i).} \qquad \square$$

Now we define a new quantity by letting $\varrho(i) = \limsup \sqrt[n]{\ell_{ii}(n)}$. Then $\varrho(i) \leq \lambda$ and $1/\varrho(i)$ is the radius of convergence of $L_{ii}(z)$. When dealing with several matrices we will use $\varrho_T(i), \lambda_T, \ell_{ii}^T(n), L_{ii}^T(z)$ to distinguish the $\varrho(i), \lambda, \ell_{ii}(n)$ and $L_{ii}(z)$ that correspond to a particular matrix T.

Lemma 7.1.23 *If $S \leq T$, with inequality occurring somewhere, and $\lambda_S = \lambda_T$ then S is transient.*

Proof. Suppose $S_{ij} < T_{ij}$. Then for some n, $\ell_{ii}^S(n) < \ell_{ii}^T(n)$. This means

$$L_{ii}^S(1/\lambda_S) = L_{ii}^S(1/\lambda_T) < L_{ii}^T(1/\lambda_T) \leq 1. \qquad \square$$

Lemma 7.1.24 *If T is a matrix where $\varrho_T(i) = \lambda_T$ for some state i, then there exists a matrix $S \leq T$, with $S_{ij} < T_{ij}$, for some j, and $\lambda_S = \lambda_T$.*

Proof. Fix a state i in T where $\varrho_T(i) = \lambda_T$. Consider the matrix $T_{(a)}$ that is defined for $a > 0$ by

$$(T_{(a)})_{jk} = T_{jk}, \quad \text{for} \quad j \neq i, \quad \text{and} \quad (T_{(a)})_{ik} = aT_{ik}.$$

Then $\ell_{ii}^{T_{(a)}}(n) = a\ell_{ii}^T(n)$ for all n. This means $\varrho_{T_{(a)}}(i) = \varrho_T(i)$. If $0 < a < 1$, $\varrho_{T_{(a)}}(i) \leq \lambda_{T_{(a)}} \leq \lambda_T = \varrho_T(i)$. Any $T_{(a)}$ has the desired property. $\qquad \square$

In the proof of Lemma 7.1.24 we defined a one-parameter family of matrices. It is interesting to examine how the recurrence properties of $T_{(a)}$ vary with a. This is done for two families in Example 7.1.31.

Lemma 7.1 25 *If T is a matrix where $\varrho_T(i) < \lambda_T$ for some state i, then T is positive recurrent.*

Proof. Fix a state i where $\varrho_T(i) < \lambda_T$. First we will prove that $L_{ii}(1/\lambda) = 1$. From Lemma 7.1.6 (iv) we know that $T_{ii}(z) = 1/(1 - L_{ii}(z))$. The radius of convergence of $T_{ii}(z)$, $1/\lambda$, is less than the radius of convergence of $L_{ii}(z)$, $1/\varrho(i)$. If $|L_{ii}(1/\lambda)| < 1$, then since $|L_{ii}(z)| \leq L_{ii}(|z|)$, $|L_{ii}(z)| < 1$ for all $|z| \leq 1/\lambda$.

This together with the fact that $1/\lambda < 1/\varrho(i)$ means that there is an $\varepsilon > 0$ so that the function $1/(1 - L_{ii}(z))$ is analytic for all $|z| < 1/\lambda + \varepsilon$. But the power series expansion for $1/(1 - L_{ii}(z))$ at the origin is $T_{ii}(z)$, a contradiction.

Next we prove that $L'_{ii}(1/\lambda) < \infty$. When $1/\lambda < 1/\varrho(i)$, $L'_{ii}(1/\lambda) < \infty$, because $L_{ii}(z)$, is absolutely convergent for all $|z| < 1/\varrho(i)$. $\qquad\square$

Example 7.1.31 (ii) exhibits a positive recurrent matrix with $\varrho(i) = \lambda$ for a state i.

Lemma 7.1.26 $L_{ii}(1/\varrho(i)) \le 1$ *if and only if* $\varrho(i) = \lambda$.

Proof. We know that $\varrho(i) \le \lambda$ and that $L_{ii}(1/\lambda) \le 1$. This means that if $\varrho(i) = \lambda$, $L_{ii}(1/\varrho) \le 1$. Conversely, if $\varrho(i) < \lambda$ then by Lemma 7.1.25 T is positive recurrent. This means $1 = L_{ii}(1/\lambda) < L_{ii}(1/\varrho(i))$. $\qquad\square$

Given two matrices S and T we define the *join of S and T at i_S and j_T*. Let $L_S = \{1_S, 2_S, 3_S, \ldots\}$ be the states of S and $L_T = \{1_T, 2_T, 3_T, \ldots\}$ the states of T. For $i_S \in L_S$ and $j_T \in L_T$ define a new matrix $S \vee_{(i_S, j_T)} T$ whose states are $L_S \cup L_T$ with i_S and j_T identified, and whose entries are

$$(S \vee_{(i_S, j_T)} T)_{kl} = \begin{cases} S_{kl} & \text{if } k, l \in L_S \\ T_{kl} & \text{if } k, l \in L_T \\ 0 & \text{otherwise.} \end{cases}$$

Lemma 7.1.27 *Suppose* $\lambda_S \le \lambda_T$ *and T is recurrent, then $S \vee_{(i_S, j_T)} T$ is positive recurrent, for all i_S and j_T.*

Proof. Let $\varrho(i_S) = \varrho_{(S \vee_{(i_S, j_T)} T)}(i_S) = \varrho_{(S \vee_{(i_S, j_T)} T)}(j_T)$. Since T is recurrent $\lambda_T < \lambda_{(S \vee_{(i_S, j_T)} T)}$ by Lemma 7.1.23. Next observe that

$$\varrho(i_S) = \limsup \sqrt[n]{\ell_{i_S i_S}^{(S \vee_{(i_S, j_T)} T)}(n)}$$

$$= \limsup \sqrt[n]{\ell_{i_S i_S}^{S}(n) + \ell_{j_T j_T}^{T}(n)}$$

$$\le \lambda_T$$

$$< \lambda_{(S \vee_{(i_S, j_T)} T)}.$$

Then by Lemma 7.1.25, $S \vee_{(i_S, j_T)} T$ is positive recurrent. $\qquad\square$

Before going further we examine a number of examples to help get a feel for the ideas we have discussed.

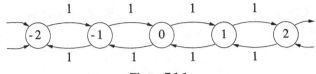

Figure 7.1.1

Example 7.1.28 This is a classical example from probability theory. It is a random walk on the integers. The adjacency matrix for $\mathbb{Z}$ is irreducible but not aperiodic. It has period 2, but we will examine it anyway, considering only $t_{ii}(2n)$ and $\ell_{ii}(2n)$.

First recall Stirling's formula for asymptotic approximations of factorials. See the notes at the end of the chapter for a reference. It shows

$$n! \sim \frac{n^n \sqrt{2\pi n}}{e^n}.$$

We use it to approximate binomial coefficients. This allows us to compute λ, since

$$t_{00}(2n) = \binom{2n}{n} \sim \frac{2^{2n}}{\pi^{1/2} n^{1/2}}.$$

So $\lambda = 2$. Next use the Ballot Theorem (Exercise 4) to compute

$$\ell_{00}(2n) = 2\left[\binom{2n-2}{n-1} - \binom{2n-2}{n}\right].$$

Then compute to see that

$$\ell_{00}(2n) = 2\left[\frac{(2n-2)!}{n!(n-1)!}\right] = \frac{2}{n}\binom{2n-2}{n-1} \sim \frac{2^{2n-1}}{\pi^{1/2} n(n-1)^{1/2}}.$$

This means $\varrho(0) = 2$. Compare the series

$$T_{00}(1/2) = \sum \binom{2n}{n}\frac{1}{2^{2n}}$$

with the series $\sum 1/n^{1/2}$ to see that $T_{00}(1/2)$ diverges with $t_{00}(2n)/2^{2n} \to 0$. By Theorem 7.1.18, T is null recurrent. It also means that

$$L_{00}(1/2) = \sum \left(\frac{2}{n}\binom{2n-2}{n-1}\right)\frac{1}{2^{2n}} = 1.$$

Example 7.1.29 These examples are also periodic of period two.
(i) The adjacency matrix for $\mathbb{N}$.

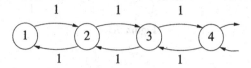

Figure 7.1.2

Then

$$\ell_{11}(2n) = \frac{1}{n}\binom{2n-2}{n-1}$$

because the number of first returns to 1 in $\mathbb{N}$ is 1/2 the number of first returns to 0 in $\mathbb{Z}$, as in Example 7.1.28. This means that $\varrho(1) = 2 \leq \lambda$. Furthermore, $\lambda \leq 2$ because T has row sum less than or equal to two so that $t_{11}(2n) \leq 2^{2n}$. From Example 7.1.28 we see,

$$L_{11}(1/2) = \sum \frac{1}{n}\binom{2n-2}{n-1}\frac{1}{2^{2n}} = \frac{1}{2}\sum \frac{2}{n}\binom{2n-2}{n-1}\frac{1}{2^{2n}} = 1/2,$$

so T is transient. Another way to see that T is transient is to observe that the row vector of all 1's is a left subinvariant (not invariant) vector for $\lambda = 2$. Then by Remark 7.1.12, T is transient. Another point is that the vector

$$v = (1, 2, 3, 4, \ldots)$$

satisfies satisfies $vT = \lambda T$ even though T is transient.

(ii) A Random walk on $\mathbb{N}$ defined by

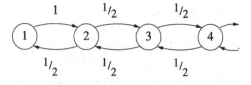

Figure 7.1.3

Use Example 7.1.28 and 7.1.29 (i) to calculate

$$\ell_{11}(2n) = \frac{1}{n}\binom{2n-2}{n-1}\frac{1}{2^{2n-1}} \sim \frac{1}{2\pi^{1/2}n(n-1)^{1/2}}$$

so $\varrho = 1$.

$$L_{11}(1) = \sum \frac{1}{n}\binom{2n-2}{n-1}\frac{1}{2^{2n-1}} = 1$$

by Example 7.1.28. This means T is recurrent. Then notice

$$t_{11}(2n) \quad \text{is less than 1/2 times} \quad \binom{2n}{n} \cdot \frac{1}{2^{2n}}$$

as computed in Example 7.1.28. So $t_{11}(2n) \to 0$ and T is null recurrent by Theorem 7.1.18.

(iii) T is defined by

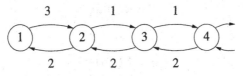

Figure 7.1.4

We see that $\lambda = 3$. The column vector, r, of all 1's is a right eigenvector for $\lambda = 3$ and the row vector $\ell = (1/3, 1/2, 1/2^2, 1/2^3, \dots)$ is a left eigenvector for λ. Then $\ell \cdot r < \infty$ and by Remark 7.1.19 T is positive recurrent. Another way to see this is to observe

$$\ell_{11}(2n) = 3 \cdot 2^n \frac{1}{n} \binom{2n-2}{n-1} \sim \frac{3 \cdot 2^{3n-2}}{\pi^{1/2} n (n-1)^{1/2}}.$$

Then

$$\varrho(1) = \lim \sqrt[2n]{\ell_{11}(2n)} = 2^{3/2} < \lambda$$

and T is positive recurrent by Lemma 7.1.25.

Example 7.1.30 Define T by

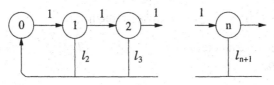

Figure 7.1.5

With $T_{i0} = \ell_{i+1} \geq 0$. Suppose $0 \leq \ell_i \leq M$ and there is a subsequence of the $\{i\}$ going to infinity whith $1 \leq \ell_i$. Then $\ell_{00}(n) = \ell_n$ and

$$\varrho(0) = \limsup \sqrt[n]{\ell_n} = 1.$$

For some N, $t_{00}(N) > 1$. That implies that $\lambda > 1 = \varrho(0)$ and by Lemma 7.1.25 that T is positive recurrent.

Example 7.1.31 (i) Define a one- parameter family of matrices $\{T_{(a)}\}, a > 0$, by the following family of graphs:

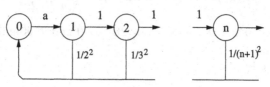

Figure 7.1.6

$$\ell_{00}^a(n) = a\ell_{00}^1(n) = \frac{a}{n^2}$$

so $\varrho_a(0) = 1$ for all a, and $L_{00}^a(z) = aL_{00}(z)$ for all z. Let

$$L^* = L_{00}^1(1) = \sum_{n=2}^{\infty} 1/n^2 < 1.$$

For $0 < a \le 1/L^*$, $L_{00}^a(1) \le 1$ and $1 = \varrho_a(0) = \lambda_a$ by Lemma 7.1.26. For $a > 1/L^*$, $L_{00}^a(1) > 1$ and $1 = \varrho_a(0) < \lambda_a$ by the same lemma. Also, $L_{00}^{a\,'}(z) = aL_{00}^{1\,'}(z)$ for all a and

$$L_{00}^{1\,'}(1) = \sum_{n=2}^{\infty} \frac{1}{n} = +\infty.$$

For $0 < a \le 1/L^*$, $L_{00}^{a\,'}(1) = \infty$. To summarize:

$$
\begin{aligned}
0 < a < 1/L^*, &\quad T_{(a)} \text{ is transitive with } L_{00}^{a\,'}(1/\lambda_a) = \infty; \\
a = 1/L^*, &\quad T_{(a)} \text{ is null recurrent;} \\
a > 1/L^*, &\quad T_{(a)} \text{ is positive recurrent.}
\end{aligned}
$$

(ii) Define a one- parameter family of matrices $\{T_{(a)}\}, a > 0$ by the following family of graphs:

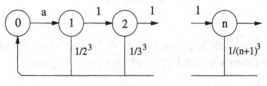

Figure 7.1.7

$$\ell_{00}^a(n) = a\ell_{00}^1(n) = \frac{a}{n^3}$$

so $\varrho_a(0) = 1$ for all a, and $L_{00}^a(z) = aL_{00}(z)$ for all z. Let

$$L^* = L_{00}^1(1) = \sum_{n=2}^{\infty} 1/n^3 < 1.$$

As in (i), $L_{00}^a(1) \le 1$ and $1 = \varrho_a(0) = \lambda_a$, for $0 < a \le 1/L_*$ and $1 = \varrho_a < \lambda_a$, for $a > 1/L^*$. But, here we see that

$$L_{00}^{a\,\prime}(1/\lambda_a) \le L_{00}^{a\,\prime}(1) = aL_{00}^{1\,\prime}(1) = a\sum_{n=2}^{\infty} \frac{1}{n^2} < \infty.$$

This gives the following information:

$0 < a < 1/L^*$, $T_{(a)}$ is transitive with $L_{00}^{a\,\prime}(1/\lambda_a) < \infty$;
$a = 1/L^*$, $T_{(a)}$ is positive recurrent with $1 = \varrho_a(0) = \lambda_a$;
$a > 1/L^*$, $T_{(a)}$ is positive recurrent with $1 = \varrho_a(0) < \lambda_a$.

(iii) Define a one- parameter family of matrices $\{T_{(a)}\}$ by the following family of graphs:

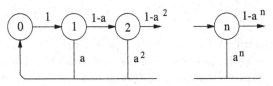

Figure 7.1.8

For $0 < a < 1$, each $T_{(a)}$ is stochastic and

$$\ell_{00}^a(n) = \prod_{k=1}^{n-2} (1 - a^k)a^{n-1}$$

so for each a, $\varrho_a(0) = a$. First we see that $0 < L_{00}^a(1) < 1$, for all a. This follows because

$$1 > 1 - L_{00}^a(1) = \prod_{k=1}^{\infty} (1 - a^k).$$

To see that

$$\prod_{k=1}^{\infty} (1 - a_k)$$

is not zero we observe that

$$\log\left(\prod_{k=1}^{\infty}(1-a_k)\right) = \sum_{k=1}^{\infty}\log(1-a^k) = -\sum_{k=1}^{\infty}\sum_{n=1}^{\infty}\frac{a^{nk}}{n} \qquad (*)$$

$$> -\sum_{k=1}^{\infty}a^k = -\frac{1}{1-a^k} > -\infty.$$

The equality (*) holds because

$$-\log(1-u) = \sum_{n=1}^{\infty}\frac{u^n}{n}.$$

By Remark 7.1.2 we see that each $T_{(a)}$ is probabilistically transitive. Secondly observe that

$$L_{00}^a(1/\varrho_a(0)) = \sum_{n=2}^{\infty}\prod_{k=1}^{n-2}(1-a^k)\frac{a^{n-1}}{a^n} = \frac{1}{a}\sum_{k=1}^{\infty}\prod_{k=1}^{n-2}(1-a^k).$$

But,

$$\prod_{k=1}^{n-2}(1-a^k) > \lim_{n\to\infty}\prod_{k=1}^{\infty}(1-a^k) = c_a > 0$$

by the first observation. Consequently $L_{00}^a(1/\varrho_a(0))$ diverges and by Lemma 7.1.26 $\varrho_a(0) < \lambda_a$. By Lemma 7.1.25 we conclude that in the terminology we are using each $T_{(a)}$ is positive recurrent. This demonstrates the confusion spoken of in Remark 7.1.2.

Example 7.1.32 Choose two sequences of positive integers. Take one of them, $\{d_k\}$, to be strictly increasing and the other, $\{r_k\}$, arbitrary. Define a matrix T by the following graph.

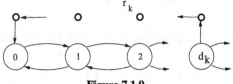

Figure 7.1.9

This means there is a single path of length r_k that goes from the vertex d_k back to the vertex 0. Then we see

$$\ell_{00}(n) = \begin{cases} 2^{d_k} & if \quad n = d_k + r_k \\ 0 & otherwise. \end{cases}$$

Compute

$$\varrho(0) = \limsup \sqrt[n]{\ell_{00}(n)} = \limsup 2^{\frac{d_k}{d_k+r_k}} = 2^\ell,$$

where $\ell = \limsup \frac{d_k}{d_k+r_k}$. Also,

$$L_{00}(z) = \sum_k 2^{d_k} z^{d_k+r_k}, \qquad \text{so} \qquad L_{00}(1/\varrho(0)) = \sum_k \frac{2^{d_k}}{2^{\ell(d_k+r_k)}}.$$

(i) Let $r_k = d_k$, so $\ell = \limsup \frac{d_k}{d_k+r_k} = 1/2$. Then compute

$$L_{00}(1/\varrho(0)) = \sum_k \frac{2^{d_k}}{2^{\ell(d_k+r_k)}} = \sum_k \frac{2^{d_k}}{2^{d_k}} > 1.$$

By Lemma 7.1.26 $\varrho(0) < \lambda$ and then by Lemma 7.1.25 T is positive recurrent.

(ii) Let $r_k = d_k + 2^k$ with $\{d_k\}$ so that $k/d_k \to 0$. Then $\ell = 1/2$.

$$L_{00}(1/\varrho(0)) = \sum_k \frac{2^{d_k}}{2^{\ell(d_k+r_k)}} = \sum_k 1/2^k = 1.$$

By Lemma 7.1.26 $\sqrt{2} = \varrho(0) = \lambda$, $L_{00}(1/\lambda) = 1$ and T is recurrent. Then

$$(1/\lambda)L'_{00}(1/\lambda) = \sum \frac{2^{d_k} + 2^k}{2^k}$$

and by choosing $\{d_k\}$ in different ways we can make T either positive or null recurrent.

(iii) Let $r_k = d_k + 4^k$ with $\{d_k\}$ so that $k/d_k \to 0$. Once again $\ell = 1/2$, but now

$$L_{00}(1/\varrho(0)) = \sum_k \frac{2^{d_k}}{2^{\ell(d_k+r_k)}} = \sum_k 1/2^{2k} < 1.$$

By Lemma 7.1.26 $\sqrt{2} = \varrho(0) = \lambda$, $L_{00}(1/\lambda) < 1$ and T is transient.

Next we turn to finite approximations of an infinite T. Let $\{V(n)\}$ be a nested family of finite subsets of the states of T. We require that their union is all of the states of T and that for each n the finite matrix defined by $(T_n)_{ij} = T_{ij}$ for $i, j \in V(n)$ is irreducible and aperiodic. Let λ_n be the Perron value for T_n.

Lemma 7.1.33 *Suppose T is a countable, nonnegative matrix. Assume T is irreducible and aperiodic. Let λ be the Perron value of T and $\{T_n\}$ and $\{\lambda_n\}$ be as just defined. Then $\lambda = \lim\limits_{n \to \infty} \lambda_n$.*

Proof. Fix a vertex i in T_1. Then for all n

$$\lambda_n = \sup_k \sqrt[k]{(T_n^k)_{ii}} \le \lambda_{n+1} = \sup_k \sqrt[k]{(T_{n+1}^k)_{ii}} \le \sup_k \sqrt[k]{(T^k)_{ii}} = \lambda$$

by Lemma 7.1.1 Let $\lambda^* = \sup_n \lambda_n$ and suppose $\lambda^* < \lambda$. Choose k so that $\lambda^* < \sqrt[k]{(T^k)_{ii}} < \lambda$. If there are only a finite number of paths from i to i of length k in T, there exists an N so that for all $n \ge N$, $(T_n^k)_{ii} = (T^k)_{ii}$. This is a contradiction because then we would have that $\lambda^* < \sqrt[k]{(T)_{ii}} \le \lambda_n$. If there are infinitely many paths in T from i to i of length k we observe by the same reasoning that for every $\varepsilon > 0$ there is an N so that for $n \ge N$, $(T_n^k)_{ii} \ge (T^n)_{ii} - \varepsilon$. This again gives a contradiction. $\qquad \square$

Lemma 7.1.34 *Suppose T is a countable, nonnegative matrix. Suppose T is irreducible, aperiodic and recurrent. Let y^n be the right Perron eigenvector for T_n, normalized, so that $y_1^n = 1$. Then $\lim_n y^n = r$, where r is the unique eigenvector for T normalized so that $r_1 = 1$.*

Proof. By Lemma 7.1.33 we know that $\lambda = \lim \lambda_n$.
Define a vector y^* by $y_i^* = \liminf\{y_i^n\} \ge 0$. Fix a state i. For every T_n that has i as a state,

$$\sum_j (T_n)_{ij} y_j^n = (T_n y^n)_i = \lambda_n y_i^n. \tag{*}$$

A simple case of Fatou's Lemma (Exercise 7.1.5) says that if

$$\left\{ \sum_j b_j^n \right\}$$

is a collection of series with $b_j^n \ge 0$ then

$$\sum_j \liminf_n b_j^n \le \liminf_n \sum_j b_j^n.$$

This follows because for every $\varepsilon > 0$ there is an N so that for $n \ge N$,

$$b_j^n - \left(\liminf_n b_j^n \right) > -\varepsilon.$$

Which means for a fixed K we can choose N large enough that

$$\sum_{j=1}^k b_j^n < \sum_{j=1}^K \left(\liminf b_j^n + \varepsilon/K \right) \le \left(\liminf \sum_{j=1}^\infty b_j^n \right) + \varepsilon.$$

We apply this to our equation (*) with

$$b_j^n = (T_n)_{ij} y_j^n, \quad \text{and} \quad \liminf_n (T_n)_{ij} y_j^n = T_{ij} y_j^*.$$

This proves that

$$(Ty^*)_i = \sum_j \liminf_n (T_n)_{ij} y_j^n \leq \liminf \sum_j (T_n)_{ij} y_j^n = \liminf \lambda_n y_i^n = \lambda y_i^*.$$

Since $y_1^* = 1$ and T is irreducible this equation tells us that each y_i^* is finite. We see that y^* is a nonnegative, nonzero vector with $Ty^* \leq \lambda y^*$. Using Lemma 7.1.11 we conclude y^* is the unique positive right eigenvector for T normalized so that $y_1^* = 1$. Next suppose there exists a state i so that $\limsup y_i^n = y_i' > y_i^*$. Choose a subsequence, $\{n_r\}$, of $\mathbb{N}$ so that $\lim y_i^{n_r} = y_i'$. Repeat the argument on this subsequence obtaining an eigenvector y' with $y_1' = 1$ but $y_i' > y_i^*$. This is a contradiction because the right eigenvector is unique up to normalization.   □

The same result clearly holds for the left eigenvector.

Next we examine matrices that are irreducible without making an assumption about the period. We will need these results when we study stochastic matrices with stationary probability vectors. With only slight modifications all the results carry over. This is exactly the same approach that is used for finite, nonnegative and irreducible matrices in Section 1.3.

Let T be an irreducible matrix. Fix a vertex i. Define the *Perron value* of T, as for an aperiodic matrix, to be $\lambda = \limsup \sqrt[n]{(T^n)_{ii}}$.

Define the functions $T_{ij}(z), L_{ij}(z),$ and $R_{ij}(z)$ exactly as in the aperiodic case and use them to define *transient, null recurrent, and positive recurrent* exactly as before.

We defined the *period of T* to be the greatest common divisor of $\{n : (T^n)_{ii} > 0\}$. This is independent of the choice of the vertex.

Remark 7.1.35 In Section 1.3 we defined the cyclic subsets of the indices for an irreducible transition matrix. Do the same for a countable, nonnegative and irreducible matrix. Let T be the matrix with period p. Fix a vertex i. The *cyclic subsets* are $C_0, \ldots, C_{p-1}$ defined by

$$C_k = \{j : (T^{np+k})_{ij} > 0 \text{ for some } n \in \mathbb{N}\}.$$

The cyclic subsets partition the indices of T. Let p be the period of T and for $0 \leq k \leq p-1$ define a matrix T_k with states in C_k by $(T_k)_{ij} = (T^p)_{ij}$, for all $i, j \in C_k$. Observe that T_k is irreducible and aperiodic with $\lambda_{T_k} = \lambda_T^p$.

Lemma 7.1.36 *Suppose T is a countable, nonnegative and irreducible matrix with period p. Let the matrices T_k be as described. Then*

(i) T is transient if and only if T_k is transient for all k.
(ii) T is null recurrent if and only if T_k is null recurrent for all k.
(iii) T is positive recurrent if and only if T_k is positive recurrent for all k.

Proof. Let i be the vertex of T that is fixed to define the cyclic subsets C_k in Remark 7.1.35. Let T_k be the irreducible and aperiodic matrices described there. Observe that

$$\ell_{ii}^T(n) = \begin{cases} \ell_{ii}^{T_0}(m) & \text{if } n = mp \\ 0 & \text{otherwise.} \end{cases}$$

This means that $L_{ii}^T(z) = L_{ii}^{T_0}(z^p)$. We use the fact that $\lambda_T^p = \lambda_{T_0}$ (Remark 7.1.35) to see that

$$L_{ii}^T(1/\lambda_T) = L_{ii}^{T_0}(1/\lambda_T^p) = L_{ii}^{T(0)}(1/\lambda_{T_0}) \qquad \text{and}$$

$$(1/\lambda_T)(L_{ii}^T)'(1/\lambda_T) = (p/\lambda_{T_0})(L_{ii}^{T_0})'(1/\lambda_{T_0}).$$

The statements follow from these equations and the fact that i was chosen arbitrarily. $\square$

Lemma 7.1.37 *(Compare to Remark 7.1.12) Let T be a countable, nonnegative and irreducible matrix with Perron value λ and period p. Also, let C_k and T_k be as as described. If T is recurrent then there exist unique strictly positive row and column vectors, ℓ and r, with $\ell T = \lambda\ell$ and $Tr = \lambda r$.*

Proof. The vectors $\ell^{(i)}$ and $r^{(j)}$ described before Lemma 7.1.9 are eigenvectors for T corresponding to λ by the same reasoning used in the proof of Lemma 7.1.9. Let ℓ and r be eigenvectors. They are unique because for each $0 \leq k \leq p - 1$ the vectors $(\ell_{j_1}, \ldots, \ell_{j_n}, \ldots)$ and $(r_{j_1}, \ldots, r_{j_n}, \ldots)$, $j_t \in C_k$, are the unique eigenvectors for T_k corresponding to λ^p. $\square$

Lemma 7.1.38 *(Compare to Remark 7.1.17) Let T be a countable, nonnegative and irreducible matrix with Perron value λ. Then T is positive recurrent if and only if there exist a nonzero row vector, $x \geq 0$, and a nonzero column vector, $y \geq 0$ with $xT = \lambda x$, $Ty = \lambda y$ and $y \cdot x < \infty$.*

Proof. For transient T apply Lemma 7.1.16 to possible row and column eigenvectors restricted to each T_k. For recurrent T apply Lemma 7.1.14 to the row and column eigenvectors from Lemma 7.1.37 restricted to each T_k. $\square$

Before Lemma 7.1.14 the mean recurrence weight, $\mu(i)$, was defined for aperiodic matrices. Make the same definition for arbitrary irreducible matrices.

Lemma 7.1.39 *(Compare to Lemma 7.1.19) If T is a countable, nonnegative and irreducible matrix with a Perron value λ and period p then:*

(i) $\lim\limits_{n\to\infty} (T^n)_{ij}/\lambda^n = 0$ *for all pairs i and j when T is transient;*

(ii)

$$\lim_{n\to\infty} \frac{(T^{np})_{ij}}{\lambda^{np}} = \begin{cases} \dfrac{L_{ij}(1/\lambda)}{p\cdot\mu(i)} = \dfrac{R_{ij}(1/\lambda)}{p\cdot\mu(j)} & \text{if } i,j \in C_k \\[2mm] 0 & \text{otherwise} \end{cases}$$

when T is recurrent.

Proof. First, for i and j not in the same C_k, $(T^{np})_{ij} = 0$. Consider i and j in the same C_k. Statement (i) follows because $(T^n)_{ij} = 0$ when $n \neq mp$. When $n = mp$ apply Lemma 7.1.19(i) to T_k. Statement (ii) follows from Lemma 7.1.19(ii) applied to T_k.   □

Exercises

1. Prove Lemma 7.1.1 (iii) when $\limsup \sqrt[n]{T^n} = \infty$.
2. Suppose T is a stochastic matrix as discussed before Remark 7.1.2. Show that the expected number of returns to a vertex is finite when the matrix is transient and infinite when the matrix is recurrent. Show that the expected return time is finite when the matrix is positive recurrent and infinite when the matrix is null recurrent.
3. Use the usual Perron-Frobenius Theorem 1.3.2 and the functions $T_{ii}(z)$ and $L'_{ii}(z)$ to show that a finite, nonnegative, irreducible and aperiodic matrix is positive recurrent.
4. Ballot Theorem: Consider the lattice $\mathbb{Z}^2$. Let $(n_0, m_0), (n_k, m_k) \in \mathbb{Z}^2$ with $n_0 < n_k$. A *walk* from (n_0, m_0) to (n_k, m_k) is a sequence of lattice points $\{(n_0, m_0), (n_1, m_1), \ldots, (n_{k-1}, m_{k-1}), (n_k, m_k)\}$ with $n_i = n_{i-1} + 1$ and $m_i \in \{m_{i-1} - 1, m_{i-1}, m_{i-1} + 1\}$ for $i = 1, \ldots k$. It is a walk on the lattice where each step moves once to the right and either moves down one level, maintains the same level or goes up a level. The Ballot Theorem states that the number of walks from $(0, 0)$ to (n, m) with $n \geq m > 0$ that do not touch the x-axis between $(0, 0)$ and (n, m) is

$$\binom{p+q-1}{p-1} - \binom{p+q-1}{p}$$

where $n = p + q$ and $m = p - q$.

(i)   Prove the theorem. Hint: The first factorial is the total number of walks between the two points and the second is the number of such walks that do touch the x-axis between the points [Fe].

(ii) Show how this applies to the computation of $\ell_{00}(2n)$ in Example 7.1.28.

5. Fatou's Lemma states that if $\{f_n\}$ is a sequence of nonnegative measurable functions which converges to a function f almost everywhere on a set E, then

$$\int_E f \le \liminf \int_E f_n.$$

See [Rd] for a reference. Show how this applies in the proof of Lemma 7.1.34.

§ 7.2 Basic Symbolic Dynamics

Let $\{1, 2, \ldots\}$ be a countable set. With the discrete topology it is a noncompact metrizable space. Form the product space, $\Sigma_{\mathcal{N}} = \{1, 2, \ldots\}^{\mathbb{Z}}$. With the product topology it is a noncompact metrizable space. As in the finite state case, the cylinder sets form a countable basis of open-closed sets. Again, the space is totally disconnected or 0-dimensional and perfect. The shift, σ, is a homeomorphism of the space to itself. The time zero partition $\mathcal{P} = \{[1]_0, [2]_0, \cdots\}$ generates the topology under the action of σ. The dynamical system $(\Sigma_{\mathcal{N}}, \sigma)$ is the *full shift on countably many symbols*. If T is a countable, zero-one matrix or a countable, nonnegative, integer matrix then as in the finite state case we use transition rules to define a shift-invariant subset of the full shift on countably many symbols. It is denoted by Σ_T. With the subspace topology it is a noncompact, metrizable space and $\sigma : \Sigma_T \to \Sigma_T$ is a homeomorphism. The dynamical system (Σ_T, σ) is the *countable state Markov shift* defined by T. The Markov shift Σ_T is *topologically transitive* if there is a point $x \in \Sigma_T$ whose orbit is dense in Σ_T. The Markov shift is *topologically mixing* if for any two open sets U and V, there is an $N \ge 0$ so that for every $n \ge N$, $U \cap \sigma^n(V) \ne \emptyset$. The next two observations are analogues of the statements about finite state Markov shifts.

Observation 7.2.1 *A countable state Markov shift, (Σ_T, σ), is topologically transitive if and only if T is irreducible.*

Proof. Suppose T is irreducible. For every pair of states i and j there is an allowable string $[k_0, k_1, \ldots, k_m]$ with $k_0 = i$ and $k_m = j$. Enumerate the cylinder sets. Then we can produce a point $x \in \Sigma_T$ whose orbit passes through each cylinder set. We do this by successively connecting the last symbol of one cylinder set to the first symbol of the next by an allowable string.

If T has a dense forward orbit then it is clear that for each pair of states i and j there is an allowable string that begins with i and ends with j. This means there is an n so that $(T^n)_{ij} > 0$.   □

As for subshifts of finite type we say a countable state Markov shift is *irreducible* if its transition matrix is irreducible.

Observation 7.2.2 *A countable state Markov shift* (Σ_T, σ) *is topologically mixing if and only if* T *is irreducible and aperiodic.*

Proof. Suppose T is irreducible and aperiodic. Lemma 7.1.1 (i) shows that for any pair of states i and j there is a k so that for every $n \geq k$, $(T^n)_{ij} > 0$. This means that if $[i_0, \ldots, i_k]_0$ and $[j_0, \ldots, j_\ell]_0$ are cylinder sets there is k so that for each $n \geq k$, $[i_0, \ldots, i_k]_0 \cap \sigma^n([j_0, \ldots, j_\ell]_0) \neq \phi$. Since every open set is a union of cylinder sets, Σ_T is topologically mixing. If (Σ_T, σ) is topologically mixing then by considering the open sets in the time zero partition it is clear that T is irreducible and aperiodic. □

We say a countable state Markov shift is *irreducible and aperiodic* when its transition matrix is irreducible and aperiodic.

A space X is *locally compact* if for every x in X there is a compact set whose interior contains x. A nonnegative matrix is *locally finite* if every row and column sum is finite. Equivalently, it is locally finite if its graph has finitely many edges entering and leaving each vertex.

Observation 7.2.3 *Let* (Σ_T, σ) *be an irreducible, countable state Markov shift.*

(i) Σ_T *is locally compact if and only if* T *is locally finite.*

(ii) Σ_T *is locally compact if and only if (any) every cylinder set is compact.*

(iii) Σ_T *is locally compact if and only if (any) every open set contains a compact subset with interior.*

(iv) Σ_T *is nonlocally compact if and only if every compact set has no interior.*

Proof. Statements (i) and (ii) are clearly equivalent. Let $[i_0, \ldots, i_k]_0$ be a cylinder set. For each ℓ we have

$$[i_0, \ldots, i_k]_0 = \bigcup [i_\ell, \ldots, i_0, \ldots, i_k, \ldots, i_{k+\ell}]_{-\ell},$$

for some collections of cylinder sets $\{[i_\ell, \ldots, i_0]\}$ and $\{[i_k, \ldots, i_{k+\ell}]\}$. The cylinder set $[i_0, \ldots, i_k]_0$ is compact if and only if the collections of cylinder sets $\{[i_\ell, \ldots, i_0]\}$ and $\{[i_k, \ldots, i_{k+\ell}]\}$ are finite for every ℓ. This is true if and only if T is locally finite. This proves (i) and (ii). Since the cylinder sets form a basis for the topology (iii) follows from (ii). For (iv) let K be a compact subset of Σ_T. For each t, $\{[i]_t : i = 1, 2, 3, \ldots\}$ is an open cover of Σ_T and so there is a finite subcover of K. If T is not locally finite K cannot contain a cylinder set. □

In Section 1.4 we defined topological entropy for subshifts of finite type by counting allowable words. Then in Section 6.2 we examined shift-invariant probability measures and defined their measure-theoretic entropy. We saw that the topological entropy is the supremum over the measure-theoretic entropies. Also, we showed that their is one measure whose entropy is equal to the topological entropy. It is a Markov measure. In the next several pages we will examine these questions for countable state Markov shifts. Because countable state Markov shifts are noncompact defining topological entropy is a problem. We will examine several definitions. No one definition produces a notion of entropy with all the desired properties. We will also look at shift-invariant, Borel probability measures and compute some of their entropies. We will prove that the supremum over the measure-theoretic entropies is equal to one of the types of entropy we define. But, only when the transition matrix is positive recurrent is there a measure whose entropy equals the supremum. In the positive recurrent case, that measure is unique. It is a Markov measure similar to the maximal measure for subshifts of finite type.

Suppose X is a compact topological space and T is a continuous homeomorphism of X to itself. We define the *topological entropy of T*, denoted by $h(X, T)$ or $h(T)$, as follows. Let $\alpha = \{\alpha_1, \dots, \alpha_k\}$ be a finite open cover of X. Define the *time n wedge of α* by

$$\overset{n-1}{\underset{}{\vee}} T^{-i}\alpha = \{\alpha_{i_0} \cap T^{-1}\alpha_{i_1} \cap \dots \cap T^{-(n-1)}\alpha_{i_{n-1}}, \text{ all } [i_0, \dots, i_{(n-1)}]\}$$

and define $N(\overset{n-1}{\vee} T^{-i}\alpha)$ to be the minimal cardinality of all subcovers of $\overset{n-1}{\vee} T^{-i}\alpha$. Next define the *topological entropy of T with respect to α* by

$$h(T, \alpha) = \lim \frac{1}{n} \log N(\overset{n-1}{\vee} T^{-i}\alpha).$$

Finally define the *topological entropy of T* by

$$h(T) = \sup_{\alpha}\{h(T, \alpha)\}$$

where the supremum is over all finite open covers of X. With this definition it is easy to see that topological entropy is an invariant of topological conjugacy.

Now suppose X (not necessarily compact) is a metrizable space and T a homeomorphism of X to itself. A metric on X is *compatible* if the topology it induces agrees with the topology on X. Let d be a metric on X that is compatible with the topology. We define one type of entropy of T which depends on d. Let K be a compact subset of X. A subset E of X is a (K, n, ε) *spanning set* if for every $y \in K$ there is an $x \in E$ so that $d(T^i(x), T^i(y)) < \varepsilon$ for $0 \le i < n$. Let $r_d(K, n, \varepsilon)$ be the minimal cardinality of a (K, n, ε) spanning set. Define the *topological entropy of T with respect to d* by

$$h_d(T) = \sup_K \lim_{\varepsilon \to 0} \overline{\lim_{n \to \infty}} \frac{1}{n} \log r_d(K, n, \varepsilon),$$

where the final supremum is over all compact subsets of X. There is also the dual formulation. A subset F of K is an (K, n, ε) *separated* set if for every x and y in F there is a k, $0 \le k < n$, so that $d(T^k(x), T^k(y)) \ge \varepsilon$. Let $s_d(K, n, \varepsilon)$ be the maximal cardinality of a (K, n, ε) separated set. Define the *topological entropy of T with respect to d* by

$$h_d(T) = \sup_K \lim_{\varepsilon \to 0} \overline{\lim_{n \to \infty}} \frac{1}{n} \log s_d(n, \varepsilon).$$

where the final supremum is again taken over all compact subsets of X. This quantity depends on the metric. Two uniformly equivalent metrics will produce the same entropy. Metrics that are equivalent but not uniformly equivalent may give different values for the entropy of T. When X is compact all metrics are uniformly equivalent and produce the same entropy. When X is a compact metric space the metric definition of topological entropy agrees with the open cover definition (Exercise 4). These entropies agree with the definition of the entropy of the shift acting on a subshift of finite type as defined in Section 1.4 (Exercise 5).

Let X be a metrizable space and T a homeomorphism of X to itself. In general, there may or may not be an invariant Borel probability measure (Exercise 7). If X is compact there is always at least one such measure. If X is an irreducible, countable state Markov shift then the Borel σ-algebra is generated by the shift acting on the time zero cylinder sets and there are always shift-invariant probability measures. Recall the definition of measure-theoretic entropy from Section 6.2.

There is a general theorem relating the entropy of a transformation with respect to metrics to the measure-theoretic entropies.

Proposition 7.2.4 *If X is a metrizable space and T is a homeomorphism of X to itself then,*

$\sup\{h(Y, T) : Y \subseteq X$ is compact and invariant$\}$

$\quad \le \sup\{h_\mu(T) : \mu$ is an invariant Borel probability measure$\}$

$\quad \le \inf\{h_d(T) : d$ is a compatible metric on $X\}$

$\quad \le \inf\{h(\widehat{X}, \widehat{T}) : \widehat{X}$ is a metric compactification of X,

$\quad \widehat{T}$ is a homeomorphism of $\widehat{X}$ and $T = \widehat{T}$ on $X\}$.

Proof. Line one is less than or equal to line two because if Y is a compact invariant subset of X then by the Variational Principle (see the notes on Section

6.2) $h(Y, T) = \sup\{h_\mu(Y, T)\}$, where the supremum is taken over all invariant Borel probability measures on Y, and the set of invariant Borel probability measures for T on Y is a subset of the set of invariant Borel probability measures for T on X. Line two is less than or equal to line three but the proof is slightly complicated. We will not prove it here but a reference can be found in the notes at the end of this chapter. For our purposes (Proposition 7.2.6) it is enough to prove that line one is less than or equal to line three and that both lines two and three are less than or equal to line four. Line one is less than or equal to line three because if Y is a compact invariant subset of X and d is a metric on X, then it restricts to a metric on Y and $h(Y, T) = h_d(Y, T) \leq h_d(X, T)$. Let $(\widehat{X}, \widehat{T})$ be as described in the fourth line. Then line two is less than $h(\widehat{X}, \widehat{T})$ because the set of invariant Borel probability measures for T is a subset of the set of invariant Borel probability measures for $\widehat{T}$. Line three is less than or equal to $h(\widehat{X}, \widehat{T})$ because a metric on $\widehat{X}$ restricts to a metric on X with $h_d(X, T) \leq h_d(\widehat{X}, \widehat{T}) = h(\widehat{X}, \widehat{T})$.   □

Let Σ_T be a countable state Markov shift. Define a metric on the alphabet by $\varrho_-(i,j) = |\frac{1}{i} - \frac{1}{j}|$. We are thinking of the alphabet as $\{1, \frac{1}{2}, \frac{1}{3}, \dots\}$. We use this metric on the alphabet to define a product metric on $\{1, 2, 3, \dots\}^{\mathbb{Z}}$, and so on Σ_T. It is explicitly given by the equation

$$d_-(x, y) = \sum_{i=-\infty}^{+\infty} \frac{|\frac{1}{x_i} - \frac{1}{y_i}|}{2^{|i|}}.$$

Remark 7.2.5 The metrics ϱ_- and d_- are totally bounded but not complete. The completion of the metric ϱ_- and the one point compactification of the alphabet are homeomorphic. We can think of both of them as $\{1, \frac{1}{2}, \frac{1}{3}, \dots, 0\}$. This means d_- extends to a metric on the compact space $\{1, \frac{1}{2}, \frac{1}{3}, \dots, 0\}^{\mathbb{Z}}$. The shift extends to a homeomorphism of $\{1, \frac{1}{2}, \frac{1}{3}, \dots, 0\}^{\mathbb{Z}}$, since the new space is defined in terms of a compactified alphabet. We have $\Sigma_T \subseteq \{1, \frac{1}{2}, \frac{1}{3}, \dots, 0\}^{\mathbb{Z}}$. Denote the closure of Σ_T in this space by $\overline{\Sigma}_T$. We now have that $\Sigma_T \subseteq \overline{\Sigma}_T$, the shift is a homeomorphism on each and $\overline{\Sigma}_T$ is compact. In fact, the completion of the metric d_- on Σ_T is homeomorphic to $\overline{\Sigma}_T$. Denote the topological entropy of the shift on $\overline{\Sigma}_T$ by $h(\overline{\Sigma}_T)$.

Proposition 7.2.6 *Let Σ_T be an irreducible, countable state Markov shift, then the following are equivalent.*

(i) $\log \lambda$, *where λ is the Perron value for T*

(ii) $\limsup_{n\to\infty} \frac{1}{n} \log(T^n)_{ii}$, *for any state i*

(iii) $\sup\{h(\Sigma_A) : \Sigma_A \subseteq \Sigma_T$ *and Σ_A is a finite state Markov shift*$\}$

(iv) $\sup\{h_\mu(\Sigma_T) : \mu$ *is an invariant, Borel probability measure on $\Sigma_T\}$*

(v) $inf\{h_d(\Sigma_T) : d$ is a compatible metric on $\Sigma_T\}$

(vi) $h_{d_-}(\Sigma_T)$

(vii) $h(\overline{\Sigma}_T)$

Proof. First observe that we have already established the following relations.

$$\log \lambda = lim_{n\to\infty} \frac{1}{n} \log(T^n)_{ii}$$
$$= \sup\{h(\Sigma_A) : A \text{ is a finite subgraph of } T\}$$
$$\leq \sup\{h(Y,\sigma) : Y \subseteq \Sigma_T \text{ is compact and invariant}\}$$
$$\leq \sup\{h_\mu(\Sigma_T) : \mu \text{ is an invariant Borel probability measure}\}$$
$$\leq \inf\{h_d(\Sigma_T) : d \text{ is a compatible metric on } \Sigma_T\}$$
$$\leq h_{d_-}(\Sigma_T)$$
$$\leq h(\overline{\Sigma}_T)$$

The equality in the second line is just the definition of the Perron value. The equality in the third line is the Finite Approximation Theorem 7.1.4. The inequality in the fourth line follows because the collection of subshifts of finite type in line three is a proper subset of the collection of all compact invariant sets. The inequalities in the fifth and sixth lines are part of Proposition 7.2.4. The inequality in the seventh line is clear. The last inequality is because the metric d_- on Σ_T extends to a metric on $\overline{\Sigma}_T$ (see Remark 7.2.5). To complete the proof we will show that

$$h(\overline{\Sigma}_T) \leq \sup\{h(\Sigma_A) : A \text{ is a finite subgraph of } T\}.$$

We need a simple coding argument. For each $n \in \mathbb{N}$ define a partition of $\overline{\Sigma}_T$, $\mathcal{P}_n = \{[1]_0, \ldots, [n]_0, [\bar{n}]_0\}$, where $[\bar{n}]_0 = \{x : x_0 > n\}$. For each n let $h(\overline{\Sigma}_T|\mathcal{P}_n)$ denote the topological entropy of $\overline{\Sigma}_T$ relative to the open cover (partition) $\mathcal{P}_n$. Since the diameter of $\mathcal{P}_n \to 0$ as $n \to \infty$, $\lim_{n\to\infty} h(\overline{\Sigma}_T|\mathcal{P}_n) = h(\overline{\Sigma}_T)$. For each n there is an m_n so that for every $k \geq m_n$ and any pair of states $i,j \in \{1,\ldots,n\}$ there is a path of length k from i to j. Choose a finite, irreducible and aperiodic subgraph, A_n of T, with the following three properties. First, it contains the states $1,\ldots,n$. Second, it contains every path of length m_n or less that begins at a state $i \in \{1,\ldots,n\}$ and ends at a state $j \in \{1,\ldots,n\}$. Third, it contains some path of length k from i to j for every $i,j \in \{1,\ldots,n\}$ and $k \geq m_n$. Let Λ_{m_n} denote the subshift of Σ_2 that consists of all sequences of zeros and ones where ones occur only in blocks of length $m_n - 2$ or more. It is easy to see that as $n \to \infty$, $m_n \to \infty$, and $h(\Lambda_{m_n}) \to 0$. Define a one block map from $\Sigma_{A_n} \times \Lambda_{m_n}$ into $\{1,\ldots,n,\bar{n}\}^{\mathbb{Z}}$ by

$$\phi(i,t) = \begin{cases} i & \text{if } i \leq n, \text{ and } t = 0 \\ \bar{n} & \text{otherwise.} \end{cases}$$

For every point $x \in \overline{\Sigma}_T$ the $\mathcal{P}_n$ description of x lies in
$\phi(\Sigma_{A_n} \times \Lambda_{mn}) \subseteq \{1, \ldots, n, \bar{n}\}^{\mathbb{Z}}$. This means that

$$h(\overline{\Sigma}_T | \mathcal{P}_n) \leqq h(\Sigma_{A_n} \times \Lambda_{mn}) = h(\Sigma_{A_n}) + h(\Lambda_{mn}).$$

Taking the limit as n goes to infinity and using statement (*iii*) we see that
$h(\overline{\Sigma}_T) \leq \sup\{h(\Sigma_A : A \text{ is a finite subgraph of } T\}$.   □

In the literature on countable state Markov shifts this entropy, which we
will denote by $h(\Sigma_T)$, is often referred to as the Gurevich or loop entropy and
denoted $h_G(\Sigma_T)$.

We carry over some of the terminology for subshifts of finite type to countable state Markov shifts. Let Σ_T and Σ_S be two countable state Markov shifts.
A continuous, onto, shift commuting map from Σ_T to Σ_S is called a *factor map*.
We say that Σ_S is a *factor* of Σ_T. If the map is a homeomorphism we say that
Σ_T and Σ_S are *topologically conjugate* or simply *conjugate* and that the map is
a *conjugacy*.

Remark 7.2.7 A continuous shift commuting map between finite state Markov
shifts is necessarily a finite block map as we saw in Chapter 1. This is not true in
the infinite state case. See Example 7.2.12. The entropy, $h(\Sigma_T)$, is an invariant
of topological conjugacy. This follows from (*iii*), (*iv*), or (*v*) of Proposition 7.2.6.
However, it is possible for Σ_S to be a factor of Σ_T but have $h(\Sigma_S) > h(\Sigma_T)$.
See Example 7.2.8 This is, of course, not possible for subshifts of finite type as
we saw in Chapter 4.

Example 7.2.8 Choose two sequences of integers: $\{d_k\}$ an increasing sequence,
and $\{r_k\}$ arbitrary, as in Example 7.1.32. Define Σ_T by the following graph.
There are two edges from state n to state $n + 1$ for each $n \in \mathbb{Z}$ and a path of
length r_k leading from state d_k to state $-d_k$ for each k. The computations to
identify the recurrence properties of the graph are just like those in Example
7.1.32, except that $\ell_{00}(2d_k + r_k) = 2^{2d_k}$. By varying the sequences we can vary
the recurrence properties while simultaneously making the entropy as small
as we like, in particular less than $\log 2$. Map Σ_T onto the full two shift by the
one block map described by the labelling on the graph.

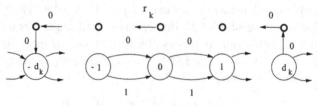

Figure 7.2.1

Next is an observation stated without proof. It is a special case of a more general theorem and is included for completeness. See the notes for references.

Observation 7.2.9 *Let Σ_T be an irreducible, countably infinite state Markov shift then*
$$\sup\{h_d(\Sigma_T) : d \text{ is a compatible metric on } X\} = +\infty.$$

Let Σ_T be a countable state Markov shift. Define a metric on the alphabet by $\varrho^*(i,j) = 1 - \delta_{ij}$, the usual discrete metric. We use this metric on the alphabet to define a product metric on $\{1, 2, 3, \ldots\}^{\mathbb{Z}}$, and so on Σ_T. It is explicitly given by the equation
$$d^*(x,y) = \sum_{i=-\infty}^{+\infty} \frac{1 - \delta_{ij}}{2^{|i|}}.$$

The metrics ϱ^* and d^* are complete but not totally bounded. Let $B(i,n) = \sum_j (T^n)_{ij}$, the number of blocks of length n that begin at state i.

Observation 7.2.10 *Let Σ_T be an irreducible, countable state Markov shift. Then*
$$h_{d^*}(\Sigma_T) = \limsup \frac{1}{n} \log B(i,n).$$

Proof. When a state of T has an infinite number of edges leaving it both values are clearly infinite. Suppose every state has a finite number of edges leaving it. Fix a state i and construct a nonempty compact subset, K, that is homeomorphic to an infinite product with all possible successor states to the right of i and some finite subcollection of the possible predecessors at each stage to the left of i. Recall Observation 7.2.3 and its proof. Observe that $h_{d^*}(\Sigma_T, K)$ and $\limsup \frac{1}{n} \log B(i,n)$ are equal. Then it is easy to see that all such K give the maximal value for $h_{d^*}(\Sigma_T)$ and that $\limsup \frac{1}{n} \log B(i,n)$ does not depend on the choice of i. $\square$

Remark 7.2.11 In the literature on countable state Markov shifts this entropy is often referred to as the Salama or block entropy and denoted $h_S(\Sigma_T)$. See the notes for references. Salama defined it and showed that for any α, β, with $0 < \alpha \leq \beta \leq \infty$ there is a countable state Markov shift with $h_{d_*}(\Sigma_T) = \alpha$ and $h_{d^*}(\Sigma_T) = \beta$. The entropy $h_{d^*}(\Sigma_T)$ is not a conjugacy invariant. This is easily seen because it depends on a particular choice of metric or state partition.

Example 7.2.12 In this example the shifts Σ_T and Σ_S are topologically conjugate but $h_{d^*}(\Sigma_T) = \log 2$ and $h_{d^*}(\Sigma_S) = \log 4$. Define Σ_T exactly as in Example 7.2.8 using the two sequences of integers $\{d_k\}$ (now chosen to be all even) and $\{r_k\}$. As observed there, by choosing the sequence $\{r_k\}$ going to infinity quickly enough we can make $h(\Sigma_T)$ as small as we want. When $h(\Sigma_T) < \log 2$, $h_{d*}(\Sigma_T) = \log 2$ because $B(0,n) = \sum_{k=0}^{n} t_{00}(k)2^{n-k}$ and so $\lim 1/n \log B(0,n) = \log 2$. As observed in Example 7.2.8.

$$\ell_{00}^T(2d_k + r_k) = 2^{2d_k}.$$

Define Σ_S similarly, but with four edges from vertex n to vertex $n+1$, and a path of length $r_k + d_k$ from state $d_k/2$ to state $-d_k/2$. Here we have

$$\ell_{00}^S(2d_k + r_k) = 4^{d_k}.$$

A similar computation as before will show that $h_{d*}(\Sigma_S) = \log 4$.

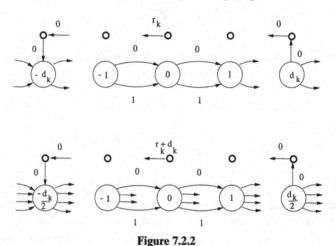

Figure 7.2.2

Now define a conjugacy, π, from Σ_S to Σ_T using the state 0 as a "marker". Recall how we used markers in Chapter 4. We think of points in Σ_T and Σ_S as infinite walks on the graphs. Consider those paths with state 0 at the decimal point (ie. an edge leading out of state 0 at time one) that never pass through state 0 again (in either forward or backward time). It is easy to see how to define a one-to-one, onto map, π_0, from those paths in Σ_T to those paths in Σ_S. We make the map so that it sends blocks of length $2n$ from Σ_T that begin (or end) at state 0 to blocks of length n in Σ_S that begin (or end) at state 0, and so that it is consistent (with itself) on shorter blocks. We will use this to make the conjugacy. First extend the map to all paths that pass through state 0 once. The image point will pass through state 0 in Σ_S at the same time. We are using the state 0 as a "marker", it tells us how to apply the map π_0. Now we define the map π on loops that begin and end at 0 without returning to 0 at an intermediate time. Such a loop from Σ_T has length $2d_k + r_k$. It begins with d_k edges leading along the integer vertices from state 0 to state d_k, takes a return path of length r_k to state $-d_k$, and ends with d_k edges leading from state $-d_k$ to state 0. We map this to a similar loop in Σ_S by applying π_0 to the first and last d_k edges and filling in the middle $r_k + d_k$ edges in the only way possible. This defines a one-to-one correspondence between the loops of length $2d_k + r_k$ in Σ_T that begin and end at state 0 and do not pass through 0 at any intermediate time and the same type of loops in Σ_S. Now we piece these together to define a conjugacy. To determine the image of a point in Σ_T under π we use 0 as a marker. The image point will pass through 0 at the same times. The images of the pieces of the point between successive occurrences of 0 are determined as just described. If there is a first or last occurrence of 0 we use the original π_0. This defines a conjugacy between the shifts. This map is far from a finite block map, it is not uniformly continuous in either direction when the two Markov shifts are equipped with the d^* metric, and it is very different from the type of conjugacy that occurs between finite state Markov shifts. It demonstrates how the noncompactness comes into play.

Next we turn to a discussion of Markov measures, measure-theoretic entropy, and maximal measures. We will use the definitions, terminology and notation from Section 6.2. We recall a few.

Let T be a countable state, irreducible, zero-one matrix. If P is a (not necessarily irreducible) stochastic matrix compatible with T, and p is a stationary probability vector for P, then the pair (p, P) defines a one-step Markov measure, μ_P, on Σ_T. It is defined on cylinder sets by

$$\mu_P([i_0, \ldots, i_\ell]_t) = p_{i_0} P_{i_0 i_1} \cdots P_{i_{\ell-1} i_\ell}.$$

This will extend to produce a shift-invariant, probability measure on Σ_T. The entry P_{ij} is the *transition probability* from state i to state j. This is exactly as in Section 6.2.

Next for an invariant, Borel probability measure μ on Σ_T we define its *one-step Markov approximation*, as before, to be the Markov measure with

$$p_i = \mu([i]) \quad \text{and} \quad P_{ij} = \frac{\mu([i,j])}{\mu([i])}.$$

Recall the definition of the support of a measure. The *support* of a measure μ, $supp(\mu)$, is the smallest closed set with measure one. Also, recall that the support of a Markov measure defined by (p, P) on Σ_T is $\Sigma_{P0} \subseteq \Sigma_T$.

Let Σ_T be a countable state Markov shift. Proposition 7.2.6 tells us that the log of the Perron value is equal to the supremum over the measure-theoretic entropies for invariant, Borel probability measures. A measure whose measure-theoretic entropy is equal to this supremum is called a *maximal measure*.

We will prove the following theorem and show the maximal measure has the same form as the maximal measure for a subshift of finite type. To do so we will make use of the entropy theory discussed in Section 6.2.

Proposition 7.2.13 *An irreducible, countable state Markov shift, Σ_T, has a maximal measure if and only if T is positive recurrent. When this maximal measure exists, it is unique and Markov.*

Remark 7.2.14 First we will give an intuitive argument for the theorem. Consider the partition of Σ_T by length n cylinder sets. The renewal theorem estimates the asymptotic number of all such cylinders that begin with a fixed i and end with a fixed j, $(T^n)_{ij}$. The Shannon-McMillan-Breiman theorem estimates the asymptotic measure of almost all such cylinders in terms of the entropy of the measure. If there is a maximal measure it must have entropy $\log \lambda$. The Shannon-McMillan-Breiman theorem says that within constant bounds almost all cylinders will have measure λ^{-n}. The total measure of all such cylinders that begin with i and end with j will be within a constant less than or equal to $(T^n)_{ij}/\lambda^n$. The renewal theorem says that this will go to zero unless T is positive recurrent. This argument doesn't actually work because the rates of convergence are too delicate.

The proof of the theorem follows the lines of the proof for a subshift of finite type, Theorem 6.2.20. The theorem will follow from the next four lemmas.

Lemma 7.2.15 *If T is an irreducible, positive recurrent matrix. Then (Σ_T, σ) has a maximal measure. It is a Markov measure defined by the pair (p, P) with*

$$p_i = \ell_i r_i \qquad P_{ij} = T_{ij} \frac{r_i}{\lambda r_j}.$$

Here, λ is the Perron value for T and ℓ and r are the left and right eigenvectors for λ, normalized so that $\ell \cdot r = 1$.

Proof. By Observation 6.2.13 the entropy of any measure is less than or equal to $\log \lambda$ where λ is the Perron value of T. If T is positive recurrent then by Remark 7.1.17, T has unique left and right eigenvectors corresponding to λ whose dot product is finite. Given these vectors we can define P and p as stated in the lemma. Observation 6.2.10 provides a formula for the entropy of a Markov measure. The same computation used for Observation 6.2.11 proves Lemma 7.2.15.   □

Lemma 7.2.16 *If μ is a maximal measure for the irreducible, countable state Markov shift, (Σ_T, σ), then it is a one-step Markov measure.*

Proof. The proof in the countable state case is the same as the proof of Corollary 6.2.19.   □

Lemma 7.2.17 *Suppose* (Σ_T, σ) *is an irreducible, countable state Markov shift. Then there is at most one one maximal measure with support equal to* Σ_T.

Proof. First we apply Lemma 7.2.16 to see that any maximal measure must be a one-step Markov measure. Next suppose there are two maximal measures μ_P and μ_Q defined by (p, P) and (q, Q) with support equal to Σ_T. Now we use the Markov weight sets on the graph of T as described in Lemma 6.2.14. We produce a new one-step Markov measure μ_R as described in the first proof of Observation 6.2.15. Either the entropy of μ_R is greater than the entropy of μ_P and μ_Q or by the assumption that both have full support $P_{ij} = Q_{ij}$ for every pair i and j with $T_{ij} > 0$.   □

There are two differences between the countable state case and the finite state case as described in the proof of Lemma 6.2.14. The first is that we have only produced a maximal measure when T is positive recurrent. The second is that to prove uniqueness in Lemma 7.2.17 we have had to assume that any maximal measure has full support.

Lemma 7.2.18 *If there is a maximal measure for the irreducible, countable state Markov shift* (Σ_T, σ) *it must have the form of the measure described in Lemma 7.2.15. This can happen only if* T *is positive recurrent.*

Proof. There are two cases to consider. First we consider the case when T is recurrent and then turn to the case when T is transient.

Suppose T is recurrent. By Observation 6.2.13 we know that the entropy of any measure is less than or equal to $\log \lambda$, where λ is the Perron value of T. Apply the Generalized Perron-Frobenius Theorem 7.1.3 to see that any maximal measure must have support equal to Σ_T. Now apply Lemma 7.2.17 to prove the uniqueness of the measure. Lemma 7.2.15 proves the existence of a maximal measure when T is positive recurrent. Next apply the method of Lagrange multipliers as in the second proof of Corollary 6.2.15. It shows that if there is a maximal measure it must have the form described in Lemma 7.2.15. When T is null recurrent Remark 7.1.17 shows that the dot product of the left and right eigenvectors for T corresponding to λ is infinite. Consequently, the maximal measure does not exist.

Suppose T is transient. The method of Lagrange multipliers shows that a maximal measure with support Σ_T cannot exist. Since T is transient there could be a maximal measure with support not equal Σ_T. Suppose such a measure exists. By Lemma 7.2.16 it is a one-step Markov measure. So it has support equal to Σ_S where S is a submatrix of T. By Lemma 6.2.13, S and T must have the same Perron value. Now apply Lemma 7.1.24 to see that S is transient and obtain a contradiction.   □

Exercises

1. Show that a countable state Markov shift with one state having infinitely many followers is not locally compact.
2. Prove that a topological space has a one point compactification if and only if it is locally compact.
3. Prove that the definitions of topological entropy using spanning sets and separated sets are equivalent.
4. Prove that the open cover definition of topological entropy and the metric definition are equivalent when the space is compact.
5. Show that for a subshift of finite type the definitions of topological entropy given in this section are equivalent to the one given in Section 1.4.
6. Give an example of a noncompact space with a homeomorphism from it to itself and two metrics where the topological entropies determined by the metrics disagree.
7. Give an example of a noncompact space with a homeomorphism from it to itself where there is no invariant, Borel probability measure.
8. Give an example of a compact space with a homeomorphism from it to itself where there is no maximal measure.

Notes

Section 7.1

Countable state Markov chains were introduced into probability theory by A. Kolmogorov in the mid 1930's. In our terminology they are countable state Markov shifts with one-step Markov measures. The measures were classified as positive recurrent, null recurrent or transient. This is discussed before and in Remark 7.1.2 and also in Exercise 2. W. Feller's book [Fe] contains a thorough discussion of the probability aspects. The Perron-Frobenius theory for countable, nonnegative matrices presented here is due to D. Vere-Jones [V-J1], [V-J2] in the 1960's. In particular, he proved the Generalized Perron-Frobenius Theorem 7.1.3 and the Finite Approximation Theorem 7.1.4. The proofs contained here are somewhat different from his proofs. E. Seneta's book [Se] also contains a discussion of this theory. Theorem 7.1.8 is a special case of the Renewal Theorem of P. Erdős, W. Feller and H. Pollard [EFP]. It was proved in 1949 and is contained in [Fe]. The definition of $\varrho(i)$ preceeding Lemma 7.1.3

as well as Lemmas 7.1.3 to 7.1.7 are due to I. Salama in the mid to late 1980's [Sl1], [Sl2], [Sl3]. Stirling's Formula and the Ballot Theorem used in Example 7.1.2 are classical combinatorial formulas [Fe].

Section 7.2

Our discussion of topological entropy begins with the open set definition due to R. Adler, A. Konheim and H. McAndrew [AKM] in 1965. Then we use the metric definition on noncompact spaces due to R. Bowen [Bo2] in 1971. There are many formulations of entropy on noncompact spaces. A discussion and comparison of several of them can be found in [HK]. In particular, Proposition 7.2.4 and Observation 7.2.9 are proved there. The metric defined before Remark 7.2.6 has been used by many people. Proposition 7.2.6 defines an entropy that is usually called the Gurevich or loop entropy. B.M. Gurevich defined and investigated this entropy in the late 1960's [Gv1]. I. Salama [Sl1] observed $h_{d_-}(\Sigma_T)$ is equal to $\log \lambda$, ((iv) of Proposition 7.2.6). Example 7.2.8 which shows the Gurevich entropy of a factor can be higher than that of the domain is due to K. Petersen [Pe2] in 1986. The Salama or block entropy discussed in Remark 7.2.11 was investigated by I. Salama in his 1984 thesis [Sl1]. There, he compared the metric entropies $h_{d_-}(\Sigma_T)$ and $h_{d*}(\Sigma_T)$. Theorem 7.2.13 which explains that only positive recurrent Markov shifts have a maximal measure is due to B.M. Gurevich [Gv] in 1970. Related issues were discussed by Y. Takahashi [Tk] and F. Hofbauer [Hb].

Observation 7.2.3 characterized locally compact Markov shifts. D. Fiebig and U. Fiebig have extended many of the results in Section 4.1 concerning embeddings and lower entropy factor maps to this setting [FF1], [FF2].

References

[AKM] R. Adler, A. Konheim and M.H. MacAndrew, *Topological Entropy*, Transactions of the American Mathematical Society no. 114 (1965).

[Bo2] R. Bowen, *Entropy for Group Endomorphisms and Homogeneous Spaces*, Transactions of the American Mathematical Society **153** (1971), 401–414.

[EFP] P. Erdős, W. Feller, H. Pollard, *A Property of Power Series with Positive Coefficients*, Bulletin of the American Mathematical Society **55** (1949), 201–204.

[Fe] W. Feller, *An Introduction to Probability Theory and Its Applications*, 3rd. edition, Wiley, 1968.

[FF1] D. Fiebig and U. Fiebig, *Topological Boundaries for Countable State Markov Shifts*, Proceedings of the London Mathematical Society, III **70** (1995), 625–643.

[FF2] D. Fiebig and U. Fiebig, *Entropy and Finite Generators for Locally Compact Subshifts*, Ergodic Theory and Dynamical Systems **70** (1997), 349–368.

[Hb]   F. Hofbauer, *On the Intrinsic Ergodicity of Piecewise Monotonic transformations with positive entropy,* Israel Journal of Mathematics **34** (1979), 213–237.

[Pe2] K. Petersen, *Chains, Entropy, Codings,* Ergodic Theory and Dynamical Systems **6** (1986), 415–448.

[Rd]   H.L. Royden, *Real Analysis,* MacMillan, 1968.

[Sl1] I. Salama, *Topological entropy and the Classification of Countable Chains,* Ph.D Thesis, University of North Carolina, Chapel Hill, 1984.

[Sl2] I. Salama, *Topological Entropy and Recurrence of Countable Chains,* Pacifac Journal of Mathematics **134** (1988), 325–341; errata Pac. J. of Math. **140** (1889) 397.

[Sl3] I. Salama, *On the Recurrence of Countable Topological Markov Chains,* American Mathematica Society Contemporary Mathematics **35** (1992), 349–360, *Symbolic Dynamics and its Applications.*

[Se]   E. Seneta, *Non-negative Matrices and Markov Chains,* Springer-Verlag, 1981.

[Tk]   Y. Takahashi, *Isomorphisms of β-Automorphisms to Markov Automorphisms,* Osaka Journal of Mathematics **10** (1973), 175–184.

[V-J1] D. Vere-Jones, *Geometric Ergodicity in Denumerable Markov Chains,* Quarterly journal of Mathematics **13** (1962), 7–28.

[V-J2] D. Vere-Jones, *Ergodic Properties of Nonnegative Matrices,* Pacific Journal of Mathematics **22** (1967), 361–386.

Bibliography

[ACH]   R. Adler, D. Coppersmith and M. Hassner: *Algorithms for Sliding Block Codes.* IEEE Transactions on Information Theory **29** (1983) 5–22

[AGW]   R. Adler, L. W. Goodwyn and B. Weiss: *Equivalence of topological Markov Shifts.* Israel Journal of Mathematics **27** (1977) 49–63

[AKM]   R. Adler, A. Konheim and M. H. MacAndrew: *Topological entropy.* Transactions of the American Mathematical Society **114** (1965)

[AM]   R. Adler and B. Marcus: *Topological entropy and equivalence of dynamical systems.* Memoirs of the American Mathematical Society **219** (1979)

[AW1]   R. Adler and B. Weiss: *Entropy, a complete metric invariant for automorphisms of the torus.* Proceedings of the National Academy of Sciences, USA **57** (1967) 1573–1576

[AW2]   R. Adler and B. Weiss: *similarity of automorphisms of the torus.* Memoirs of the American Mathematical Society **98** (1970)

[ArM]   M. Artin and B. Mazur: *On periodic points.* Annals of Mathematics **81** (1965)

[Ay1]   J. Ashley: *Marker automorphisms of the one-sided d-shift.* Ergodic Theory and Dynamical Systems **10** (1990) 247–262

[Ay2]   J. Ashley: *Bounded-to-1 factors of an aperiodic shift of finite type are 1-to-1 almost everywhere factors also.* Ergodic Theory and Dynamical Systems **10** (1990) 615–625

[Ay3]   J. Ashley: *Resolving factor maps for shifts of finite type with equal entropy.* Ergodic Theory and Dynamical Systems **11** (1991) 219–240

[Ba]   K. Baker: *Strong shift equivalence of* 2×2 *matrices of non-negative integers.* Ergodic Theory and Dynamical Systems **3** (1983) 501–508

[Be]   K. Berg: *On the conjugacy problem for K-systems.* Ph. D Thesis University of Minnesota, 1967

[BDK]   P. Blanchard, R. Devaney and L. Keen: *The dynamics of complex polynomials.* Inventiones Mathematicae **104** (1991) 545–580

[Bo1]   R. Bowen: *Markov partitions for axiom A diffeomorphisms.* American Journal of Mathematics **92** (1970) 725–747

[Bo2]   R. Bowen: *Entropy for group endomorphisms and homogeneous spaces.* Transactions of the American Mathematical Society **153** (1971) 401–414

[BF]   R. Bowen and J. Franks: *Homology for zero-dimensional nonwandering sets.* Annals of Mathematics **106** (1977) 73–92

[BL]   R. Bowen and O. Lanford: *Zeta functions of restrictions of the shift transformation,* in: Global Analysis, Proceedings of Symposia in Pure and Applied

Math XIV (S-S. Chern and S. Smale, eds.). American Mathematical Society 1970, pp. 43–49

[By1]   M. Boyle: *Lower entropy factors of sofic systems.* Ergodic Theory and Dynamical Systems **4** (1984) 541–557

[By2]   M. Boyle: *Constraints on the degree of sofic homomorphisms and the induced multiplication of measures on unstable sets.* Israel Journal of Mathematics **53** (1986) 52–68

[By3]   M. Boyle: *Factoring factor maps.* Preprint

[BF]   M. Boyle and U.-R. Fiebig: *The action of inert finite order automorphisms on finite subsystems of the shift.* Ergodic Theory and Dynamical Systems **11** (1991) 413–425

[BFK]   M. Boyle, J. Franks and B. Kitchens: *Automorphisms of one-sided subshifts of finite type.* Ergodic Theory and Dynamical Systems **10** (1990) 421–449

[BKM]   M. Boyle, B. Kitchens and B. Marcus: *A note on minimal covers for sofic systems.* Proceedings of the American Mathematical Society **95** (1985) 403–411

[BK1]   M. Boyle and W. Krieger: *Periodic points and automorphisms of the shift.* Transactions of the American Mathematical Society **302** (1987) 125–149

[BK2]   M. Boyle and W. Krieger: *Automorphisms and subsystems of the shift.* Journal für die reine und angewandte Mathematik **437** (1993) 13–28

[BLR]   M. Boyle, D. Lind and B. Rudolph: *The automorphism group of a shift of finite type.* Transactions of the American Mathematical Society **306** (1988) 71–114

[BMT]   M. Boyle, B. Marcus and P. Trow: *Resolving maps and the dimension group for shifts of finite type.* Memoirs of the American Mathematical Society **377** (1987)

[CP1]   E. Coven and M. Paul: *Endomorphisms of irreducible subshifts of finite type.* Mathematical Systems Theory **8** (1974) 167–175

[CP2]   E. Coven and M. Paul: *Sofic systema.* Israel Journal of Mathematics **20** (1975) 165–177

[CP3]   E. Coven and M. Paul: *Finite procedures for sofic systems.* Monatshefte für Mathematik **83** (1977) 265–278

[CK1]   J. Cuntz and W. Krieger: *A class of C^*-algebras and topological Markov Chains.* Inventiones Mathematicae **56** (1980) 251–268

[CK2]   J. Cuntz and W. Krieger: *Topological Markov chains with dicyclic dimension groups.* Journal für die reine und angewandte Mathematik **320** (1980) 44–50

[DGS]   M. Denker, C. Grillenberger and K. Sigmund: *Ergodic theory on compact spaces.* Lecture Notes in Mathematics, vol. 527. Springer 1976

[EFP]   P. Erdős, W. Feller, H. Pollard: *A property of power series with positive coefficients.* Bulletin of the American Mathematical Society **55** (1949) 201–204

[FTW]   D. Farmer, T. Toffoli and S. Wolfram, editors: *Cellular automata.* North-Holland 1984

[Fa]   P. Fatou: *Sur les equations fonctionnelles.* Bulletin de la Société Mathématique de France **47** (1919) 161–247

[Fe]   W. Feller: *An introduction to probability theory and its applications,* 3rd. edn. Wiley 1968

[FF1]   D. Fiebig and U. Fiebig: *Topological boundaries for countable state Markov Shifts.* Proceedings of the London Mathematical Society, III **70** (1995) 625–643

[FF2]   D. Fiebig and U. Fiebig: *Entropy and finite generators for locally compact subshifts.* Ergodic Theory and Dynamical Systems **17** (1997) 349–368

[Fg]   U. Fiebig: *Symbolic dynamics and locally compact Markov Shifts.* Habilitationsschrift der Universität Heidelberg (1997)

[Fi1]   R. Fischer: *Sofic systems and graphs.* Monatshefte für Mathematik **80** (1975) 179–186

[Fi2]   R. Fischer: *Graphs and symbolic dynamics.* Proceedings of the Colloquia Mathematical Society János Bolyai, Information Theory **16** (1975) 229–244

[Fa]   P. Fatou: *Sur les equations fonctionnelles.* Bulletin de la Société Mathématique de France **47** (1919) 161–247

[Fz]   P. Franaszek: *Sequence state methods for run-length limited codes.* IBM Journal of Research and Development **14** (1970) 376–383

[Fk]   J. Franks: *Flow equivalence of subshifts of finite type.* Ergodic Theory and Dynamical Systems **4** (1984) 53–66

[Fd]   D. Fried: *Finitely presented dynamical systems.* Ergodic Theory and Dynamical Systems **7** (1987) 489–507

[Fn]   J. Friedman: *On the road coloring problem:* Proceedings of the American Mathematical Society **110** (1990) 1133–1135

[G]   W. Gantmacher: *The theory of matrices,* vol. 1. Chelsea Publishing Co. 1959

[Gu]   H. Gutowitz, editor: *Cellular automata: theory and experiment.* MIT Press, 1991

[Hd]   J. Hadamard: *Les surfaces à courbures opposées et leurs lignes géodésiques.* Journal de Mathématiques Pures et Appliquées **4** (1898) 27–73

[Ha]   M. Hall, Jr.: *The theory of groups.* Chelsea Publishing Co. 1976

[Hs]   P. Halmos: *On automorphisms of compact groups.* Bulletin of the American Mathematical Society **49** (1943) 619–624

[H1]   G. A. Hedlund: *Transformations commuting with the shift.* Topological Dynamics (J. Auslander and W. Gottschalk, eds.). Benjamin 1968

[H2]   G. A. Hedlund: *Endomorphisms and automorphisms of the shift dynamical system.* Mathematical Systems Theory **3** no. 4 (1969) 320–375

[HR]   E. Hewitt and K. Ross: *Abstract harmonic analysis.* Academic Press and Springer 1963

[Hb]   F. Hofbauer: *On the intrinsic ergodicity of piecewise monotonic transformations with positive entropy.* Israel Journal of Mathematics **34** (1979) 213–237

[HU]   J. E Hopcroft and J. D. Ullman: *Frmal languages and their relation to automata.* Addison-Wesley Publishing Co. 1969

[Ju]   G. Julia: *Iteration des applications fonctionnelles.* Journal de Mathématiques Pures et Appliquées **8** (1918) 47–245, reprinted in *Oeuvres de Gaston Julia,* Gauthier-Villars, vol. I, pp. 121–319

[Kh]   A. Khinchin: *Mathematical foundations of statistical mechanics.* Dover Publications 1949

[KR1]   K. H. Kim and F. Roush: *Decidability of shift equivalence.* Dynamical Systems: Proceedings, University of Maryland 1886–87 (J. C. Alexander, ed.). Springer 1988, pp. 374–424

[KR2]   K.H. Kim and F. Roush: *William's conjecture is false for reducible subshifts.* Journal of the American Mathematical Society **5** (1992) 213–215

[KR3]   K. H. Kim and F. Roush: *On the automorphism groups of subshifts.* P.U.M.A Series B **1** (1990) 203–230

[KR4]   K. H. Kim and F. Roush: *On the structure of inert automorphisms of subshifts.* P.U.M.A Series B **2** (1991) 3–22

[KRW1]   K. H. Kim, F. Roush and J. Wagoner: *Automorphisms of the dimension group and gyration numbers.* Journal of the American Mathematical Society **5** (1992) 191–212

[KRW2]   K. H. Kim, F. Roush and J. Wagoner: *Inert actions on periodic points,* preprint

[K1]   B. Kitchens: *An invariant for continuous factors of Markov Shifts.* Proceedings of the American Mathematical Society **83** (1981) 825–828

[K2]   B. Kitchens: *Expansive dynamics on zer dimensional groups.* Ergodic Theory and Dynamical Systems **7** (1987) 249–261

[KMT]   B. Kitchens, B. Marcus and P. Trow: *Eventual factor maps and compositions of closing maps.* Ergodic Theory and Dynamical Systems **11** (1991) 857–113

[Ko]   A. Kolmogorov: *A new metric invariant for transient dynamical systems.* Academiia Nauk SSSR, Doklady **119** (1958) 861–864 (Russian)

[Kg1]   W. Krieger: *On topological Markov Chains.* Astérisque **50** (1977) 193–196

[Kg2]   W. Krieger: *On dimension for a class of homeomorphism groups.* Mathematische Annalen **252** (1980) 87–95

[Kg3]   W. Krieger: *On dimension functions and topological Markov Chains.* Inventiones Mathematicae **56** (1980) 239–250

[Kg4]   W. Krieger: *On the periodic points of topological Markov Chains.* Mathematische Zeitschrift **169** (1979) 99–104

[Kg5]   W. Krieger: *On subsystems of topological Markov Chains.* Ergodic Theory and Dynamical Systems **2** (1982) 195–202

[Kg6]   W. Krieger: *On sofic systems I.* Israel Journal of Mathematics **48** (1984) 305–330

[Kg7]   W. Krieger: *On sofic systems II. Journal of Mathematics* **60** (1987) 167–176

[Li1]   D. Lind: *Entropies and factorizations of topological Markov Shifts.* Bulletin of the American Mathematical Society **9** (1983) 219–222

[Li2]   D. Lind: *The entropies of topological Markov Shifts and a related class of algebraic integers.* Ergodic Theory and Dynamical Systems **4** (1984) 283–300

[MKS]   D. Lind and B. Marcus: *Symbolic dynamics and coding.* Cambridge University Press 1995

[Mc]   B. McMillan: *The basic theorems of information theory.* Annals of Mathematical Statistics **24** (1953) 196–219

[MKS]   W. Magnus, A. Karrass and D. Solitar: *Combinatorial group theory.* Dover Publications 1976

[Mg]   A. Manning: *Axiom a diffeomorphisms have rational zeta functions.* Bulletin of the London Mathematical Society **3** (1971) 215–220

[Mr1]   B. Marcus: *Factors and extensions of full shifts.* Monatshefte fü Mathematik **88** (1979) 239–247

[MH]   M. Morse and G. A. Hedlund: *Symbolic dynamics.* American Journal of Mathematics **60** (1938) 815–866

[Nu1]   M. Nasu: *Uniformly finite-to-one and onto extensions of homomorphisms between strongly connected graphs.* Discrete Mathematics no. 39 (1982) 171–197

[Nu2]   M. Nasu: *Constant-to-one and onto global maps of homomorphisms between strongly connected graphs.* Ergodic Theory and Dynamical Systems **3** (1983) 387–413

[Nu3]   M. Nasu: *An invariant for bounded-to-one factor maps between transitive sofic subshifts.* Ergodic Theory and Dynamical Systems **5** (1985) 85–105

[O'B]   G. L. O'Brien: *The road coloring problem.* Israel Journal of Mathematics **39** (1981) 145–154

[Or]   D. Ornstein: *Bernoulli Shifts with the same entropy are isomorphic.* Advances in Mathematics **4** (1970) 337–352

[P1]   W. Parry: *Intrinsic Markov Chains.* Transactions of the American Mathematical Society **112** (1964) 55–66

[P2]   W. Parry: *Symbolic dynamics and transformations of the unit interval.* Transactions of the American Mathematical Society **122** (1966) 368–378

[P3]   W. Parry: *A finitary classification of topological Markov Chains and sofic systems.* Bulletin of the London Mathematical Society **9** (1977) 86–92

[PS]   W. Parry and D. Sullivan: *A topological invariant for flows on one-dimensional spaces.* Topology **14** (1975) 297–299

[PT]   W. Parry and S. Tuncel: *Classification problems in ergodic theory.* London Mathematical Society Lecture Series, 67. Cambridge University Press 1982

[Pe]   K. Petersen: *Ergodic theory.* Cambridge Press 1983

[Pe2]   K. Petersen: *Chains, entropy, codings.* Ergodic Theory and Dynamical Systems **6** (1986) 415–448

[Rn]   J. Rotman: *An introduction to the theory of groups,* 3rd. edn. Allyn and Bacon, Inc. 1984

[Rd]   H. L. Royden: *Real analysis.* MacMillan 1968

[Ru]   D. J. Rudolph: *Fundamentals of measurable dynamics.* Clarendon Press 1990

[Ry1]   J. P. Ryan: *The shift and commutivity.* Mathematical Systems Theory **6** (1973) 82–85

[Ry2]   J. P. Ryan: *The shift and commutivity II.* Mathematical Systems Theory **8** (1975) 249–250

[Sl1]   I. Salama: *Topological entropy and the classification of countable chains.* Ph.D Thesis University of North Carolina, Chapel Hill 1984

[Sl2]   I. Salama: *Topological entropy and recurrence of countable chains.* Pacific Journal of Mathematics **134** (1988) 325–341; errata Pac. J. of Math. **140** (1889) 397

[Sl3]   I. Salama: *On the recurrence of countable topological Markov Chains.* Contemporary Mathematics, vol. 35: Symbolic Dynamics and its Applications. American Mathematical Society 1922, pp. 349–360

[Sd]   K. Schmidt: *Dynamical systems of algebraic origin.* Birkhäuser 1995

[Se]   E. Seneta: *Non-negative matrices and Markov Chains.* Springer 1981

[S]   C. Shannon: *A mathematical theory of communication.* Bell System Technical Journal (1948) 379–473 and 623–656, reprinted in *The Mathematical Theory of Communication* by C. Shannon and W. Weaver, University of Illinois Press, 1963

[Si1]   Y. Sinai: *Construction of Markov Partitions.* Funkcional'nyi Analiz i Ego
          Prilozheniya **2** no. 3 (1968) 70–80 (Russian); English transl. in Functional
          Analysis and Its Applications **2** (1968) 245–253

[Si2]   Y. Sinai: *On the concept of entropy for a dynamical system.* Academiia Nauk
          SSSR, Doklady **124** (1959) 768–771 (Russian)

[Sm]   S. Smale: *Diffeomorphisms with many periodic points, Differential and Com-
          binatorial Topology.* Princeton University Press, 1965, pp. 63–80

[Tk]   Y. Takahashi: *Isomorphisms of β-automorphisms to Markov Automorphisms.*
          Osaka Journal of Mathematics **10** (1973) 175–184

[To]   R. C. Tolman: *Principles of statistical mechanics.* Oxford, Clarendon 1938

[Tw1]   P. Trow: *Resolving maps which commute with a power of the shift.* Ergodic
          Theory and Dynamical Systems **6** (1986) 281–293

[Tw2]   P. Trow: *Degrees of finite-to-one factor maps.* Israel Journal of Mathematics
          71 (1990) 229–238

[V-J1]   D. Vere-Jones: *Geometric ergodicity in denumerable Markov Chains.* Quar-
          terly Journal of Mathematics **13** (1962) 7–28

[V-J2]   Vere-Jones: *Ergodic properties of nonnegative matrices.* Pacific Journal of
          Mathematics **22** (1967) 361–386

[vN]   J. von Neumann: *Theory of self-reproducing automata.* (A. W. Burks, ed.).
          Univerity of Illinois Press 1966

[Wa1]   J. Wagoner: *Markov Partitions and K_2.* Publications Mathématiques IHES
          65 (1987) 91–129

[Wa2]   J. Wagoner: *Triangle identities and symmetries of a subshift of finite type.*
          Pacific Journal of Mathematics **144** (1990) 181–205

[Wa3]   J. Wagoner: *Eventual finite order generation for the kernel of the dimension
          group representation.* Transactions of the American Mathematical Society
          317 (1990) 331–350

[Wa4]   J. Wagoner: *Topological Markov Chains, C^*-algebras and K_2.* Advances in
          Mathematics **71** (1988) 133–185

[Wa]   P. Walters: *An introduction to ergodic theory* GTM. Springer 1981

[We]   B. Weiss: *Subshifts of finite type and sofic systems.* Monatshefte für Mathe-
          matik **77** (1973) 462–474

[Wi1]   R. F. Williams: *Classification of one-dimensional attractors,* in Global Anal-
          ysis, Proceedings of Symposia in Pure and Applied Math. (S-S. Chern and
          S. Smale, eds.). American Mathematical Society 1970, pp. 341–361

[Wi2]   R. F. Williams: *Classification of subshifts of finite type.* Annals of Mathematics
          98 (1973) 120–153; *Errata* **99** (1974) 380–381

[Wm]   S. Williams: *Lattice invariants for sofic shifts.* Ergodic Theory and Dynamical
          Systems **11** (1991) 787–801

Name Index

The chapter numbers after each name refer to the *Notes* at the end of the corresponding chapters.

Subject Index

Printing: Mercedesdruck, Berlin
Binding: Buchbinderei Lüderitz & Bauer, Berlin

H.-C. Hege, K. Polthier (Eds.)

Visualization and Mathematics

Experiments, Simulations and Environments

1997. XIX, 386 pp. 230 figs., 43 in color
Hardcover DM 138,-
ISBN 3-540-61269-6

Visualization and mathematics have begun a fruitful relationship, establishing links between problems and solutions of both fields. In some areas of mathematics, like differential geometry and numerical mathematics, visualization techniques are applied with great success. However, visualization methods are relying heavily on mathematical concepts.
Applications of visualization in mathematical research and the use of mathematical methods in visualization have been topic of an international workshop in Berlin in June 1995. Selected contributions treat topics of particular interest in current research. Experts are reporting on their latest work, giving an overview on this fascinating new area. The reader will get insight to state-of-the-art techniques for solving visualization problems and mathematical questions.

O. Moeschlin, E. Grycko, C. Pohl, F. Steinert

Experimental Stochastics

1997. CD-ROM with booklet, approx. 200 pp. 52 figs.
Softcover DM 130,-
ISBN 3-540-14619-9

The electronic monograph Experimental Stochastics deals with the generating and testing of artificial random numbers and demonstrates their applications in practice. Artificial random experiments allow to treat a large variety of problems. The software is organized according to areas of applications: stochastic models, stochastic processes, evaluation of statistical methods. The examples chosen range from the control of traffic lights at road bottlenecks to Maxwell's law on the velocity distribution of gases. The presentation of the material is done via appropriate computer visualization, and the theory behind the experiments is given in both the electronic component and the accompanying booklet. This product, presenting many well-prepared simulations, will be very useful for instructional purposes.

Springer

Springer-Verlag, P. O. Box 31 13 40, D-10643 Berlin, Germany

Y.G. Sinai

Probability Theory

An Introductory Course

1992. VIII, 138 pp. 14 figs.
(Springer Textbook)
Softcover DM 48,-
ISBN 3-540-53348-6

Sinai's book leads the student through the standard material of Probability Theory, with stops along the way for interesting topics such as statistical mechanics, not usually included in a book for beginners. The text is divided into sixteen lectures, each covering a major topic. The introductory notions and classical results are included, of course: random variables, the central limit theorem, the law of large numbers, conditional probability, random walks, etc. Sinai's style is accessible and clear, with interesting examples to accompany new ideas. Besides statistical mechanics, other interesting, less common topics found in the book are: percolation, the concept of stability in the central limit theorem and the study of probability of large deviations. Little more than a standard undergraduate course in analysis is assumed of the reader. Notions from measure theory and Lebesgue integration are introduced in the second half of the text. The book is suitable for second or third year students in mathematics, physics or other natural sciences. It could also be used by more advanced readers who want to learn the mathematics of probability theory and some of its applications in statistical physics.

Cucker, M. Shub (Eds.)

Foundations of Computational Mathematics

Selected Papers of a Conference Held in Rio de Janeiro, January 1997

1997. XV, 441 pp.
Softcover DM 128,-
ISBN 3-540-61647-0

This book contains articles corresponding to some of the talks delivered at the Foundations of Computational Mathematics (FoCM) conference at IMPA in Rio de Janeiro in January 1997. FoCM brings together a novel constellation of subjects in which the computational process itself and the foundational mathematical underpinnings of algorithms are the objects of study. The conference was organized around nine workshops: systems of algebraic equations and computational algebraic geometry, homotopy methods and real machines, information based complexity, numerical linear algebra, approximation and PDE's, optimization, differential equations and dynamical systems, relations to computer science and vision and related computational tools. The book gives the reader an idea of the state of the art in this emerging discipline.

Please order from
Springer-Verlag Berlin
Fax: + 49 / 30 / 8 27 87- 301
e-mail: orders@springer.de
or through your bookseller

Prices subject to change without notice.
In EU countries the local VAT is effective.

Springer

Springer-Verlag, P. O. Box 31 13 40, D-10643 Berlin, Germany